Ecological Studies, Vol. 174

Analysis and Synthesis

Edited by

M.M. Caldwell, Logan, USA
G. Heldmaier, Marburg, Germany
Robert B. Jackson, Durham, USA
O.L. Lange, Würzburg, Germany
H.A. Mooney, Stanford, USA
E.-D. Schulze, Jena, Germany
U. Sommer, Kiel, Germany

Ecological Studies

Volumes published since 1998 are listed at the end of this book.

David W. Roubik   Shoko Sakai
Abang A. Hamid Karim

Editors

Pollination Ecology and the Rain Forest

Sarawak Studies

With 76 Illustrations, 12 in Full Color

 Springer

David W. Roubik
Smithsonian Tropical Research Institute
Ancon, Balboa
Republic of Panama

Shoko Sakai
Center for Ecological Research
Kyoto University
Kamitanakami Hiranocho
Otsu 520-2113, Kyoto
Japan

Abang A. Hamid Karim
Department of Agriculture
Menara Pelita
Petra Jaya, 93050 Kuching
Malaysia

Cover illustration: Concepts of coevolution, ecological fitting, and loose niches, applied to ecological interactions among plants and pollinators. Adapted from the island biogeographic model of MacArthur and Wilson, 1963.

ISSN 0070-8356
ISBN 0-387-21309-0   Printed on acid-free paper.

Printed in the United States of America.   (WBG/EB)

9 8 7 6 5 4 3 2 1   SPIN 10958995

springeronline.com

Preface

Rain Forest Biology and the Canopy System, Sarawak, 1992–2002

The rain forest takes an immense breath and then exhales, once every four or five years, as a major global weather pattern plays out, usually heralded by El Niño–Southern Oscillation. While this powerful natural cycle has occurred for many millennia, it is during the past decade that both the climate of Earth and the people living on it have had an increasing influence on the weather pattern itself, with many biological consequences. In Southeast Asia, as also in most of the Neotropics, El Niño accompanies one of the most exuberant outpourings of nature's diversity. After several years of little activity, the incredibly diverse rain forests suddenly burst into flower—a phenomenon referred to as General Flowering in Asia. Plant populations are rejuvenated and animals are fed, but the process involves a delicate and complex balance.

When the canopy access system was under construction at Lambir Hills National Park in the early 1990s, it made use of an underlying technology that was already in place: bridges. For centuries, bridges have spanned the natural chasms over rivers. This existing network of bridges and the people who built and use them produced the technology we needed to gain access to the canopy. Bridge builders were our natural allies in the quest for biological knowledge of the high canopy. We saw the two massive tree towers take shape, then the walkways between them, all in a setting that would make any naturalist or explorer dizzy with excitement, if not vertigo. Studies at the top of the living envelope of forest were finally to gain a firm footing and would soon be incorporated with the more traditional, earthbound observations. Professor Tamiji Inoue recognized that the special environment of the rain-forest canopy held the future for tropical scientific exploration.

Now, over a decade later, technology has placed at our disposal a new canopy access system—an immense construction crane towering 80 meters high, with a jib reaching 75 meters across the surrounding forest, and a remote-controlled gondola that can travel from the ground to well above the canopy. This represents a revolution in the study of tropical rain forests. It may also represent a final frontier in natural history studies, in one of the most important, but little known, biomes on Earth.

Students of the rain forest strive to see the entire forest and its denizens, across both space and time. Of the 367 species of mammals, birds, reptiles, and frogs at Lambir Hills National Park, the disturbed or open habitat species are increasing, while forest animals such as hornbills and primates are in decline or have disappeared (Shanahan and Debski 2002). An unusually severe drought and an El Niño in 1997 and 1998 increased tree mortality by seven times (Nakagawa et al. 2000) and led uniformly to local extinctions of mutualistic insects (Harrison 2000). Also following that event was an outbreak of certain insect herbivores (Itioka et al. 2003). Many changes and dynamics continue apace.

Similar themes are emerging elsewhere. At the other side of the world, in Costa Rica, a gathering to commemorate the fortieth anniversary of the Organization for Tropical Studies recognizes a worldview with particular resonance for the tropics. One of the speakers is Dr. Edward O. Wilson, a spokesman for, and well-known pioneer of, themes about the rain forest that have captured attention with their urgency; for example:

- In 1988, the term *biodiversity* was introduced, yet even today, 90% of the world's species remain undescribed and unappreciated. Half of them live only in the tropical forests.
- The second-greatest block of rain forest on the planet is in Borneo. It is representative of what remains on Earth in the standing tropical forests, now diminished from 12% to 6% of the planet's surface, since the precipitous advance of human populations.
- In the small but biodiverse region of Costa Rica, national parks and preserves now include 37% of all land, an increase from 20% a short time ago. Why? One reason is purely economical, because the water provided by forest is more valuable than one of its popular economic alternatives—beef cattle that would be produced on land cleared of its natural vegetation.
- Currently, the poor outnumber the rest of humanity by about 75:1, and almost 100 million people live in absolute poverty. However, future generations will pay the heaviest price. It will stem from the loss of biodiversity and the services, quality of life, culture, and potential for development that biodiversity provides.

- Our collective retirement funds lie, now and in the future, in the sustained partnership of people and their environment, not in the short-term profit taking that leads to erosion of all that is valued by society.

Even though the pessimists seem to outnumber the optimists, we still agree with Dr. Wilson and the participants of that tropical conference in the Americas. We need to act, we need to reason, and we need to understand. From a tract of rain forest in the north of Borneo, the information given here brings us a little closer to seeing the scientific reality of the rain forest. We are striving to keep in step with the race to realize our potential before the great forests are taken away, for, as Professor Inoue once remarked, these places are the windows in which we can behold the entire history of life on Earth. As presented in the closing chapter of this work, expressed by our friend the late Professor Inoue, who died tragically during the Sarawak studies, there is enduring relevance in rain-forest research. Maintaining the human birthright—the preservation of nature's masterpieces while fulfilling the true goals of our lives and histories—is still the primary purpose of science.

David W. Roubik
Shoko Sakai
Abang A. Hamid Karim

Acknowledgments

This book was compiled after more than 10 years of extensive ecological studies in lowland dipterocarp forest at Lambir Hills National Park, Malaysia. The work was conducted by cooperative projects that included the Forest Department Sarawak, Malaysia; the Japan Science and Technology Agency; Ehime University; Harvard University; Kyoto University; Osaka City University; Smithsonian Tropical Research Institute; and other universities and organizations from the United States and Japan.

Sarawak is endowed with true splendors of nature and recognized as one of the world's centers of species diversity. Like almost all tropical forests, those of Sarawak are threatened and may even disappear under strong economic pressures. The authorities of this Malaysian state have made serious and strenuous efforts to enlarge protected areas and to conserve biodiversity, as symbolized by the state emblem of the rhinoceros hornbill. As biologists, we greatly appreciate opportunities to wander, as Beccari did a century ago, the Great Forests of Borneo. We wander even farther now, to climb up to the canopy and conduct studies in the Lambir Hills forest, which is blessed with an amazingly high biodiversity and a safe system of canopy access, permitting biologists to go where none have gone before.

Rather than attempting to cover all the topics studied so far in this part of Borneo's rain forest, the present volume highlights interactions between plants and animals in the context of dynamic natural environments. Encouraged by recent attention given to the significance and real inspiration provided by biodiversity and biological interactions, we climbed to the canopy to observe first-hand the flowering, fruiting, sexual recombination, predation, fighting, and parasitism that occur there, and in the forest below. In addition, long-term monitoring of insects and plants has revealed that the forest's biological activities are very dynamic, with a cycle of more than one year under mild, uniform climate conditions with little seasonality. The tight links in regeneration of dipterocarp forests and rhythms of the global climate, related to El Niño, are exciting to recognize as major factors in rain-forest biology; at the same time, such links are cautionary signs, indicating the sensitive and fragile nature of the ecosystem. We hope our studies can contribute to the conservation of tropical forests by emphasizing that pollination and diversity are truly partners, and that they have been understudied or, unfortunately, altogether neglected not only in schemes for conservation, but also in research on forest ecology.

This book owes much to many people. First, we would like to apologize that it is impossible to list everyone who contributed to studies in Lambir Hills and to this book. Ms. Lucy Chong and many other staff members from the Forest Department, Sarawak supported fieldwork and management of the projects. The local people at Lambir have taken us to many interesting sites while providing fascinating knowledge about the forest. A large number of researchers and students conducted studies and contributed to Lambir projects, but they do not appear in this book as authors. In particular, Dr. Lee and professors Ogino and Yamakura played leading roles to establish our field site. We acknowledge the support and editorial assistance of Dr. Rhett Harrison. We also are grateful to Springer for unfailing support for this book. The studies were funded by various sources, including the Ministry of Education, Science, Sports, and Culture (nos. 04041067, 06041013, 09NP1501) and Japan Science, Technology Corporation (Core Research for Evolutionary Science and Technology Program) and the Research Institute for Humanity and Nature Project (P2-2) in Japan. This publication was supported by the Japan Society for the Promotion of Science (Grant 165296).

Special thanks are also due to Mrs. Eiko Inoue, who shared our enthusiasm for the forest and our research, and assisted our research activities in many ways. She also gave permission to translate and reprint part of a book written by Professor Tamiji Inoue and to use his beautiful photographs.

David W. Roubik
Shoko Sakai
Abang A. Hamid Karim

Contents

Contributors

Peter S. Ashton

Organismic and Evolutionary Biology,
Harvard University, USA, and
Royal Botanic Gardens, Kew, UK

Stuart J. Davies

Center for Tropical Forest Science—
Arnold Arboretum Asia Program,
Harvard University, USA

Abang A. Hamid Karim

Department of Agriculture,
Menara Pelita, Petra Jaya,
Kuching, Malaysia

Rhett D. Harrison

Smithsonian Tropical Research Institute,
Ancon, Balboa, Republic of Panama

Tamiji Inoue (deceased)

Center for Ecological Research,
Kyoto University, Otsu, Japan

Takao Itino

Department of Biology, Faculty of
Science, Shinshu University,
Nagano, Japan

Takao Itioka

Graduate School of Human and
Environmental Studies,
Kyoto University, Kyoto, Japan

Makoto Kato

Graduate School of Human and
Environmental Studies,
Kyoto University, Kyoto, Japan

James V. LaFrankie

Center for Tropical Forest Science—
Arnold Arboretum Asia Program,
Smithsonian Tropical Research Institute;
c/o National Institute of Education,
Singapore

Kuniyasu Momose

Department of Agriculture,
Ehime University, Ehime, Japan

Hidetoshi Nagamasu

The Kyoto University Museum,
Kyoto, Japan

Teruyoshi Nagamitsu

Hokkaido Research Center,
Forestry and Forest Products Research
Institute, Hokkaido, Japan

Michiko Nakagawa

Research Institute for Humanity and
Nature, Kyoto, Japan

Tohru Nakashizuka

Research Institute for Humanity and
Nature, Kyoto, Japan

Matthew D. Potts

Institute on Global Conflict and
Cooperation, University of California,
California, USA

David W. Roubik

Smithsonian Tropical Research Institute
Ancon, Balboa, Republic of Panama

Shoko Sakai

Center for Ecological Research,
Kyoto University, Otsu, Japan

Mike Shanahan

Centre for Biodiversity and
Conservation, School of Biology,
University of Leeds,
UK

Sylvester Tan   Forest Department, Sarawak,
                     Kuching, Malaysia

Takakazu Yumoto   Research Institute for Humanity and
                     Nature, Kyoto, Japan

1. Large Processes with Small Targets: Rarity and Pollination in Rain Forests

David W. Roubik

1.1 Ecological Interactions Among Plants, Animals, Microbes, and Fungi

Perhaps nowhere on Earth has there been such a remarkably long period of uninterrupted tropical forest evolution, some 36 million years (Morley 2000), as within the old forest in Borneo. An example of tropical forest ecology from this area is Lambir Hills National Park, Sarawak, shown in Plates 1–12.

For studies of terrestrial ecology in forests to be realistic they must consider the movement of organisms and turnover of populations. At the base of the food chain, plants are fixed in space; the fungi that grow with them are also immobile. Their reproductive propagules, however, exhibit impressive mobility. Animals locate and harvest their food as they explore the forest and feed on fungi, roots, wood, sap, dung, leaves, fruit, nectar, pollen, seeds, or flowers. In turn, the predators that follow such prey include the human hunters, and a large, forest-wide cycle is created. The cycle depends on very small ecological targets: flowers, fruits, seeds, pollen grains, the sites in which seeds, microbes, or fungi can grow, and the receptive stigmata of flowers.

On a grand scale, the forest displays periodic migrations within its bounds. Feeding groups of several hundred white-lipped peccaries *Tayassu*, which follow the fruit drop of palms along waterways in the Amazon basin, are matched by the movement of bearded pigs *Sus*, moving in number to find patches of fruit on the ground, during a heavy fruiting year in Southeast Asia. Preceding such

consumer migrations, there is always a burst of flowers opening and petals dropping to the ground, and the noisy commotion of pollinators high in the trees.
Yet the forest canopy in Southeast Asia may remain relatively silent for years,
because most of the fruiting and flowering occurs in a supra-annual fashion,
generally once every four or five years (Inoue and Nakamura 1990; Inoue et al.
1993). One wonders if the intensity of those rare events is greater than the
flowering peaks and annual glut of fruits taking place each year in the more
predictably seasonal forests of Asia, Africa, or the Neotropics. Most observers
who have witnessed both phenomena believe that the annual peaks in flowering
and the resulting fruit are more intense in such seasonal forests than in their
counterparts in the rain forest of Southeast Asia, although not lasting as long.

Why is the Lambir Hills National Park, Sarawak, which is located in the
floristically rich north of Borneo, extremely valuable when left intact? The giant
trees in the ocean of forest have often been measured in terms of their economic
value or the ways in which plantations can be made by selecting certain species
(Panayotou and Ashton 1992; Appanah and Weinland 1993; Guariguata and
Kattan 2002; Okuda et al. 2003). Such forests lay outside the experience of most
people, even biologists, yet few natural environments are so rich in detail and
offer such great potential for insight. Lambir Hills yields insights that further
the development of classical theory or concepts, as seen in the physical sciences,
art, or music. We certainly have theories that address biology, culture, and many
other disciplines, but tropical field biologists primarily begin their work by observing a concrete, physical world—one that is often full of surprises. When
the studies are concluded, we are closer to understanding the forests and their
component species; often we come away with concepts and perspectives that
we had never before imagined.

What shapes the lives and evolution of living things in the rain forests? In
terms of interactions (see Fig. 1.1), consider three guiding principles: *coevolution* (Janzen 1980), *ecological fitting* (Janzen 1985), and *loose niches* (Roubik
1992; Roubik et al. 2003). The first implies tight and sustained interactions over
many generations, as part of the general process known as *adaptive radiation*.
The interacting populations are affected genetically in significant ways.

For instance, pollinating *fig wasps* or beetles have the right size and physiological traits to fly to their host plants and to pollinate them, for which they
must live their lives in synchrony with the highly specialized flowers. The flowers often have only one important pollinator, which they sustain by providing
food and access to flowers. In contrast, in *ecological fitting* there is no coevolution, but interactions can be subtle and complex. The organisms may come
from different places, having evolved their characteristics in other circumstances,
but now combine to form an ecological relationship. The third process, the loose
niches, derives from population cycles, with the strength of interaction tied to
the changing abundance of participants. Modern participants may have a coevolutionary history or not, but the modern interactions often demand behavioral
adjustments by the animals. The three types of relationships combine in highly
diverse communities, where the highest proportions of coevolutionary relation-

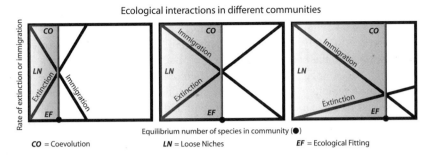

Figure 1.1. Concepts of coevolution, ecological fitting, and loose niches—applied to ecological interactions among plants and pollinators. Empirical data indicate loose pollination niches include 50% of plant species (Roubik et al. 2003); other interaction categories are complementary (shown by shaded triangles). Differing extinction and immigration rates determine local species richness; the richest community has the largest proportion of coevolved interactions (adapted from the island biogeographic model, MacArthur and Wilson 1963).

ships may exist (see Fig. 1.1). Undeniably, all such matters concern the weather, changing climate, geomorphology, continental drift, sea level, and oceans—not just life in and under the rain forest canopy. Such variables affect the origin, presence, and extinction of players in the game. The biological setting is traditionally known, thanks to G.E. Hutchinson, as the 'ecological theater' and the 'evolutionary play.'

In the rain forest, there is a relentless dynamic centering on events that can be as explosive as a volcanic eruption. An individual tree, group of plants, or entire population bursts into flower, dispensing pollen and nectar. As they drop the last of their flowers, the plants begin to sprout offspring in the form of seeds and fruit, which are afterwards dropped or carried away. Consumers, certainly including humans and animals of all kinds, come in as though filling a vacuum. They have taken their cue for the localized event from its coincident weather patterns or, if from nothing else, the colors or fragrances of flowers or fruit.

The major consumers in tropical forest include folivores and plant pathogens, which are not strictly tied to reproductive botany. Their dynamics are similar to animals that use the fruit, flowers, and seeds, but they seem to operate on a much smaller spatial scale. They are not, after all, moving to and from objects that are *designed* to be attractive. Quite the contrary, herbivores using particular leaves or small seeds often find them by searching the appropriate habitats, seeking a chance encounter with their small targets. While dispersal of seeds to forest openings or gaps seems rarely to involve a distance greater than 100 meters (Dalling et al. 2002; Levey et al. 2002; see Higgins and Richardson 1999), the dispersal of pollen by pollinators to flowers can easily cover distances of several to dozens of kilometers. Fungal spores or microbes that can infest seeds or growing plants are transmitted by wind, water, or animals, while invertebrates in pursuit of host plants may walk, crawl, or fly a moderate distance.

Consumers that are not feeding on leaves—the pollinators, frugivores, and granivores—may require areas encompassing tens to hundreds of kilometers: the scale that is ultimately important to Lambir Hills. Particularly in a forest with so many species, the canopy and understory both share the all-important environmental and ecological factor of rarity. Ecological and evolutionary processes that cause or maintain rarity are clear, and constitute the flip side of species richness and biological diversity. The second unifying theme is the double standard of the rain forest. Large-scale events, like general flowering or a severe drought, are uncommon, while the normal, annual flowering of certain trees and understory plants in a warm and humid environment has taken place consistently for millions of years.

1.2 Pollen, Seeds, and the Red Queen

Because of their relatively slow evolutionary rates, long-lived plants' best chances for keeping up with the evolutionary advances of natural enemies include diversifying offspring and maximizing seed and pollen dispersal to favorable sites. Within the lifetime of an individual plant, generations of insects or pathogens may produce new genetic combinations that allow toxic or unpalatable foliage to be eaten and digested. Not to be forgotten is the fact that immigrant species may arrive from other communities, providing a chance for ecological fitting (Janzen 1985). Such community building is complementary to evolution, or, coevolutionary fitting between a particular host and mutualist (see Fig. 1.1). A functioning community is a product of biogeography, ecology, behavior, and genetics. Under the *Red Queen* hypothesis, genetic dynamics are not all that pertain to unequal life spans. For plants, the evolution of a breeding system and pollination ecology are among consequences that can be traced to the Red Queen. An invertebrate, fungus, or microbe may, as natural enemies, genetically overcome any conceivable defense of the trees (Summers et al. 2003; Arnold et al. 2003; Normark et al. 2003). The Red Queen hypothesis rests on this premise (Hamilton 2001; Summers et al. 2003). A further consideration is the population buildup in small, fast-breeding insects (Itioka et al. 2003), which can go through multiple generations even during a single flowering or fruiting event. Pollination ecology helps plants to persist.

Once they have located their target resource, insects or pathogens sometimes consume almost all its seeds or leaves. Even though they may not kill a reproductively mature host, they diminish its potential reproduction (Strauss 1997). But, if they repeatedly cause extensive damage, they threaten their own survival and propagation. One may reasonably expect them to follow options to the evolutionary arms race. One of the most attractive is mutualism (if you can't beat them, join them). That selective pressure, in particular, may be a basis for the evolution of rather unusual pollination systems—wherein pollination is by species that use flowers or seeds as breeding sites or consume foliage when no flowers are present—and the existence of plants that do not participate in the general, community-wide flowering peak emphasized in this book (Itioka et al. 2003; Momose et al. 1998c). Exceptions involve ecological fitting or coevolution.

Fungi and bacteria not only feed the trees, but also kill their offspring. Mutualist fungi upon which the root systems of many tropical trees depend for nutrient acquisition (Turner 2001) or for defense of the foliage (Arnold et al. 2003) might have a starting point similar to that of herbivores that, over evolutionary time, have been converted into pollinators. Even some pathogenic fungi have been found playing a role in pollination in the Lambir Hills environment (Sakai et al. 2000). The transition from pathogen or herbivore to mutualist seems prevalent among the Dipterocarpaceae and their pollinators, root or seed associates. Because this plant family is so abundant at Lambir Hills, possessing by far the greatest stem area in the forest, and because a plant's natural enemies tend to evolve feeding specializations that are most effective on related host species (Janzen 2003), the evolutionary ecology of the Lambir Hills plant community is bound to the biology of abundant families maintaining a large biomass, like the euphorbs and dipterocarps.

Perhaps for a hardwood tree like Belian, *Eusideroxylon*, deaths from drought, fire, or specialist natural enemies are equally important. Woody plants with extremely hard wood and capable, especially among dipterocarps, of countering the breach of an insect mandible with copious resin (Langenheim 2003), are far from defenseless. It is no surprise that highly *eusocial* bees, most of the genus *Trigona*, are both abundant and ecologically diverse in Borneo, because they exploit the dipterocarp resin to build nests and defend their colonies (Plate 9F, G). Lodged within cavities in the dipterocarp trees, the bees obtain much pollen and nectar from their flowers, while also serving as pollinators.

Nonetheless, it is instructive to consider that millions upon millions of seeds are produced in order to maintain a tree population by contributing a single reproductive individual. Extremely large tropical trees make numerous tiny flowers, often dominated by social bees (Whitmore 1984; Momose et al. 1998c; Roubik et al. 2003), but these flower visitors are not prone to disperse pollen among trees (Roubik 1989). If no other individual is flowering within a short distance, in most cases not a single seed is produced (Ghazoul et al. 1998). This is largely because the mature seeds are derived only from non-self pollen in more than 85% of all tropical trees that have been investigated (Bawa 1990; Loveless 2002). Contrary to agricultural and domesticated plants, in which *outcrossing* and genetic diversity in seeds decrease fitness of the parents (Richards 1997), differences at the genetic level are strongly favored in tropical trees and become accentuated with rarity (Shapcott 1999; Loveless 2002). Loveless indicates, from studies of 176 tropical tree populations and nearly 100 species, average *heterozygosity* per locus is relatively high: 53%. Selection for inbreeding and uniformity among progeny would produce levels close to 0%.

If the entire lifetime of a tree could be witnessed, we would observe, in slow motion, behavior like that of a highly intelligent animal as it escapes from natural enemies and propagates its genes. Although it may stand rooted in the ground, a tree with a seed crop more than 40 m from the forest floor can disperse its seeds far by wind. Trees in varied tropical forests show 8% to 30% of species disperse seeds in this manner (Regal 1977; Mori and Brown 1994). Most seed dispersers consume the fruit or seeds (thereby not killing them), but some passively carry the seeds (Levey et al. 2002). Some ovipositing seed predators are used as pollinators (Pellmyr 1997) and some pollinators are also used as seed dispersers (Dressler 1993; Wallace and Trueman 1995). Such cases imply that natural selection and evolution have forged a beneficial relationship from a one-sided detrimental one. On the other hand, an adult tree may buy time. Its options for success include making seeds have as wide a variety of pollen-donating parents as possible and dropping developing seeds that have not received sufficiently diverse pollen (Willson and Burley 1983; Kenta et al. 2002). Many cohorts of seeds and pollen may be made over many years; trees also are paying dispersers to carry seeds to favorable sites, where species-specific pathogens or insects are unlikely to find them. Last but not least, because wind is inadequate and self-pollinated seeds usually do not survive, animals must accomplish outcrossing pollination. Flowering trees and other plants reward pollinators, both for bringing in and for dispersing pollen, with some extremely rare or important floral resources. These include oviposition sites, antimicrobial floral resins, sweet nectar, high-quality protein in pollen, and emblems of foraging prowess that impress choosy females (Roubik and Hanson 2004).

At the base of this remarkable chain of life are tiny capsules containing genes. The currency in plant reproduction is pollen, one of the smallest natural forest materials. Pollen is protein for pollinators, but it carries genetic information that includes capacity for reproduction, the avoidance of natural enemies, and colonization ability. Exactly the same qualities apply to seeds, except that they result from maternal ovules combined with paternal pollen nuclei.

We believe that every seed has a micro-site where mutualists and the physical qualities of soil, nutrients, mutualist fungi, microbes, water, heat, and light are optimal. Such a site has much in common with a *conspecific* stigma needed for successful pollination in a forest of more than 1000 different plant species. Spatially, the odds are great that a pollen grain or seed will fail. Furthermore, the intricacies of compatibility between pollen and ovule show that the quality of pollinator ecology is key to the success of plant reproduction (Wiens et al. 1987). We also believe that the fate of either a pollen grain or a seed depends on the rareness or distribution of its enemies (Janzen 1983; Bawa 1994; Wright 2002; Terborgh et al. 2002; Olesen and Jordano 2002; Ricklefs 2003; Degen and Roubik 2004). Seeds are normally destroyed, either on the mother plant or on the way to another site, by insects or pathogens. Of course many are consumed by larger animals, which either defecate or drop them where they can grow, or digest them as food. Pollen grains, in parallel, most often nourish pollinator offspring (Thomson 2003), but sometimes they are taken by nonpollinating flower visitors and consumed in situ by thrips, microbes, or larger consumers, both invertebrate and vertebrate. Only rarely does a pollen grain experience mortality after reaching its germination site, although it often is outcompeted by other pollen grains in fertilizing the target ovule; most ovules fail to produce a seed (Mulcahy 1979; Wiens et al. 1987; Thomson 1989).

Because plants are fixed in space, every natural enemy strikes twice with a single blow. Not only is an individual plant affected, so are its neighbors and progeny. Few plants escape herbivores, and these have remarkably precise defenses, chemical, intrinsic (e.g., Arnold et al. 2003) and mutualistic. Among the most impressive defoliators are caterpillars, which normally are adept at circumventing the defenses of a small number of plant species (Janzen 1984, 2003). When a pest outbreak occurs, the caterpillars spread between plants, or the next generation of adults lays its eggs on those plants nearest to the former host. Moreover, like their host plants, when the herbivores are hyperabundant, their natural enemies, including faster-reproducing viruses, locate and then decimate their populations.

To date, the root cause of diversity in an ecological community does not seem to fit the expectations of any single model (see Fig. 1.2 and below); there are too many exceptions, not enough data, and knotty problems with the application of both statistics and theoretical models (Leigh 1999; Hubbell 2001; Wright 2002; Terborgh et al. 2002; Uriarte et al. 2004).

The processes of extinction and colonization, which generate community richness in species, are tied to regional and local conditions (Fig. 1.1; Ricklefs 2004). While the Red Queen provides support for the well-known Janzen-Connell hypothesis, neither is established as the sine qua non of tree diversity in hyperdiverse forests (Condit et al. 1992; Gilbert et al. 1994; Wright 2002; Delissio et al. 2003; Normark et al. 2003; Uriarte et al. 2004). In addition, no convincing evidence exists that the number of tree species drives the species richness of herbivores (Odegaard 2003). The knowledge gap widens considerably when either the history of colonization or the relative tendencies for extinction or speciation are considered (Colinvaux 1996; Morley 2000; Dick et al. 2003; Ricklefs 2003). Nonetheless, the Red Queen demonstrates why it is important that seed and seedling mortality seem highest near the mother tree (Givnish 1999; Leigh 1999). After mortality occurs, surviving seed and seedling density still remain relatively high near a parent tree (Hubbell 1980; Condit et al. 2000). The density-dependence of tree mortality has been clearly demonstrated in data from Malaysia and Panama (Peters 2003). It is appealing to apply so-called negative density-dependent models to populations, because as any city-dweller is already aware, every outbreak has a focus. Diseases, like other natural enemies, are broadcast from their points of origin. Sedentary organisms depend much on the sites to which they are attached, making the distributions of individual species naturally aggregate in space, thus perpetuating the Red Queen and other phenomena dominated by spatiality. Another phenomenon of equal importance concerns the distribution of pollinators and flowers.

1.3 Flowering in the Face of Adversity

Flowers form the basis for plant populations to both purge *lethal mutations* and increase their fund of *genetic variation* available for short-term opportunities or necessities. Those necessities generally involve escape from natural enemies.

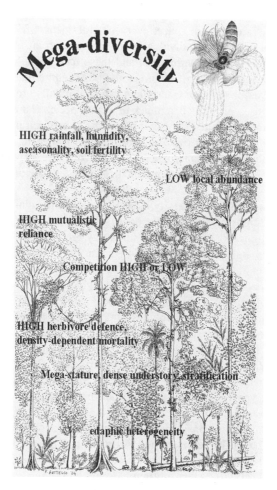

Figure 1.2. The mega-diversity phenomenon, viewing major factors that promote the astonishing richness of life in the ever-wet tropical forests of Borneo (adapted from Givnish 1999; original drawing by F. Gattesco).

Moreover, flowers and their parts represent a commitment in sexual reproduction. Without adverse conditions, and with no genetic mutations, it can be argued that plants would be better served by maintaining a single, female sex that would clonally produce its seeds or offshoots. The *cost of sex* hypothesis raises these points for all organisms (e.g., Kumpulainen et al. 2004). As already mentioned, outbreeding is advantageously avoided in flowering crops (Richards 1997). If asexual breeding or clonal reproduction were favored by natural selection, then flowers and pollen could be done away with altogether. That is certainly not the case for tropical trees, nor for wet tropical forests in particular. For example, our study area at Lambir and a similarly biodiverse area called Yasuní National Park in eastern Ecuador have roughly one-third of their tree species totally committed to sexual reproduction (Valencia et al. 2004b). The male flowers or the female flowers are on different individual trees. No selfing is allowed!

To be at least somewhat rare, or to be dioecious (Bawa 1980), seems an integral part of tropical plant life. Wind pollination will not work in this setting, unless the plant is a grass or gap specialist. Such plants may grow in high densities where there are intermittent winds—a condition also found in secondary growth trees like Neotropical *Cecropia* and Paleotropical *Macaranga,* which grow along river banks and, now, roadways. In these special cases, mutualist ants may be essential, to protect trees from herbivores which easily locate them (Chapters 13–15). A mutualism between ants and *Macaranga* has been traced to seven million years of coevolution (see Chapter 14).

Rarity, in contrast, brings special problems for maintaining beneficial relationships with mutualists, whether as defensive agents, nutritional suppliers, or dispersers of pollen and seeds. In light of the data presented in this book, it would seem that in the case of flowers and seeds rareness in space is characteristic of the understory, or of the non-emergent vegetation (except gap specialists). Rareness in time, often in addition to scarcity, is more common in flowering and fruit production among trees. Considering pollinators, resource rareness in time seems to promote generalization and diversity in interactions (ecological fitting and loose niches), while rareness in space favors specialization and sometimes tight coevolution (see Chapters 4, 9–12).

The *pest pressure hypothesis,* or *escape hypothesis*, (Gillet 1962; Losos and Leigh 2004) has been the basis for much discussion of why so many plant species coexist in a single tropical forest. Its key argument is provocatively simple: Rarity is a product of specialized natural enemies, which frees up space for competitors. The direct complement, although often neglected, is that intelligent or abundant pollinators *permit* plant rarity in general, both in space and in time (Janzen 1970; Regal 1977; Roubik 1993). There may be an added benefit for the plant in a synergism that naturally follows rarity, encouraging animal pollination and plant rareness to evolve together; and yet pollination occasionally involves *exaptations* that, incorporated as pollinator rewards, become less effective as herbivore deterrents (Armbruster 1997). The key concept is also a simple one, found in sexual selection models for animals (West-Eberhard 2003) and plants (Willson and Burley 1983). There must be considerable economic or ecological superiority in an individual that can send its pollen grain, or attract pollen to its stigma, over the many vicissitudes of weather, time, space, and interactions. A 'spatial filter' helps to select the mate, causes genetic remixing among parental gametophytes in seeds and the spreading of new alleles, and encourages rapid and diversified evolution of interactions. The patterns in evolution of flower shape, size, color, smell, and other varied features (Regal 1977; Endress 1994) benefit from the synergism that increased rarity creates. Pollinators are thus selected for spatial and temporal memory, color vision, and olfactory acuity (Dobson 1987; Chittka et al. 1994; Lunau 2000). The predicted results can be considered both from aspects of flowering phenology (see Chapters 3–5) and from qualities of the flowers that permit successful interaction with pollinators (see Chapters 6–12).

When seeds and flowers are both attacked intensely by herbivores, flowers like those of dipterocarps may evolve to be large and fleshy, thereby becoming attractive as oviposition sites or feeding sites for some insects. Most thrips that visit flowers of tropical trees are not their pollinators, and most beetles consume flowers or leaves rather than pollinate them, but both of these animals are important pollinators among dipterocarps and other plants at Lambir Hills. The seeds evolve nutritionally attractive arils or fleshy fruit, and repellents or deterrents, to ensure dispersal by the right animal. However, when the two consumer groups constantly overwhelm tree fecundity, the evolutionary result is thought to be masting, or making resources for the natural enemy populations particularly scarce for long periods of time (but see Herrera et al. 1998). If successful, the masting plant will have to contend with ever more generalized seed predators, which it may then attempt to satiate. We are just beginning to discern whether seed predators are relatively specialized to host trees while the prediction that pollinators often tend to be generalists compared to their flowers (Olesen and Jordano 2002; Roubik et al. 2003) seems upheld. Why should this be so?

1.4 Patterns in Mutualist Biodiversity

The mutualists that are given special domiciles as well as food in plants are products of a long and sustained evolutionary history, well documented in the ant genus *Crematogaster* and the pioneer plant *Macaranga*, in the fig genus *Ficus* and the agaonid wasps, and in other insects and flowers. The diversity and range of different mutualisms demonstrate how finely resources such as a mutualistic genus can be divided among plants, fungi, or organisms receiving the benefits. Often, the mutualism is largely specific to participating species. In the forest of Lambir Hills, our accumulated studies reveal more varied pollinators than are known for any other rain forest, yet variations among interacting species are largely unknown. The understory holds an unusually wide array of organisms in the plant-pollinator mutualism, from fungi to slugs to cockroaches, and from dung beetles to hordes of stinging bees, moths, butterflies, beetles, fruit bats, and squirrels. In contrast, the forest canopy does not display this diversity. Although reduced coevolution between flower visitors and hosts is likely when the host has flowers only once every four or five years, and loose pollination niches beget generalist associations, an ecological fitting seems more likely when the pollinators of the same dipterocarp trees are thrips in Peninsular Malaysia but beetles on Borneo (Sakai et al. 1999b). Many other canopy flowers are visited extensively and seem pollinated by the perennial, colonial stingless bees, or honey bees.

The most abundant tropical forest bees are the eusocial, perennial colonies. There are more than 60 local species of stingless bees in some Neotropical forest, about three times as many as in Lambir Hills (and five times as many genera). In addition, there are up to 50 species of long-tongued traplining bees (most are euglossines) in the same Neotropical forests (Roubik 1990, 1998; Roubik and Hanson 2004), compared to less than a dozen at Lambir Hills, although four genera are found in each. Why is Lambir Hills poor in these key forest pollinators, both solitary and social? The perennial bees possessing large colonies, and the proportion of flowering angiosperm species visited by them, seem comparable in Southeast Asia and the Neotropics (Roubik 1990; Roubik et al. 2003; Wilms et al. 1996). The proportions of different animal pollinators do not differ appreciably in the two of the best-studied tropical wet forests: La Selva, Costa Rica (Kress and Beach 1994), and Lambir Hills, Sarawak.

Nonetheless, if one compares the tall, dense forest at Lambir Hills with outstanding examples of mature, Neotropical forests (see Chapter 16) there are obvious differences in mutualists that pollinate. Regardless of differences in forest stature, total annual rainfall, or its seasonality, there are far fewer pollinating bee species in the Asian tropical forests, and generally far fewer in the tropics than in many warm, temperate areas (Roubik 1989, 1990, 1996; Michener 2000). The understory of the forest at Lambir Hills is packed with immature, emergent trees, so that relatively few individuals are ever in flower there. Flower abundance is low. The sheer numerical and temporal dominance of the honey bees in Asia and the stingless bees is impressed upon all tropical field biologists. These bees are scavengers in forests, especially those with periodic flowering. In equatorial Africa, Asia, and America, bees are extremely fond of sodium and concentrate on removing it from vertebrate skin or the carcasses and feces left by predators. Bees that are stinging or biting pests are so practiced at locating sodium, which has no smell, that they use other vertebrate products to find its source. Their olfactory senses and exploratory behavior are strongly developed, and they rapidly locate floral and other potential resources, especially in the canopy of large forest trees. As alternative resources in times of scarcity, the colonies can use the colony food stores, or forage for non-floral food (Roubik 1989).

The explanations for social forager dominance in tropical forests hinge on floral scarcity. Social bees are generalized in flower choice, are good competitors, and can bring nest mates to resources at any level above the ground, thereby dominating many flowering plants and possibly curtailing evolution of more specialized or seasonal competitors (see Roubik 1996b, 2002; Roubik et al. 1999). An added feature is that the extremely tall emergent trees in Sarawak constitute a foraging environment absent in other tropical forests (Richards 1952; Allen 1956; Momose et al. 1998a). Momose and collaborators advanced a mathematical model stressing the importance of rapid reproduction for plants of small stature in the understory and in gaps, often having specialized mutualists then predicted that the generalist pollinators, honey bees, and stingless bees would be favored for emergent trees. The argument that a protracted periods without flowers further reduces development of a rich bee community in the tropics of Southeast Asia (Roubik 1990) is compatible with this view. However, the negligible response to increased flowering during a general flowering by forest bees specializing on understory flowers has further implications (see Chapter 11). More data on flower use and further attempts at realistic models are needed. There are at least three alternative hypotheses to account for unexpected low species richness: a *lack* of evolution of specialization and de-specialization

(Thompson 1994), a *lack* of dynamic-refugia (Colinvaux 1996), and *isolation* from adequate source populations (see Fig. 1.1; Ricklefs and Schluter 1993).

A key factor in the pollination ecology of the forests of Southeast Asia is thought to be the large interval between general flowering periods and the predominance of *Apis,* usually two abundant species. The prominent migratory honeybees, giant *Apis dorsata* may escape predation from the sun bear *Helarctos* by nesting in *Koompassia*, a legume that is the tallest emergent tree. Honeybees generalize on many flower species. Their migratory pattern of populating the rain forest is testimony to the many loose niches that exist there, promoted perhaps most strongly by the periods of non–General Flowering that may last for 20 years (Wood 1956). These migratory flower visitors may compensate for, and take advantage of, the local poverty of pollinators created by an intermittent, general flowering phenomenon, and a low abundance of flowers in the understory.

Comparative studies can be used to assess the impact of social bee dominance on rain forest pollination ecology. Studies in the Neotropics pointed to the role of stingless bees (where formerly there were no *Apis*) in molding plant breeding systems. Dioecy would evolve to avoid inefficiency of pollen capture and wastage of pollen and ovules (Bawa 1980). The wet forest of Sri Lanka, at Sinharaja, has many tree species with mixed breeding systems (Stacy and Hamrick 2004). That is, self-compatibility and hermaphroditic flowers are not rare, even though honey bees are common. However, stingless bees, having only one species, are rare (Karunaratne 2004). Does the low incidence of dioecy imply *Apis* provides more reliable outcrossing services than small Meliponini?

In summary, consider the pollinators that allow mega-events like the general burst of plant reproduction to occur, and then consider the general rarity of the flowers and plants that are involved. Like the dipterocarp trees that now dominate the forest, there is an intrinsic difference in tropical ecology where General Flowering occurs. Pollinators range from the very large to the very small, with high abundance being roughly compensatory for relatively poor ability to reach a rare or distant target. Fig wasps are extraordinarily small, most less than 1.5 mm; thrips are similar; and the smallest meliponine bees and beetles are slightly larger. Those groups can, however, be spectacularly numerous. Being the only pollinators of many plants, they and other small animals will often determine which plants and animals survive in tropical forests. In contrast, the long-distance bat, bird, or certain insect pollinators, while never very abundant, are equipped, instead, to reach distant flowers and to locate them efficiently (Gribel and Griggs 2002). Both are master chefs of gene combinations. Frugivores, although often not as specialized as pollinators in their interactions, must provide mutualistic services for seeds, virtually all of which need transport away from the parent. Ecology in the rain forest of Southeast Asia functions as it does because the flowers are as rare as their pollinators are scarce, opportunistic, tiny, or specialized. The Borneo forest at Lambir Hills, Sarawak, demonstrates loose niches and limited coevolution, side by side with highly diverse and specific relationships that derive only from extensive coevolution, in a usually benign yet dynamic physical environment that will persist as long as we allow it to do so.

2. The Canopy Biology Program in Sarawak: Scope, Methods, and Merit

Takakazu Yumoto and Tohru Nakashizuka

The mixed dipterocarp forests in Sarawak are among the richest tropical rain forests in the world with almost 1200 species of trees known in an area of only 52 hectares (see Condit el al. 2000). We can easily imagine that the reproductive system, from flowering through pollination and seed dispersal, plays a crucial role in maintaining the rich diversity of plants. However, relatively little study has explored this topic at the community level in the tropical rain forests of Southeast Asia.

Tropical rain forests in this region are also known for the *general flowering* or *mass-flowering* phenomenon in the community (Ashton et al. 1988; Appanah 1993). More than 80% of the canopy trees species bloom during a period of up to 6 months at irregular intervals of 2 to 10 years, with a mean of once every 4 to 5 years. In a general flowering, or GF, so many species bloom in such a short period that a pollinator shortage inevitably occurs for some length of time. Effective pollinators must provide populations that can quickly meet the needs of millions of flowers and hundreds of species.

Previously, tropical rain forests in Southeast Asia were believed to exist in a stable, warm, humid climate throughout the year. However, it has recently been found that rainfall in this region changes from month to month, and with various long-term rhythms (Inoue and Nakamura 1990; Inoue et al. 1993). Among these repeated cycles, the most dominant occurs every 4 to 5 years. Its cause is known as the El Niño-Southern Oscillation (ENSO). This relatively dry period may last for a few to 10 months. An early hypothesis was that drought triggered general flowering in mixed dipterocarp forests in the Malay Peninsula and Borneo Island (Ashton et al. 1988; Ashton 1993; Appanah 1993), but the physiological mechanism that produces flower induction in spite of little environmental change is still under study. How the insects that are the predominant pollinators, and others that consume flowers or seeds, respond to such an unpredictable and drastic change of food availability is one of a number of topics currently being investigated. To understand patterns of community dynamics in a changing environment, a systematic program to monitor plant phenology and insect abundance for at least one episode, from one general flowering event to another, was founded in Sarawak.

The Canopy Biology Program in Sarawak (CBPS), led by the late Professor Tamiji Inoue, Kyoto University (see Plate 1A), began as part of an international cooperative project known as the Long-Term Forest Ecology Research Project at Lambir Hills National Park, Sarawak, organized in 1992 by the Forest Department of Sarawak, Harvard University, Ehime University, Osaka City University, and Kyoto University, and financially supported by the Japanese Ministry of Culture, Sports, Science, and Technology, and by other sources.

The first goal of CBPS was to clarify how unpredictable environmental changes at the global level influence phenology and reproductive systems of forest plants, from flowering through pollination and flower/seed-predation by herbivores to seed dispersal, in the mixed dipterocarp forest. A second aim was to understand how the animals that build mutualistic relationships with plants (pollinators, seed dispersers, and ant-mutualists) as well as antagonistic relationships (flower or seed predators, and herbivores) are affected by the same environmental changes, directly or indirectly, mediated by plant phenology.

The importance of such studies is increasingly clear, but technical difficulties tend to inhibit their progress. The canopy of the mixed dipterocarp forests reaches up to 75 meters aboveground, making any sustained work there quite difficult. Nonetheless, forest tree reproduction occurs mostly in the canopy; CBPS therefore established a canopy observation system that consisted of tree towers and walkways. Using this canopy access system, we conducted bimonthly continuous censuses of both plants and insects along a fixed route in the canopy, aimed at a period long enough to securely encompass at least one GF.

2.1 Location and Vegetation

Lambir Hills National Park is located about 30 kilometers south of Miri, the capital of the Fourth Division, Sarawak, Malaysia, at 4°20'N, 113°50'E (see Fig. 2.1A). The park covers an area of approximately 6,949 hectares. The highest peak in the park is Bukit Lambir, at 465 meters (see Plate 1A). The vegetation can be classified as typical *lowland mixed dipterocarp* forest (Ashton and Hall 1992), dominated by Dipterocarpaceae in the emergent canopy layer. The habitat is further typified by an extraordinarily rich diversity of tree species (Lee et al. 2002).

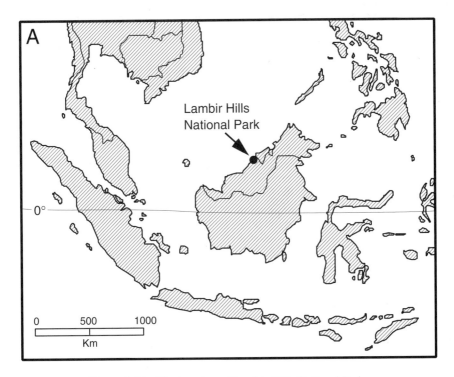

Figure 2.1A. The location of Lambir Hills National Park.

In 1992 and 1993, we established an 8 hectare (200 by 400 m) plot, or Canopy Biology Plot (CBP), at an elevation of 150 meters to 200 meters above sea level (see Fig. 2.1B). The plot includes *humult* and *udult* soils (sandy clay, light clay, or heavy clay in texture), several ridges and valleys, a closed stand (mature-stage forest), and canopy gaps. At the central part of the plot we made two tree towers on neighboring ridges and connected them by walkways of approximately 300 meters long.

In March 2000, a canopy crane was installed in a nearby forest with comparatively flat topography. A permanent plot, or, Crane Plot (CP), was established at 400 meters to 500 meters northeast of the CBP. The soil in CP is poorer and trees are lower than at CBP; together, these facilities greatly increase the accessibility of the forest canopy.

2.2 The Canopy Observation System

In the early 1980s, pioneer work in tropical rain forests revealed that the canopy is the center of most plant activities (Sutton et al. 1983; Whitmore 1984), both in production and reproduction. Animal abundance (mainly insects) also displays

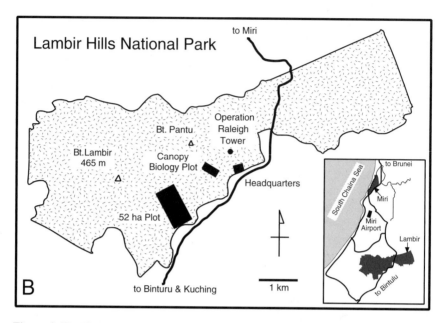

Figure 2.1B. The site of the canopy biology plots in Lambir Hills National Park (*inset* Miri and surrounding area).

its peaks there (Erwin 1983, 1988; Stork 1987a, 1987b, 1988a, 1988b; Rees 1983), regulated by factors such as distribution of food and shelter. These findings necessitated a concerted effort to approach the high canopy, at least 40 meters aboveground. Several access methods to the canopy have been developed: for example, ascent by tower and walkway (see Plate 3A, C-E, G; Mitchell 1982), by crane (see Plate 3B, F; Illueca and Smith 1993; Joyce 1991), by rope (see Plate 4D; Mitchell 1982; Perry 1978, 1984; Dial and Tobin 1994), and by raft (Hallé and Pascal 1992).

In CBP, we combined tree towers and aerial walkways for the long-term observation described above. An existing technology, such towers and walkways are constructed in various places in Southeast Asia (Pasoh, Peninsular Malaya; Poring, Sabah; Semengoh, Sarawak among others). We modified previous methods to accommodate long-term use and integrated devices in order to cover a wide area of canopy. We constructed two tree towers with heights of 50 meters and connected them by nine spans of aerial walkway that pass through various canopy layers. The total length of walkways spanned 300 meters, making the total system one of the largest in the world.

Tree tower 1 (T1) was constructed around an emergent tree of *Dryobalanops lanceolata* (Dipterocarpaceae), 70 meters in height and 1.5 meters in diameter at breast height on a gently sloping ridge (see Plate 3A). Eleven wooden platforms and 11 flights of stairs are set around the trunk of the tree, and observers walk in a spiral up to the top wooden platform 33 meters above the ground.

Pillars are made with ironwood, or, Belian (*Eusideroxylon zwageri*, Lauraceae). Above the top platform of T1 we made three emergent platforms among the branches at 45, 55, and 65 meters aboveground, to which aluminum ladders were connected. Tree tower 2 (T2) was constructed to one side of a canopy tree, *Dipterocarpus pachyphyllus* (Dipterocarpaceae), at 48 meters high and 1.36 meters in diameter (see Plate 3C). The top platform of T2 is 16 meters higher than that of T1. The top platform higher than the neighboring trees provides a clear view in all directions. To reduce the total weight of the tower we use aluminum ladders instead of wooden steps to climb from one platform to the next. The two tree towers are connected by nine spans of aerial walkway (see Plate 3G and Fig. 2.2, 2.3).

We used trunks of emergent or canopy trees as piers. The length of a span ranges from 25 meters to 54 meters, depending on the distribution of pier trees. The walkways pass through canopy layers from 15 meters to 35 meters above the ground. The structure of one span of a standard walkway with a length of 30 meters is shown in Fig. 2.3. Observers walk on wooden boards set on horizontal aluminum ladders, suspended by steel cables from the two carrying cables. The two handrail cables are also connected to aluminum ladders. A safety cable is fixed on one side of the handrails, so that users connect at least one safety line of the two on their harness while on walkways. Carrying and handrail cables are anchored, not directly on tree trunks, but on shock-absorbing wooden buttresses (5 by 5 by 50 cm) set around the trunk, to reduce harmful effects on

Figure 2.2. Walkways through the canopy layer of the forest.

Figure 2.3. A researcher observing plants from the walkway.

the trees. Platforms are made on pier trees to allow movement from one walkway to the next.

One fatal weak point in the use of walkways is the risk of a tree falling. In October 1993, a tree trunk (40 cm diameter, 10 m length, 5 ton estimated weight) hit the walkway. Fortunately only the tip of the falling tree struck, and that only loosened some cables. If larger trunks fall, there is a possibility that the whole walkway system could collapse. In addition, during the severe drought in 1998 one of the pillar trees died in place. Since we could not predict when the dead tree would fall, we removed the two spans of the walkway tied to the tree. We could not prevent the construction of a walkway from causing some damage to the trees.

The canopy crane is 80 meters tall (to the base of the observer's gondola), with a jib length of 75 meters (see Plate 3B, F), made in Germany. Since the tallest tree in the CP is about 55 meters, the gondola can be much higher than the canopy. This extreme height causes some difficulty in control of the gondola for ecological observations, although there are advantages when it is used at full height to verify remote sensing data, take certain samples, or make other measurements. We have two kinds of gondolas of different sizes: one for three people and another for one person. The smaller one is used to dive into small canopy gaps. The crane itself provides observation stages at three levels (20, 40, and 60 m aboveground) along the crane tower, and it has an elevator to reach the control cabin (see Plate 3B). Usually, researchers in the gondola control the operation of the crane. Support for the canopy crane was provided by a part of the CREST (Core Research for Evolutionary Science and Technology) Project of JST (Japan Science and Technology Corporation).

2.3 What Have We Done?

More than 1000 tree species were expected to coexist at relatively low densities in the mixed dipterocarp forests. Although inventory work is usually difficult because flowers are not easily obtained, owing both to canopy height and an unpredictable flowering tempo, the canopy access system and the long-term project enabled us to reliably collect plant and insect specimens. The plant specimens are maintained at the herbarium of the Sarawak Forest Department and are distributed to herbaria at the Kyoto University Museum in Japan, Kew Botanical Garden in the United Kingdom, and other locations. Many new species of plants and insects, including one new plant genus, have been revealed in the course of canopy work.

The GF phenomenon in Southeast Asia has provided a framework for studying many scientific questions. Which environmental cues induce the general flowering, how many plants join the general flowering, and how do the pollinators respond? Other studies concern why all plants in rain forests do not bloom during ENSOs, and why different plants show different phenological patterns. To answer these questions we have studied temporal changes in the forests and carefully recorded various plant-animal interactions. Using the tree towers, the walkways, and the canopy crane, we have monitored plant phenology twice a month for more than 10 years. The data give us a complete picture of general flowering, which is one of the most spectacular phenomena in the tropics. We constructed some new hypotheses based on the new data and examined several already well-known ideas concerning the GF. Traps set to monitor insect population dynamics revealed fluctuations in the population size of some insects in response to general flowering, but no such changes in others. Some seed predators were observed only in a GF. In addition, species composition and diversity seem to differ among flowering events of sequential GFs.

Although pollination systems of the whole forest have rarely been documented in tropical regions, we succeeded in identifying general pollinator-plant relationships at the community level (Momose et al. 1998). In addition, because little research activity has pursued the pollination ecology in this region, we immediately obtained several original findings on pollination by cockroaches (Nagamitsu and Inoue 1997a), by beetles (Nagamitsu et al. 1999a; Sakai et al. 1999b), especially dung-beetle pollination (Sakai and Inoue 1999), by a gall midges (Sakai et al. 2000), by fig-wasps (Harrison et al. 2000), and by birds (Yumoto et al. 1997; Yumoto 2000). Certain specific groups of plants were documented intensively (*Gnetum*: Kato and Inoue 1994; Kato et al. 1995b; Zingiberaceae: Sakai et. al. 1999a; Loranthaceae: Yumoto et. al. 1997; *Durio*: Yumoto 2000). These findings revealed specific pollinator-plant interactions that had never been documented. Among such pollinating animals, several intensive studies have been done for bees (Roubik et al. 1995, 1999; Nagamitsu and Inoue 1997b, 1998; Nagamitsu et al. 1999b; Itioka et al. 2001a).

Early in our project we gave priority to an overreaching theme, seeking to clarify how periodic environmental change influences plants and animals. This was measured directly through the simultaneous observation of environmental changes, plant phenology, and animal seasonality. We also confirmed that low temperatures in 1996 might act as the local trigger of general flowering (Sakai et al.1999c), and even induce the migration of giant honeybees (*Apis dorsata*) from considerable distances (Itioka et al. 2001a) and bring about other changes among anthophilous beetles and other insects (Kato et al. 2000). On the other hand, a severe drought brought by ENSO caused different kinds of disturbances, including death of many forest trees (Nakagawa et al. 2000) and local extinction of populations of mutualist figwasps (Harrison 1999b).

The interaction between the atmosphere and the forest canopy is becoming a hot issue among certain scientists (Ozanne et al. 2003). Such studies using a canopy crane have been focused on gas exchange, in addition to GF. Since the forest in northern Borneo Island is truly aseasonal, without clear dry seasons, we could expect higher biological productivity than in other tropical forests. The canopy activity in gas exchange is also expected to be high, although few observations have been made in this area. Measurements on carbon CO_2 and H_2O have continued since 2001. Their flux observations will be compared with estimates made from measurement of tree growth.

2.4 The Present and Future of Lambir Hills

International networks for canopy studies have developed in recent years. In cooperation with more than 10 crane sites, comparative studies among different forests are being integrated as activities of the Global Canopy Program (Mitchell et al. 2002). We can expect several cross-comparisons of tropical forests, with Lambir Hills established as a reference for the most humid and aseasonal conditions.

In 2002, Lambir Hills was selected as one of the four research sites of the project known as Sustainability and Biodiversity Assessment on Forest Utilization Options, organized by the Research Institute for Humanity and Nature in Japan. That project considers what effects human activities have on forest ecosystems and biodiversity; it places emphasis on the development of sustainable management systems. Canopy processes are of importance in this project because they include mechanisms crucial to maintaining sound ecosystems. There is increasing global awareness and concern that these mechanisms might be lost with an accelerating loss of biodiversity. Therefore, the research areas have been broadened to include secondary forests under different intensity of human disturbances, logged forests, plantations, villages, and nearby cities. By cooperation with social scientists and anthropologists, the project aims to elucidate the social, economic, and cultural factors that are responsible for the recent changes in forest-use patterns on regional as well as global scales. A goal is now to establish ecological and economic models for sustainable forest use and physical planning, in much the same vein as urban planning. As for the interactive function between the canopy and Earth's atmosphere, more projects are forthcoming. Those projects will help to evaluate the effect of climatic fluctuations on tropical rain forests in Southeast Asia, and the canopy facilities will be extremely useful.

3. Soil-Related Floristic Variation in a Hyperdiverse Dipterocarp Forest

Stuart J. Davies, Sylvester Tan, James V. LaFrankie, and Matthew D. Potts

A 52 hectare permanent research plot was established in Lambir Hills National Park to enable long-term study of factors controlling the origin and maintenance of tree diversity. In this chapter we summarize some of our recent work on the relationships between floristic variation and edaphic heterogeneity in the Lambir forest. First, we provide a general description of the floristic composition of this hyperdiverse forest. Second, we use a detailed survey of soil chemistry to test whether floristic composition changes in relation to edaphic characteristics. We also assess the extent to which individual species have non-random distributions in the forest with respect to edaphic heterogeneity. Finally, to investigate the influence of habitat variation on floristic diversity we compare our results from the heterogeneous forest at Lambir Hills with a more homogeneous forest in Peninsular Malaysia.

3.1 Introduction

The lowland forests of Northwest Borneo are among the most floristically diverse forests in the world (Davies and Becker 1996; Lee et al. 2002). Individual hectares of forest in this area often have in excess of 275 species that have a diameter equal to or greater than 10 cm at breast height (dbh), a species richness that is matched only by forests in western Amazonia (Turner 2001).

Within diverse tropical forests there is growing evidence that variations in soil nutrients, soil water, and topographic position constrain the distribution of tree species and may thereby contribute to the coexistence of large numbers of species (Duivenvoorden 1996; Clark et al. 1998; Potts et al. 2002). In diverse lowland forests in Borneo, a series of studies dating back to the 1960s have demonstrated significant spatial variation in forest composition (Ashton 1964; Austin et al. 1972; Newbery and Proctor 1984; Baillie et al. 1987; Davies and Becker 1996). Several of these studies have suggested that variation in floristic composition is more strongly related to soil nutrient availability, particularly phosphorus (P) and magnesium (Mg), than to other habitat features, such as topographic effects on water availability (Baillie et al. 1987; Potts et al. 2002).

3.2 Study Description

Lambir Hills National Park, Sarawak, includes lowland mixed dipterocarp forest and kerangas forest. The lowland forest at Lambir is the most diverse forest in tree species recorded for the Palaeotropics (Ashton and Hall 1992; Davies and Becker 1996; Lee et al. 2002). Lambir receives approximately 3000 mm of rainfall per year, with all months averaging greater than 100 mm (Watson 1985), but periodic short-term droughts may have a significant impact on vegetation in this region (Becker 1992; Delissio and Primack 2003; Potts 2003).

The soils and geomorphology of the Lambir research plot are described in more detail in Lee et al. (2002) and in Chapter 17. In brief, the Lambir hills consist of a series of cuestas comprised of Neogene sediments, dominated by sandstone (Liechti et al. 1960). These soft erodable sediments overlie the calcareous Setap shale formation of the lower Miocene, which is exposed along the southern boundary of the park. The soils of Lambir are derived from these inter-bedded sandstone and shale parent materials. Sandstone-derived soils are humult ultisols, with a surface horizon of loosely matted and densely rooted raw humus, low nutrient status, and low water retention capacity (Ashton 1964; Baillie et al. 1987; Ashton and Hall 1992; Davies et al. 1998). Shale-derived soils are easily crumbled, relatively fertile, clay-rich udult ultisols with high water-holding capacity, with a shallow leaf-litter layer on top. These two ultisols represent extremes in the range of lowland soils overlying sediments in northwest Borneo. Based on a qualitative assessment of the soils of the 52 ha plot, Davies et al. (1998) estimated that shale-derived soils covered about 25% of the plot, mostly in the low-lying gullies. Humult ultisols occur on slopes and ridges. In this study we conducted a quantitative assessment of the soil chemistry in the 52 ha plot. Due to the possibility that many soil factors might be correlated with topography, and consequently water availability (Yamakura et al. 1995, 1996), we include topographic variation in our analyses of habitat-related floristic variation.

In 1991, a research project was initiated in Lambir to monitor all woody plants that are equal to or greater than 1 cm dbh in 52 ha of forest. The methods for this project followed similar studies coordinated by the Center for Tropical Forest Science, performed on Barro Colorado Island, Panama (Hubbell and Foster 1983), and Pasoh Forest Reserve, West Malaysia (Manokaran et al. 1990; Ashton et al. 1999). All trees in the specified dbh range (excluding palms) were tagged, mapped, identified, and measured at their diameters (Condit 1998). A full description of the floristic composition and stand structure of the forest is presented in Lee et al. (2002). The complete stand tables for all species encountered in the plot have recently been published (Lee et al. 2003) and are summarized below.

Soil analyses were performed with 501 samples, each a composite of three randomly collected cores of 5 cm to 15 cm deep from a single location. One sample was taken from within each 40 m by 40 m area of the plot (N=338). An additional 163 samples were taken along transects positioned to traverse apparently abrupt transitions in soils (areas of high topographic heterogeneity) and in the area known to represent the transition between the principal soil types within the plot.

Soils were analyzed at the Agriculture Research Center, Semengoh, Sarawak, following the methods of Chin (1993): air dried and then ground to pass through a 2 mm mesh sieve. Total soil C was analyzed using a dry combustion technique. Total N was determined using Kjeldahl digestion. Total soil P was determined following extraction with perchloric and sulfuric acids. Exchangeable soil P concentrations were determined following extraction with ammonium fluoride and hydrochloric acid (Bray-2 method). Exchangeable cation (K, Ca and Mg) concentrations were determined following extraction with neutral ammonium acetate. Total and extractable nutrient concentrations were measured on an inductively coupled plasma spectrophotometer.

Multivariate analyses were conducted to investigate the relationships between spatial variation in floristic composition and habitat variation. Two characteristics of habitat were assessed: soil chemistry, as described above, and topographic position as measured by mean quadrat elevation.

The coarse scale soils data were kriged using universal kriging on a 20 m^2 grid to produce estimates of soil nutrient values across the plot (Cressie 1991). The kriged soils data and existing elevation data were then standardized and normalized. K-means clustering was then used to identify four distinct habitat classes (see Fig. 3.1; Ihaka and Gentleman 1996). The significance of species' association with these four habitat classes was then tested using the Poisson cluster method (Diggle 1983; Plotkin et al. 2001).

The relationships between floristic composition and habitat (soil chemistry and topography) were investigated using ordination with canonical correspondence analysis (CCA) on 200 0.25 ha quadrats (McCune and Mefford 1999). Habitat values for the 200 samples were the means of smaller scale samples. Mantel and Partial Mantel analyses were conducted to test the relationships between soil chemistry and elevation, and floristic composition among 200, 0.25 ha quadrats (Legendre and Legendre 1998). Mantel analyses involved computing separate distance matrices for soil chemistry, mean elevation, and floristics data (Casgrain and Legendre 2001). Partial Mantel analysis was used to test for the relative strength of the relationship between two of the distance matrices (flo-

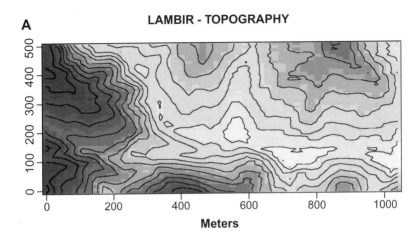

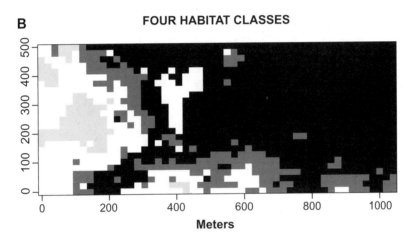

Figure 3.1. (**A**) Topographic maps of the 52 hectare plot at Lambir Hills National Park (partly adapted from Yamakura et al. 1995). Contours are at 10 meter intervals. (**B**) Map of four habitat classes derived from the K-means cluster analysis of soil chemical properties and topographic elevation. Habitat classes from Table 3.3 are humult soils (2 lightest areas) and udult soils (2 darkest areas).

ristics and soils) while controlling for the distances in the third matrix (elevation). Significance of the relationships was tested by bootstrapping the data.

3.3 Floristic Diversity

The 52 ha plot included approximately 356,000 trees having a dbh of equal to or greater than 1 cm (mean=6856 trees/ha). There were 1173 species in the plot

Table 3.1. Floristic composition of the 52 ha plot at Lambir, listing five families and genera with the greatest total basal area, number of species, and tree number. Please refer to Lee et al. (2002) for complete lists.

Family	Basal area m²(%)	Family	Species N (%)	Family	Trees N (%)
Dipterocarpaceae	918.4 (41.6)	Euphorbiaceae	125 (10.7)	Dipterocarpaceae	54089 (15.6)
Burseraceae	146.5 (6.6)	Dipterocarpaceae	87 (7.4)	Euphorbiaceae	51556 (14.9)
Euphorbiaceae	144.8 (6.6)	Lauraceae	78 (6.6)	Burseraceae	23118 (6.7)
Anacardiaceae	133.1 (6)	Rubiaceae	59 (5.0)	Anacardiaceae	19381 (5.6)
Myrtaceae	99.64 (4.5)	Annonaceae	54 (4.6)	Rubiaceae	17417 (5.0)

Genus	Family	Basal area m² (%)	Genus	Family	Species N (%)	Genus	Family	Tree number
Shorea	Dipterocarpaceae	467.8 (21.2)	*Shorea*	Dipterocarpaceae	55 (4.7)	*Shorea*	Dipterocarpaceae	23813 (6.9)
Dipterocarpus	Dipterocarpaceae	213.5 (9.7)	*Syzygium*	Myrtaceae	49 (4.2)	*Dryobalanops*	Dipterocarpaceae	11453 (3.3)
Dryobalanops	Dipterocarpaceae	164.03 (7.4)	*Diospyros*	Ebenaceae	34 (2.9)	*Dacryodes*	Burseraceae	11252 (3.2)
Santiria	Burseraceae	60.67 (2.8)	*Litsea*	Lauraceae	29 (2.5)	*Diospyros*	Ebenaceae	9471 (2.7)
Gluta	Anacardiaceae	60.03 (2.7)	*Xanthophyllum*	Polygalaceae	25 (2.1)	*Vatica*	Dipterocarpaceae	8889 (2.6)

representing 81 families and 286 genera. Total basal area for all trees in the plot was approximately 2250 m^2 with a mean of 43.30 m^2/ha. Almost 80% of the trees were less than 5 cm dbh, and there were only 1372 trees (~26 trees/ha) greater than 60 cm dbh.

The Dipterocarpaceae dominated the composition of the plot with 54,089 trees (15.6% of the total) and total basal area of 918.41 m^2, or, 41.6% of the total (see Table 3.1). The 87 species of dipterocarps made it the second-richest family after the Euphorbiaceae (including Phyllanthaceae and Putranjivaceae) with 125 species. The Euphorbiaceae with almost 52,000 trees contributed almost as many trees as the Dipterocarpaceae, but considerably less basal area (6.6%). The Lauraceae, Rubiaceae and Annonaceae were also exceptionally species-rich with equal to or greater than 54 species. In total, 21 families had equal to or greater than 20 species in the plot. The Burseraceae and Anacardiaceae were among the four most important families both in tree number and in basal area contribution, having 5% to 7% in each category (see Table 3.1).

Shorea was the dominant genus in the plot, as listed in Table 3.1, in terms of number of species (55 species, 4.7% of all species), number of stems (23,813 trees, 6.9% of all trees), and total basal area (467.8 m^2, 21% of total basal area). *Syzygium, Diospyros, Litsea* and *Xanthophyllum* were also exceptionally species-rich with equal to or greater than 25 species, and over 20 genera had more than 10 species in the plot. Two other dipterocarp genera, *Dipterocarpus* (9.7%) and *Dryobalanops* (7.4%), were the second- and third-most important genera in terms of basal area contribution; *Dryobalanops* had the second-greatest number of trees (3.3%). Several genera of small or subcanopy trees had substantial numbers of trees in the forest (e.g., *Diospyros, Vatica* and *Macaranga*).

The ten most important species by total basal area and stem numbers are listed in Table 3.2. The emergent dipterocarp, *Dryobalanops aromatica*, was the most important tree species, with over 10,000 trees (3.0%) and basal area of 152.8 m^2 (6.9%). *Dipterocarpus globosus* also contributed significantly to basal area, and together with *D. aromatica* accounted for greater than 13% of total basal area. The 6 species with the largest contribution to basal area were all dipterocarps, as were 13 of the top 20 basal area contributors. Individual dipterocarp species were less dominant, when measured by stem density (see Table 3.2). There were several very common understory non-dipterocarps, the most important of which was the legume, *Fordia splendidissima*.

3.4 Variation Related to Edaphic Heterogeneity

The plot includes almost 140 meters of elevational change from the highest to the lowest point (see Fig. 3.1); soil and topographic variation revealed four habitat classes. Nutrient-poor humult soils are in the upper northern sections of the plot (habitats A and B), and the relatively nutrient-rich udult soils derived from shale are in the lower parts of the plot (habitats C and D). The four habitat classes differ significantly in topographic position and soil chemical concentra-

Table 3.2. The 10 most important tree species in the Lambir Hills 52 ha plot.

Tree Abundance

Genus, species & author	Family	Frequency N (%)
Dryobalanops aromatica Gaertn. f.	Dipterocarpaceae	10503 (3.0)
Allantospermum borneense Form.	Simaroubaceae	7368 (2.1)
Vatica micrantha V. Sl.	Dipterocarpaceae	6261 (1.8)
Fordia splendidissima (Bl.) Buijsen	Fabaceae	3717 (1.1)
Gluta laxiflora Ridl.	Anacardiaceae	3646 (1.1)
Whiteodendron moultonianum (W.W. Sm.) v.	Myrtaceae	3387 (1.0)
Shorea beccariana Burck	Dipterocarpaceae	3361 (1.0)
Shorea laxa V. Sl.	Dipterocarpaceae	3328 (1.0)
Dipterocarpus globosus Vesq.	Dipterocarpaceae	3311 (1.0)
Dacryodes expansa (Ridl.) Lam	Burseraceae	3287 (1.0)

Basal Area

Genus, species & author	Family	Basal area m^2(%)
Dryobalanops aromatica Gaertn. f.	Dipterocarpaceae	152.75 (6.9)
Dipterocarpus globosus Vesq.	Dipterocarpaceae	137.91 (6.3)
Shorea beccariana Burck	Dipterocarpaceae	59.55 (2.7)
Shorea laxa V. Sl.	Dipterocarpaceae	47.96 (2.2)
Shorea acuta Ashton	Dipterocarpaceae	41.12 (1.9)
Shorea smithiana Sym. complex	Dipterocarpaceae	39.23 (1.8)
Allantospermum borneense Form.	Simaroubaceae	34.14 (1.6)
Whiteodendron moultonianum (W.W. Sm.) v.	Myrtaceae	31.83 (1.4)
Shorea curtisii Dyer	Dipterocarpaceae	27.86 (1.3)
Elateriospermum tapos Bl.	Euphorbiaceae	22.03 (1.0)

tions (see Table 3.3). Total soil P concentrations, and extractable Ca and Mg were significantly higher on the udult soils, and pH was significantly lower on udult soils.

Canonical correspondence analysis (CCA) maximizes the correlation between floristic and environmental variation. The first three axes of the CCA explained 16.4% of floristic variation. Total P and extractable Mg and Ca concentrations were strongly positively correlated with the first ordination axis, while mean elevation was strongly negatively correlated with that axis (see Fig. 3.2 and Table 3.4). Cation concentrations were moderately negatively correlated with the second ordination axis.

Species prominent in the two extremes of the habitat gradient are illustrated in the species bi-plot resulting from the CCA ordination (see Fig. 3.3). The species common on humult soils included the dipterocarps, *Dryobalanops aromatica* and *Dipterocarpus globosus* (the two most common species in the plot), *Whiteodendron moultonianum* and *Allantospermum borneense* (two species characteristic of humult soils throughout northwest Borneo). The common udult

Table 3.3. Mean elevation, total phosphorus, pH and extractable Mg and Ca concentrations for the four habitat classes defined by the K-means cluster analysis in Fig. 3.1; analysis based on 1300 quadrats of 20 × 20 m. Values in parentheses are standard errors based on N quadrats per cluster; significant differences among habitat variables indicated by different letters after standard errors.

Habitat Cluster	N	Total P	pH	Mg	Ca	Mean Elevation (m)
A	766	43.7 (0.7) d	4.64 (0.00) a	0.12 (0.00) d	0.21 (0.00) c	193.8 (0.7) a
B	184	66.5 (2.3) c	4.41 (0.01) b	0.15 (0.01) c	0.22 (0.00) c	183.1 (1.4) b
C	270	103.3 (2.2) b	4.32 (0.01) c	0.19 (0.01) b	0.3 (0.00) b	152.6 (1.4) c
D	80	133.6 (4.1) a	4.43 (0.01) b	0.7 (0.01) a	0.52 (0.01) a	138.8 (1.4) d

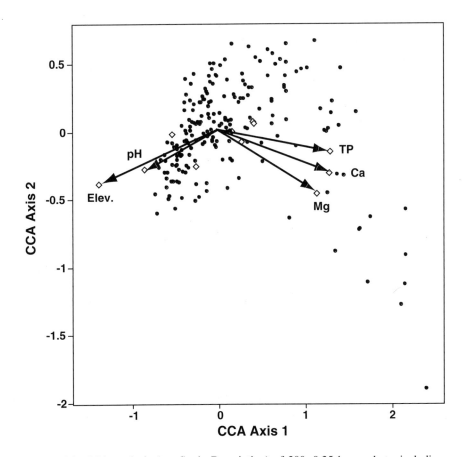

Figure 3.2. CCA analysis (see Study Description) of 200, 0.25 ha quadrats, including 776 species with equal to or greater than 50 trees. Arrows indicate direction and relative influence of habitat variables on floristic variation. TP indicates total P concentrations. Habitat scores were multiplied by 3.

Table 3.4. Intra-set correlations of habitat variables with the first three axes of the CCA, presented in Fig. 3.2. Intra-set correlations reflect relative importance of habitat variables in explaining plot floristic variation. TP and AP are total and available phosphorus, respectively.

Variable	CCAA Axis1	CCA Axis2	CCA Axis3
pH	−0.508	−0.278	−0.547
N	0.238	0.06	0.02
C	−0.157	−0.256	0.212
TP	0.765	−0.147	0.423
AP	0.23	0.076	0.392
Ca	0.758	−0.307	−0.248
Mg	0.673	−0.461	−0.384
K	0.09	0.002	−0.179
Na	0.156	−0.075	−0.312
Mean elevation	−0.823	−0.385	0.279

soil species included the dipterocarps *Dryobalanops lanceolata* and *Hopea dryobalanoides*, and the legume, *Millettia vasta*.

Because soil chemistry and topographic position were strongly correlated within the plot, we used Mantel and Partial Mantel tests to investigate their separate influence on floristic composition (Table 3.5). The analysis indicated that effects of soil chemistry were marginally stronger.

3.5 Species Spatial Aggregation on the Edaphic Gradient

The plot included 764 species with equal to or greater than 50 individuals. Of those species, 663 (86.8%) had spatial distributions significantly biased with respect to the habitat gradient in the plot. Tests revealed significantly higher or lower abundance than expected in one or more of the four habitat classes defined above (see Table 3.6). Only 101 species (13.2%) had distributions that were not biased with respect to the four habitats.

3.6 Habitat Heterogeneity and Diversity

Because Lambir forest is among the most diverse forests in the world, a single forest plot includes more species than any of the 14 other plots in the Center for Tropical Forest Science network of large-scale plots (Condit et al. 2003). In our 52 ha study plot there were 356,501 trees including 1173 tree species. To summarize, the Euphorbiaceae (125 species) and the Dipterocarpaceae (87 species) were the most diverse families. Dipterocarpaceae dominated the forest

Humult Soil Species:

Allantospermum borneensis	*Shorea acuta*
Dacryodes rostrata	*Shorea laxa*
Dipterocarpus globosus	*Shorea quadrinervis*
Dryobalanops aromatica	*Vatica micrantha*
Elateriospermum tapos	*Whiteodendron moultonianum*
Gluta laxiflora	

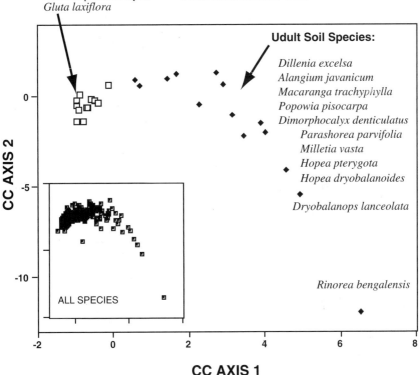

Udult Soil Species:

Dillenia excelsa
Alangium javanicum
Macaranga trachyphylla
Popowia pisocarpa
Dimorphocalyx denticulatus
Parashorea parvifolia
Milletia vasta
Hopea pterygota
Hopea dryobalanoides

Dryobalanops lanceolata

Rinorea bengalensis

Figure 3.3. Species biplots for the first two axes of the CCA; full figure shows common, characteristic species of two plot extremes; inset includes all species.

Table 3.5. Summary of Mantel and Partial Mantel tests for plot relationships between floristic composition, soil chemistry and mean elevation; composition based on stems in 0.25 ha quadrats for 777 species with more than 50 stems. Mean includes all measured elevations within each 0.25 ha quadrat. Values above the diagonal are pairwise standardized Mantel statistics (r_M scores); in bold below are Partial Mantel statistics.

	Species Compositi	Soil Chemistry	Mean Elevation
Species Compositio	—	0.559**	0.539**
Soil Chemistry	**0.444****	—	0.401**
Mean Elevation	**0.415****	**0.143****	—

Table 3.6. Association of 764 species with four habitats defined in Fig. 3.1. Species were tested for significant habitat associations using Poisson cluster tests. Analyses included species with equal to or greater than 50 trees. The list gives the number and percent of species significantly positively associated with habitats A through D; Negative = no positive associations, Neutral = no positive or negative habitat associations.

Habitat Association	Species N (%)
Neutral	101 (13.2)
Positive A	286 (37.4)
Positive B	57 (7.5)
Positive C	107 (14)
Positive D	45 (5.9)
Positive A and B	3 (0.4)
Positive B and C	29 (3.8)
Positive B and D	1 (0.1)
Positive C and D	40 (5.2)
Negative A	6 (0.8)
Negative B	13 (1.7)
Negative C	12 (1.6)
Negative D	52 (6.8)
Negative B and D	1 (0.1)
Negative C and D	11 (1.4)

with 42% of the basal area and 16% of the trees, and *Shorea* was the most important genus with 55 species and the highest basal area and stem number. Our results demonstrated that habitat diversity may contribute significantly to the coexistence of such an extraordinary species number in the Lambir Hills forest.

The northwest Borneo flora, including Sarawak, Sabah, and Brunei, includes approximately 5000 tree species (E. Soepadmo and P.S. Ashton, personal communication), thus the Lambir plot included almost one quarter of this estimated regional tree flora. The conservation value of this area is further enhanced by the considerable number of species from Lambir that are endemic to the region known as the Riau pocket. This region includes the area of Borneo north of a line between Pontianak and Kota Kinabalu (which includes Lambir), coastal Perak, the Riau Archipelago, and the east coast of Peninsular Malaysia (Corner 1960; Ashton 1995; Morley 2000). Many species restricted to the humult ultisols in Lambir are endemic to this area, including *Dipterocarpus globosus*, *Shorea acuta*, *S. laxa* and *Gluta laxiflora*.

Floristic gradients within the Lambir plot were correlated with and possibly determined by both soil chemistry and variation in topographic position. This result is similar to earlier work on floristics and soils in northwest Borneo (Ashton 1964; Baillie et al. 1987; Potts et al. 2002). Although the Partial Mantel tests suggested that floristic variation was more strongly correlated with soil chemistry than with elevation, the difference was relatively small. The multitude of possible casual effects on tree distributions, indicated by low explanatory power of the first three axes in the CCA model, suggests that correlative studies need to be supplemented with mechanistic experimental studies on factors influencing the spatial distributions of these tree species. In the Lambir context, that might include a test of the hypothesis that species have performance advantages on the habitats in which they are aggregated. Palmiotto et al. (unpublished manuscript) grew seedlings of six tree species on humult and udult soils. Species either were naturally aggregated on the low-fertility sandy humult soils (*Dryobalanops aromatica, Shorea laxa,* and *Swintonia schwenkii*), aggregated on the moderate-fertility udult ultisols (*Dryobalanops lanceolata* and *Hopea dryobalanoides*), or distributed across both soils (*Shorea balanocarpoides*). Thus the hypothesis that species have performance advantages specific to the habitat on which they aggregate was supported in four of the five cases. Furthermore, to test whether nutrient availability influences performance, P was added to seedlings on both soils, but had no effect. Additional studies with more experimental variables are needed to better understand what controls spatial distributions of Lambir's tree species.

Table 3.7. Comparison of species richness of large tree genera in two large forest plots in Malaysia. The Lambir plot has high habitat heterogeneity and 1173 species; the Pasoh plot has low heterogeneity and 822 species.

Genus	Lambir (N species)	Pasoh (N species)	Difference
Lambir Hills			
Shorea	55	14	41
Xanthophyllum	25	10	15
Litsea	29	14	15
Madhuca	14	2	12
Diospyros	34	23	11
Santiria	17	7	10
Macaranga	15	6	9
Ficus	21	12	9
Calophyllum	16	8	8
Gonystylus	8	1	7
Palaquium	13	6	7
Hydnocarpus	6	0	6
Lophopetalum	7	1	6
Urophyllum	8	2	6
Durio	9	3	6
Garcinia	23	17	6
Pasoh			
Clerodendrum	0	3	3
Psydrax	0	3	3
Trivalvaria	0	3	3
Mangifera	4	13	9

More than 85% of the more abundant species in the Lambir plot had spatial distributions biased with respect to soil chemistry and topographic position. Although the results of Poisson cluster tests are not directly comparable among plots with different habitat distributions (R. Condit, personal communication), this estimate of habitat association is substantially higher than was estimated for the more homogeneous forest on Barro Colorado Island in Panama (Harms et al. 2001).

A comparison of diversity patterns in the Lambir plot and a similar 50-ha plot at Pasoh Forest Reserve in Peninsular Malaysia provides further insight on possible effects of habitat heterogeneity on floristic diversity. The heterogeneous Lambir plot with 1173 species was 30% more diverse than the relatively homogeneous Pasoh forest with 818 species (Davies et al. 2003a). The great excess of Lambir species was due largely to a few genera with many species (see Table 3.7). Greater habitat heterogeneity has no clear correlation with richness at higher taxonomic levels (see APGII, 2003). Pasoh included the same number of Orders (35) as Lambir, 2% more families (88), and 4% more genera (296). Only one genus, *Mangifera*, was substantially more speciose in the Pasoh forest. In contrast, 16 genera had at least 6 more species in Lambir than in Pasoh. *Shorea* was the most striking example of the difference between the plots with 41 more species in Lambir.

Detailed autecological studies of several trees at Lambir have been undertaken to investigate factors potentially influencing spatial distributions and coexistence. Those investigated include *Aporosa* (Debski et al. 2002), *Dryobalanops* (Itoh et al. 1997; Itoh et al. 2003), *Scaphium* (Yamada et al. 1997), *Macaranga* (Davies 1998; Davies et al. 1998), and *Ficus* (Harrison et al. 2003). In all cases there is strong evidence that congeneric species use different edaphic microhabitats.

4. Plant Reproductive Phenology and General Flowering in a Mixed Dipterocarp Forest

Shoko Sakai, Kuniyasu Momose, Takakazu Yumoto,
Teruyoshi Nagamitsu, Hidetoshi Nagamasu,
Abang A. Hamid Karim, Tohru Nakashizuka, and Tamiji Inoue

This chapter discusses the flowering patterns at Lambir Hills observed by the Canopy Biology Program in Sarawak (CBPS), in the current perspective of tropical phenology. We begin with a review of phenological studies, mostly from seasonal forests having dry seasons, in the Neotropics. Next, the flowering phenology of lowland dipterocarp forests, characterized by general flowering (GF), is described, comparing and contrasting the flowering tempo of this forest to that found in other forests. Then, the ultimate and proximate causes of flowering phenology are reviewed. A discussion of future directions and challenges concludes the chapter.

4.1 Introduction

Phenology is the study of the periodicity or timing of recurring biological events, in relation to short-term climatic change. In the case of plants, phenological events involve flowering, fruiting, leaf flushing, and seed germination (Leith 1974). The timing of these events can profoundly affect survival and reproductive success. Not only abiotic environmental factors such as temperature and humidity, but also biotic elements including herbivory, competition, and pollination (through pollinators and flowering phenology of other conspecifics) can be selective agents for patterns of plant phenology. Germination, flowering, or leaf production at the wrong time cause low survivorship of seedlings (Tevis

1958), low seed production (Augspurger 1981), and high predation rates (Aide 1992), respectively. At the same time, plant phenology can greatly affect animals that use young leaves, flowers, seeds and mature or immature fruits (van Schaik et al. 1993), and resource cycling in the forest. Plant phenology is thus of fundamental importance for monitoring, managing, and conserving ecosystems. Most studies of plant phenology in tropical forests have been conducted to describe resource availability for consumer animals (e.g., Frankie et al. 1974; Croat 1975; Putz 1979; Opler et al. 1980; Foster 1982; Koptur et al. 1988; Murali and Sukumar 1994; Justiniano and Frederickens 2000; Morellato et al. 2000). Other studies emphasize physiological release mechanisms (e.g., Augspurger 1981; Reich and Borchert 1982) and synchronization within populations (e.g., Augspurger 1980, 1983; Primack 1980) from a perspective of plant reproductive success by monitoring the focal plants in more detail at the population level for rather short time periods. Further aspects of plant phenological studies are reviewed in Rathcke and Lacey (1985), Primack (1987) and van Schaik et al. (1993).

One central characteristic of phenology in tropical forests may be high diversity, which has two important aspects (Gentry 1974; Janzen 1978; Bawa 1983; Sarmiento and Monasterio 1983; Newstrom et al. 1994a, b). First, phenology patterns may be quite different among individuals of a given species, thus the flowering or fruiting pattern of individual plants may differ from the mean of the population and community. For example, flowering of *Boesenbergia grandifolia* (Zingiberaceae) in Borneo has irregular sub-annual or annual flowering patterns at the individual level but continuous flowering at the population level (Sakai 2000). Second, various flowering patterns are found among plants in the local community. Gentry (1974) was among the first to draw attention to the high diversity in phenology in tropical forests, compared to forests in the Temperate Zone. He qualitatively classified flowering phenology of the Bignoniaceae in four flowering types based on duration, frequency, and amplitude and discussed the ecological significance of such differences in relation to pollination. His work demonstrated the great potential of tropical phenological studies for exploring selective pressures and their evolutionary significance.

4.2 Annual Cycles at the Community Level in Seasonal Forests

Climate in tropical rain forests is characterized by continuous humid or warm conditions, which potentially allow most organisms to remain active throughout the year. Thus, one prominent theme in tropical community studies is the degree of periodicity or regularity of biological activities. In the temperate region, regular rhythms in temperature, day length, and winter, which limits all biological activities, impose clear annual cycles. In contrast, in the low latitudes the difference between the shortest and the longest day of the year is small: about 70 minutes at 10° latitude. The annual range of mean temperature is much smaller than changes during a day. The nights are the winter of the tropics. However, a periodic change in rainfall caused by movements of the intertropical convergence zone, a seasonal event in the tropics, rather than temperature and day length, plays an important part in controlling proximate and ultimate factors for tropical plant phenology (van Schaik et al. 1993). Dry seasons within an annual cycle occur in most tropical regions, and many studies have shown a correlation between phenology and rainfall (Augspurger 1981; Borchert 1983; Reich and Borchert 1984). Most Neotropical forest communities that have been studied show flowering and fruiting peaks near the end of the dry season (Janzen 1967; Croat 1975; Foster 1982; Frankie et al. 1974; Hilty 1980; Opler et al. 1980; Bullock and Solis-Magallanes 1990; Justiniano and Fredericksen 2000). The pattern may be due to high insolation and photosynthesis in dry seasons. Alternatively, or in addition, it may enhance germination and seedling survival by adjusting fruiting to precede the beginning of the wet season (van Schaik et al. 1993).

Although the effect of rainfall pattern are predominant even in wet forests without a clear dry season, detailed examination at the species and population levels can reveal wide variation in flowering phenology. At La Selva, in Costa Rica, most trees have a sub-annual flowering pattern (55% of 254 species flower more than once a year, often irregularly), and only 29% of trees show an annual flowering pattern (see Fig. 4.1; Newstrom et al. 1994b). This forest is wet and lacks a severe dry season; monthly precipitation never drops lower than 100 mm (Sanford et al. 1994).

While comparative data are not available from other Neotropical forests, a higher proportion of annual flowering species may occur in forests with stronger seasonality. Wright and Calderon (1995) analyzed flowering phenology of 217 species with 230 seed traps for five years on Barro Colorado Island (BCI) in Panama. They found that flowering was highly concentrated in time for most species, and mean flowering dates of species were concentrated in February and March, which are the driest months of the year, and in April and May when the wet season begins. In addition, year-to-year variation in intensity of plant reproduction may also be related with rainfall fluctuation. As one example, in the moist forest of BCI, an infrequent famine was shown to be linked to an unusually small fruit crop during a La Niña year (moderate dry season) that followed an El Niño-Southern Oscillation event (Wright et al. 1999).

Although many studies have reported the clear correlation between rainfall patterns and phenological events, results of irrigation experiments are not always positive. Some biologists have succeeded in manipulating flowering phenology by watering plants (Augspurger 1981; Reich and Borchert 1982; Wright and Cornejo 1990a, b; Wright 1991; Tissue and Wright 1995). However, a large-scale irrigation experiment (2.25 ha) in BCI, with a strong seasonal pattern in rainfall, showed that irrigation had no effect on the timing of leaf fall, leaf flush, flowering, or fruiting for most species of canopy trees (Wright and Cornejo 1990a,b). Deep-rooting canopy trees possibly do not experience a water deficit even in dry seasons (Steinberg et al. 1989). The mechanisms for synchronized flowering are still unknown, and little is known about consequences.

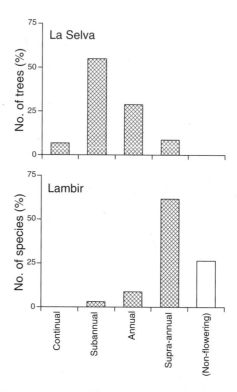

Figure 4.1. Proportions of sub-annual, annual, supra-annual and continual flowering types among trees at La Selva, Costa Rica (254 species, Newstrom et al. 1994b) and Lambir, Malaysia (187 species, Sakai et al. 1999c); modified from Sakai 2002. The General Flowering at Lambir is included in supra-annual. Note La Selva is on an individual tree basis, Lambir on a species basis; Newstrom et al. (1994b) state the two are similar at La Selva.

4.3 Flowering Phenology in Southeast Asia

In a large portion of the Asian tropical forests from Sumatra to the Philippines, there is generally no clear annual cycle of rainfall (Yasunari 1995; McGregor and Nieuwolt 1998). It is here that flowering at multi-year intervals—called GF, general flowering or mass-flowering—is known from lowland dipterocarp forests (Ashton et al. 1988). A GF usually occurs every 2 to 10 years. During GF, many trees, including most dipterocarps and other families, flower for months, yet some flowering occurs in non-GF periods (Sakai et al. 1999c; Sakai 2002). The longest records of GF comes from statistics of exports, since GF brings about a huge crop of *illipe nuts* (fruits of *Shorea* section *Pachycarpae*), an important commercial item of the region for export (Blicher 1994) and thus has a strong effect on the local economy.

Although the importance and uniqueness of the GF have been stressed by other authors (Janzen 1974; Appanah 1985, 1993; Ashton 1989; Ashton et al. 1988), there are few detailed studies that accurately describe a GF at the community level, or that examine the prevalence of the phenomenon among species of different life forms, the pollination mode, or the fruit dispersal. Records of gregarious flowering in most studies are restricted to the Dipterocarpaceae (Burgress 1972; Ng 1977; Yap and Chan 1990) or inferred by examination of herbarium specimens (Cockburn 1975). A few studies on GF have recorded reproductive phenology of plant species other than Dipterocarpaceae, but they include only a small number of individuals or species (Medway 1972; Yap 1982) and a much shorter period than one GF to non-GF cycle (Corlett 1990).

One of the major purposes of CBPS concerned the causes and consequences of GF (Chapter 2). To accomplish this study, in 1992 the CBPS began monitoring phenological events among 576 plants of 305 species. Observations were comprehensive within the study site, using tree towers and aerial walkways constructed in an 8 ha permanent plot (Sakai et al. 1999c; Inoue et al. 1995; Yumoto et al. 1996). When the censuses were initiated, the forest was at a fruiting peak following the GF of 1992. From 1993 to 1995, the proportion of flowering plants was very low, around 3%. However, in May 1996, the proportion increased dramatically to reach 17% and 20% for individuals and species, respectively. Thus, this GF was observed from its beginning (see Fig. 4.2; Sakai et al. 1999c). To our surprise, the proportion of plants flowering had two peaks in 1996, and GFs were also observed during the following two years.

The percentage of plants in flower was generally quite low in Lambir Hills, compared with other tropical regions (see Table 4.1). In most seasonal lowland forests, the proportions of species flowering average 15% to 20%. In tropical forests at higher elevations, the proportion can be higher. In a forest with a severe dry season the number of flowering species often drops to zero for a few very dry months each year, but at other times it is over 10% and sometimes exceeds 60% (Murali and Sukumar 1994). The maximum proportion recorded at Lambir Hills so far, 22%, is also much lower than the maxima observed in other forests. Medway (1972) reported similar figures to those of Lambir Hills from a lowland dipterocarp forest in Peninsular Malaysia.

Sakai et al. (1999c) analyzed the phenology data up until December 1996 to describe plant reproductive phenology and GF in 1996 at Lambir Hills, and they concluded that the low percentage of flowering individuals was mainly due to low flowering frequency and the concentration of reproductive activities in GF periods, only at multi-year intervals. They classified species into flowering types using the flowering data of individual plants for the 43 months from June 1993 to December 1996. The first is a GF type, which flowers only in the GF period. Three additional categories were based on flowering frequency: *supra-annual* (flowered once or twice in 43 months), *annual* (flowered three or four times), and *sub-annual* (flowered more than four times). When a species included individuals that displayed more than one flowering type, the majority represented the species. Species in which reproduction was not observed during the 43 months were tentatively categorized as *non-flowering* (see Fig. 4.1).

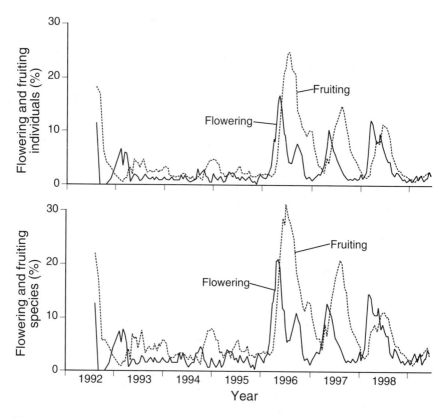

Figure 4.2. Changes in percentage of flowering or fruiting species and individuals observed from tree towers and walkways (237 spp., 428 individuals, Sakai et al. 1999c, Sakai et al., unpublished data).

In contrast to the plants at La Selva, Costa Rica, more than half of the species we observed were supra-annual and GF species, which flower once in two or more years on average. A continuous flowering pattern (extended flowering with short interruptions) was rarely found (see Fig. 4.1). Causes of this difference have scarcely been explored, although a poor nutrient level in the soil was suggested to be a factor (Janzen 1974; Inoue 1997). Out of 527 flowering events observed during 43 months from July 1993 to December 1996, 57% occurred in 10 months of GF from March to December 1996. Among species that flowered at least once in the 43 months, 85% reproduced during the GF period. Most species showed strict synchronization within species, and major flowering periods of species were usually less than one month long. During a GF, the flowering of related species tended to be aggregated in time (see Fig. 4.3).

Participation in GF was observed among various plant groups, which confirmed that GF was a general phenomenon, operating at the community level.

Table 4.1. Proportion of species recorded as flowering in various tropical forests

Study site	Forest type	Flowering spp. %	References
Neotropics			
La Selva (Costa Rica)	Wet, lowland	Overstory trees 9–30 Understory trees 17–38	Frankie et al. (1974)
Comelco (Costa Rica)	Dry, lowland	7–28	Frankie et al. (1974)
Monteverde (Costa Rica)	Montane	20–60	Koptur et al. (1988)
Alto Yunda (Clombia)	Premontane	25–40	Hilty (1980)
Lomerio (Bolivia)	Dry, lowland	8–41	Justiniano and Fredericksen (2000)
São Paulo (SE Brazil)	Wet, lowland	3–33	Morellato et al. (2000)
	Premontane	3–24	Morellato et al. (2000)
Africa			
Nyungwe (Rwanda)	Montane	14–47	Sun et al. (1996)
Asia			
Mudumalai (S India)	Dry, lowland	0–60	Murali and Sukumar (1994)
Dipterocarp forests in Asia			
Lambir (Borneo)	Wet, lowland	Non-GF period 0–3 GF–22	Sakai et al. (1999)
Ulu Gombak (Peninsular Malaysia	Wet, lowland	Non-GF period 0–7[a] GF–35[a]	Medway (1972)

[a] Proportion of individuals

As many as 35% of 257 species at Lambir were of the GF type. These comprised plants of different families and life forms, from epiphytic orchids to emergent dipterocarp trees (see Fig. 4.4). Supra-annual and annual species also reproduced more actively during a GF period than during non-GF years (see Fig. 4.5). Therefore, GF is the preeminent reproductive pattern at Lambir.

4.4 Ultimate Factors

Van Schaik et al. (1993) showed that peaks in irradiance are accompanied by peaks in leaf flushing and flowering. These authors reviewed phenological studies from all the three major tropical regions. Their work strongly suggested a major role of climate as a determinant of phenology. They proposed several explanations, including the *high radiation* hypothesis. Since it is energetically most efficient to transfer *photosynthates* directly into growing organs, rather than store them for later translocation (Chapin et al. 1990), it is advantageous for plants to produce leaves and flowers during the most productive season. The hypothesis assumes that plant production is mostly limited by *insolation* and *irradiation* which usually have a peak in the tropical dry season, mainly because of less cloudiness (van Schaik et al. 1993). This general rule was supported by

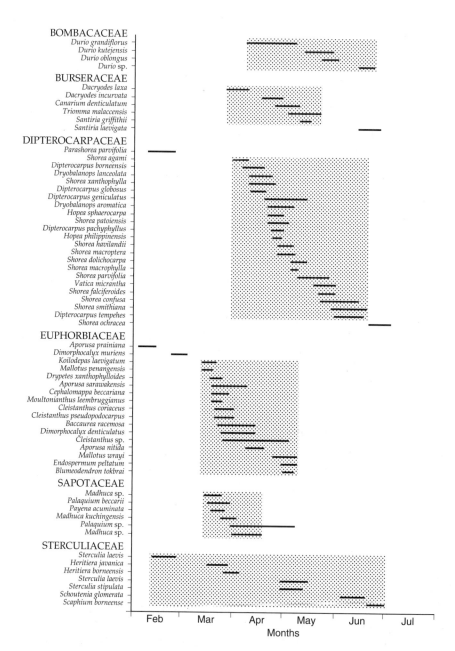

Figure 4.3. Flowering periods of different taxonomic groups at Lambir during 1996 GF (Momose et al. unpublished data).

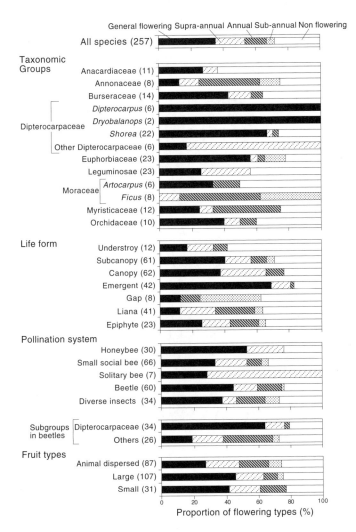

Figure 4.4. Flowering types (sub-annual, annual, and general flowering) and non-flowering species (species not flowering during the observation period) among all species observed, and taxonomic groups, life-form types, pollination systems, and fruit type; N species in parentheses (Sakai et al. 1999c).

the fact that peaks of flowering and flushing occur in the months of most intense and sustained sunshine. Besides, flowering in dry periods and fruit dispersal in the following rainy season may be adaptive, considering water conditions are critical for seed germination and survival in the tropics and that seeds of many tropical plants do not have dormancy. Sakai (2002) suggests that predictable rainy periods of the supra-annual cycle caused by ENSO (El Niño Southern

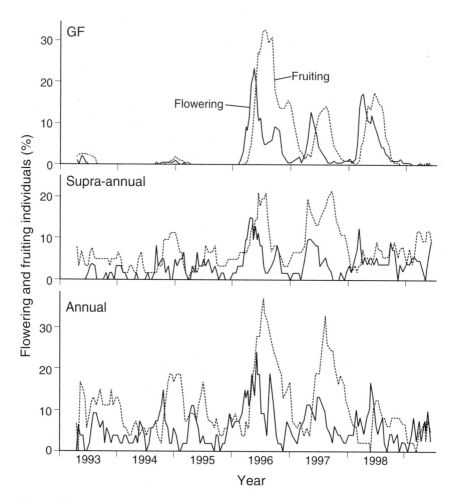

Figure 4.5. Changes in percentage flowering and fruiting individuals for GF, supra-annual and annual flowering types—193, 43, 50 individuals for GF, supra-annual, annual species, respectively (Sakai et al., unpublished data).

Oscillation) may also promote synchronized flowering triggered by drought in Southeast Asia, where rainfall pattern lacks clear annual regularity.

Among biotic factors, predator satiation has been considered to be the most important in explaining supra-annual reproduction in plants. It asserts that synchronized fruiting at long intervals is an effective means of starving the predators in low seed years, or surpassing their needs and satiating them in high years (Janzen 1971b; Silvertown 1980). Although the hypothesis is supported by some field data from temperate forests showing lower per-capita seed predation in mast years (e.g., Sork 1993; Crawley and Long 1995; Kelly and Sullivan 1997),

Kelly (1994) accurately pointed out that the validity of the hypothesis depended on the functional response of predators to crop size (Ims 1990). He suggested two possible scenarios: that predator populations or losses of seeds are limited by small crop size during non-mast years, and that predator populations are generally limited by factors other than crop size. If seed predators are specialists, predator populations determined by small crop size in non-masting years are much smaller than would be maintained by constant seed production. The occurrence of consecutive mast years can be evidence against this particular hypothesis. The other case assumes predators are generalists, thus their populations are not limited by crop size of a particular plant species. Plants may only limit predation by producing a crop far beyond the potential amount of predator consumption in a masting year, using resources accumulated in normal years. In the latter case, consecutive masting can function to satiate predators.

Predator satiation has been treated as the most important evolutionary factor for GF (Janzen 1971b). The predators in this case are birds and mammals which consume fruits of a wide range of plants, since satiation of specialized predators such as most insect fruit parasites does not explain synchronized flowering among plants of different genera and families found in GF. The hypothesis is supported by studies from Kalimantan showing the high predation rate of dipterocarp fruits in a minor flowering year, compared to a large one (Curran and Leighton 2000; Curran and Webb 2000). However, Inoue (1997) and Sakai et al. (1999c) indicated that predator satiation does not explain concentration of flowering in GF periods of non-dipterocarp species, including orchids and others with tiny fruits or seeds, which are usually neglected by birds and mammals.

Instead, the above authors suggested that the promotion of pollination, through temporal aggregation of flowering, is a strong evolutionary factor that promoted GF. The idea is contrary to a well-known concept called the "shared pollinator hypothesis," which has received particular attention for tropical forest plants. That hypothesis predicts that plant species sharing common pollinators should separate their flowering somewhat to minimize interspecific overlap in flowering times, and thus minimize ineffective pollination and/or competition for pollinators (Stiles 1977; Ashton et al. 1988). Stiles (1977) documented that Neotropical Costaceae and Heliconiaceae had clear annual rhythms in flowering and suggested strong intraspecific synchronization and temporal segregation among species sharing pollinators. Ashton et al. (1988) showed sequential flowering of *Shorea* species (Dipterocarpaceae), which significantly segregated flowering periods during a few months of GF. However, most experimental or field studies (Wheelwright 1985; Murray et al. 1987; Wright and Calderon 1995) have produced negative results. At Lambir Hills, many ginger species (Zingiberaceae and Costaceae), the most important herbaceous constituents on the forest floor, are pollinated by birds or solitary bees and showed irregular, sub-annual flowering patterns (Sakai et al. 1999a,c; Sakai 2000). Their flowering periods could be both synchronized and unsynchronized among conspecific individuals. Thus no temporal segregation among species sharing common pollinators occurred (Sakai 2000). Other studies suggested that synchronized flowering of different species could facilitate pollination through an increase of resource density and local pollinator attraction (Schemske 1981). Aggregation of flowering in time may be due to pollinator availability in a particular season. In dry deciduous forest, Frankie (1975) found that a disproportionately large number of moth-pollinated plants flower in the wet season. He suggested that this aggregation of flowering time might be related to moth population density, which was controlled by the abundance of new foliage: the larval food (Frankie 1975). Flowering may, however, be completely out of phase with pollinator abundance (Zimmerman et al. 1989).

The "promotion of pollination hypothesis" proposed by Inoue (1997) and Sakai et al. (1999c) assumed a higher pollination success in GF periods than non-GF periods. For this principle to operate, the number of available pollinators relative to flowers must increase rapidly. Aggregated flowering of various species sharing common pollinators may activate pollinators and result in higher pollination success than isolated flowering: an increase of floral resources increase the density of flower visitors through immigration, population growth and feeding. The idea is supported by higher fruit set in GF than in non-GF periods (Yap and Chan 1990; Sakai et al. 1999c). An increase in population density or activities in GF has been observed in some pollinators. Giant honeybees, important pollinators in GF (Momose et al. 1998c), have an ability to migrate long distances, up to 200 km. They are thought to immigrate into dipterocarp forests when GF begins, likely from secondary and montane forests, where some flowers are usually available. Giant honeybees pollinate the two most abundant dipterocarp species (*Dryobalanops lanceolata* and *D. aromatica*) as well as other minor plant species at Lambir Hills (see Chapters 6, 8). Therefore, the minor plants may receive benefits from synchronized flowering with abundant dipterocarps, since flowering of a single rare species cannot induce immigration of giant honey bees. Many pollinators, including many resident bees, become abundant through population growth in GF.

Appanah and Chan (1981) reported that *Shorea* (Dipterocarpaceae) were pollinated by thrips, tiny insects which feed on pollen and floral tissue of a variety of plant species. Because of low specialization to plant species and short generation time, thrips build up large populations quickly at the beginning of GF and serve as pollinators of different *Shorea* spp. On the other hand, Sakai et al. (1999a, b) found that small beetles were the main pollinators of *Shorea* (Dipterocarpaceae) at Lambir Hills. Interestingly, some of the pollinators are herbivores as well, feeding on new leaves of dipterocarp trees and possibly others during non-GF periods without dipterocarp flowers. In this case, a rapid increase of the pollinating beetles is unlikely because they do not migrate or reproduce as rapidly as do thrips. More detailed studies of life history of the pollinator beetles (copulation and breeding sites), and also the host (both for flowers and leaves) are required to clarify relationships between the pollinators and GF.

Momose et al. (1998a) addressed differences in flowering intervals among plants belonging to different forest strata using a theoretical approach. The model assumed that the flowering intervals of trees maximize visits by pollinators, including opportunist and social bees, throughout their lifetimes after they reach mature size. The model also assumed that larger displays attract more opportunist (social) pollinators per flower, while the number of other pollinators per flower is constant irrespective of display size. Social foragers recruit colony members once a display exceeds a minimum size.

When productivity is an increasing function of plant size, trees in the highest canopy layers enjoy high productivity and low mortality. Their low mortality enables them to wait long intervals between flowering, and their high productivity allows them to display heavily and attract many opportunist pollinators. By contrast, the canopy or subcanopy species cannot wait as long between reproductive episodes because of higher mortality. For these trees it is better to frequently produce smaller displays to attract pollinators. The higher proportion of social-bee-pollinated plants in the canopy and subcanopy trees than in emergent trees supports this idea, except for plants pollinated by giant honeybees (*Apis dorsata*), which respond only to extraordinarily large floral resources associated with GF (Itioka et al. 2001a).

A few studies have focused on differences in flowering patterns among plants with different pollination systems (Gentry 1974; Frankie 1975; Momose et al. 1999a). At Lambir Hills, Sakai et al. (1999c) found a correlation between flowering types and pollination systems, which may be related to characteristics of their pollinators. Because flowers of GF plants are available only during GF, their flower visitors and pollinators should use a wide range of resources in terms of foraging area, resource type, and/or plant species. In contrast, plants with high host-specificity, such as beetle-pollinated Annonaceae and fig-wasp-pollinated *Ficus* tend to flower more frequently (see Fig. 4.4).

Studies from tropical forests and other regions indicate that in addition to ecological factors, flowering phenology is under strong *phylogenetic constraint* (Kochmer and Handel 1986; Johnson 1992; Ollerton and Lack 1992; Wright and Calderon 1995). Plants sharing the same pollinators often show synchronous flowering, simply because they are closely related. The fact is often cited to reject the shared pollinator hypothesis, which depends upon segregated flowering among plants pollinated by the same animals. In the GF period at Lambir Hills, aggregation of flowering periods was found among species of the same taxonomic groups (see Fig. 4.3). Strong phylogenetic constraints detected by the above studies, however, do not necessarily indicate absence of adaptation in phenology. The diversity of tropical flowering phenologies should be guided by such phylogenetic perspectives. This means, for example, that the GF in Southeast Asia has an historical component. We may have to direct more attention to synchronization of flowering among species of different families than primarily within a family or Dipterocarpaceae, or to dipterocarps that do not share GF phenology patterns.

4.5 Proximate Factors

At the mechanistic level, flowering can be thought to be under the control of both internal and external factors. Internal factors include plant developmental stage (immature or mature) and stored resources. External variables are environmental, such as humidity, temperature, or day length. All of these factors are interrelated and function in different ways in different plant species (Bernier 1988). Little is known about which external factors function in observed flowering patterns with high diversity, in the relatively equable climate of the tropical rain forest, except for a few studies on annual-flowering species in the seasonal tropical forests (Reich and Borchert 1982; Augspurger 1981; Rivera and Borchert 2001). Here, we limit our discussion to comment on sub-annual flowering and GF patterns.

Newstrom et al. (1994b) suggested that flowering patterns of irregular, sub-annual flowering could be viewed as due to inhibiting factors, rather than inducing ones, although almost no information exists on mechanisms controlling sub-annual flowering. They reported that some trees of sub-annual species never flowered in a certain month at La Selva, Costa Rica, possibly because certain inhibiting factors occurred annually. Since non-flowering months were different among species, the putative inhibiting factors might also be varied (Newstrom et al. 1994b).

The same may be true for flowering phenology of gingers and *Macaranga* at Lambir Hills. Some ginger species flower intermittently, while flowering of other gingers was synchronized within species, although the synchronization was far less, compared with GF species. An increase of flowering intensity in the GF period was not observed (Sakai 2000). On the other hand, flowering intensity of *Macaranga hosei* (Euphorbiaceae), categorized as a sub-annual species, increased in GF in 1992 and 1996 (Sakai et al. 1999c; Davies and Ashton 1999). Eleven sympatric *Macaranga* species have a single yearly flowering peak, and most of their reproductive activities were limited to several months within a year, except for two continuously flowering species. Their flowering periods were synchronized among species, and flowering intervals were not strictly constant. Davies and Ashton (1999) argued that these *Macaranga* species responded to a common flowering cue. An increase in flowering intensity might be related to increased irradiance levels associated with drought periods, which are likely linked with GF.

The environmental trigger of GF is still somewhat controversial. An association between GF and severe drought is often reported from different forests, and important roles of prolonged drought or increased photoperiod have been suggested repeatedly (Wood 1956; Burgress 1972; Medway 1972; Janzen 1974; Whitmore 1984; Appanah 1985; van Schaik 1986; Kiyono and Hastaniah 1999). One argument is that if reproduction is limited by photosynthesis, it is reasonable that plants may only reproduce in years when they can accumulate more energy and reserves through photosynthesis. The correlation between ENSO and GF was significant, especially in eastern Peninsular Malaysia (Ashton et al. 1988) and western Kalimantan (Curran et al. 1999; Curran and Leighton 2000), and El Niño usually brings about diminished rainfall in that region (Leighton and Wirawan 1986; Salafsky 1994; McGregor and Nieuwolt 1998).

However, there is doubt that drought itself induces flowering (Ashton et al. 1988). Such skepticism exists because correlations between flowering intensity and local geography, or water availability, have not been found. If water shortage directly induces flowering, flowering should be affected by local topography, soil types, altitudes, and so on. In addition, the relationship between rainfall seasonality and timing of a GF is obscure. For example, in eastern Peninsular Malaysia and SW Borneo the driest month is often January, although GF in eastern Peninsular Malaysia occurs from February to July, while in western Borneo it is from August through November. Ng (1977) suggested that a longer photoperiod was an alternative trigger, not affected by soils or local topography. It remains uncertain whether an increase of hours of direct sunshine, caused by less cloudiness (rather than by longer day length) can provide an effective cue for synchronized flowering of dipterocarp species, when the flowering of single tree lasts only 2 to 3.5 weeks (Ashton et al. 1988). A decrease in photoperiod by some 30 minutes is now thought sufficient stimulus to cause flower bud formation in some tropical trees (Rivera and Borchert 2001).

Apparently, Wycherley (1973) was the first to propose an abnormal temperature was the cue for GF. Based on an analysis of meteorological records for 11 years, Ashton et al. (1988) suggested this condition was a decrease in the minimum temperature. Supporting that hypothesis, reductions of minimum temperature were observed about one month before the onset of GF at Lambir in 1996 and 1997 (see Fig. 4.3), and at Pasoh Forest Reserve in Peninsular Malaysia in 1996 (Yasuda et al. 1999). However, GF occurred without a preceding temperature drop in Singapore in 1987, in Danum, Sabah, in 1987 and in Gunung Palung NP, West Kalimantan, in 1987 and 1991 (Corlett and La Frankie 1998). It is often difficult to identify the direct trigger from simple observation, because many meteorological factors, such as temperature, rainfall, humidity, and solar radiation, are closely related, and never change independently. Flowering also depends on the internal conditions of plants. Therefore, the same climatic conditions do not always bring about the same plant responses. An experimental approach is needed to evaluate the possible triggers of flower production.

4.6 Directions of Future Research

Clearly, long-term monitoring of plant phenology is more important now that global environmental change and global warming are recognized as critical issues. Climate change affects the ecosystem through plant and/or animal behavior, plant-animal interactions, and their biodiversity (Reich 1995; Corlett and LaFrankie 1998; Visser and Holleman 2000; Both and Visser 2001; Chuine and Beaubien 2001; Penuelas and Filella 2001). Harrison (2000b) reported that severe drought in 1997–98 associated with El Niño caused a substantial break in the production of inflorescences on dioecious figs and led to the local extinction of their pollinators at Lambir Hills. It brought about absences of the fig crops that were essential for the survival of mammals. The global climate change is thought to strengthen effects of El Niño and drought in the region.

At the same time, strong biological seasonality provided by GF is a very interesting and important theme to study in ecology. As we have seen, lowland dipterocarp forests with GF have a flowering phenology quite different from that in the Neotropics. The differences raise many other questions. Are there any differences in regeneration ecology of trees compared with other tropical forests? Are there fewer birds and mammals feeding on fruits and seeds in dipterocarp forests than in others? Are seed predating insects less specialized? Do supra-annual flowering plants invest in reproduction as much as annual flowering species? We do not have clear answers to these questions, and we are still at the beginning of GF studies. LaFrankie (Chapter 16) discusses the higher seedling density of canopy species in Malaysian forests compared to Neotropical forests. That, also, may be related to differences in the regeneration habit of canopy tree species, but we do not understand the mechanisms. Sakai (2002) compared mammal and bird biomass and their consumption of fruits in a Malaysian and Neotropical forest from the literature and could not find any significant difference. Moreover, because most dipterocarp seeds are dispersed by gravity and wind, the proportion of animal-dispersed plants is generally low in dipterocarp forests. Vertebrate pollinators are less common, and the diversity of nectarivorous birds is lower at Lambir than at La Selva, Costa Rica. One possible cause is that vertebrates have difficulty maintaining their populations using only floral resources. The work by Nakagawa et al. (2003) on insect seed predators of dipterocarps revealed a rather broad diet and large overlap in hosts used by different insect seed predators. More surprisingly, the dominant insect group changed dramatically among GF years. Loose pollination niches and pollinator generalization seem involved. The correlation between flowering habit and specificity of seed predators has still not been examined.

Other thematic problems are related to material cycling in the ecosystem. No studies have examined whether forests with GF, in which most large trees reproduce infrequently, produce on average less fruit or reproductive tissue, according to their biomass, than do the trees of forests dominated by sub-annual and annual flowering species. Our seed trap surveys, initiated in 2002, will provide an answer to that and other questions in the next GF event. We are ignorant of consequences from large fluctuations in the amount of input from trees to the ground, in terms of biomass, or considering the amount of carbon, nitrogen, and minerals, due to the GF cycles. Synchronized flowering of many canopy species once in several years may also change photosynthetic activities of the forest, and thus even affect Earth's atmosphere.

5. A Severe Drought in Lambir Hills National Park

Rhett D. Harrison

Drought can affect the ecology of forests in several ways—via fire, plant mortality, and plant phenology—which then affect the timing and amount of resources available to herbivores, pollinators, and seed dispersers. In this chapter, I consider the incidence of drought and describe its effect on the forest and faunistic elements in turn; I conclude by outlining implications for the maintenance of biodiversity in the region.

5.1 Introduction

The antiquity and stability of a perennially humid equatorial climate was once thought sufficient to explain evolution of the enormous number of plant species found in the forests of Borneo (Whitmore 1984). However, the impact of major climatic and sea-level fluctuations, especially during the Pleistocene glaciations, has also been recognized (Whitmore 1981; Flenley 1998). The similar importance of catastrophic disturbance, such as drought, fire, typhoons, and landslides, has been noted (Whitmore 1984; Ashton 1993), and such rare but severe disturbances clearly have a major influence on forest ecology. Yet our knowledge on the impact of large-scale, severe disturbances remains fragmentary and is generally restricted to studies of plant population dynamics. As the forest is reduced to ever smaller and more isolated patches, through the expansion of human populations and the intensification of agriculture and logging, animal populations suffer from the loss of refugia and corridors connecting surviving forests. Furthermore, catastrophic disturbance may hinder the recovery of essential groups such as pollinators, seed dispersers, and parasitoids, and a gradual disintegration of the community is possible (Whitmore 1998).

Lambir Hills National Park (LHNP) is a small fragment of primary forest set in a matrix of palm plantations and slash-and-burn cultivation. In 1998 a very severe drought induced by the 1997–98 El Niño-Southern Oscillation (ENSO) event occurred. Our studies in the park, therefore, enabled us to observe some immediate results of this disturbance, and over time we will be able to follow its long-term impact.

5.2 Drought in Lambir Hills National Park

Central Southeast Asia has a tropical maritime climate with monsoon rains from the Pacific Ocean in winter and from the Indian Ocean in summer. It is thus one of the wettest and most aseasonal climates of any tropical region (Whitmore 1984). However, brief dry periods are not infrequent. They often occur between the monsoons and may become extended if the variable movements of the intertropical convergence zone delay the onset of the next monsoon (Seal 1957; Brunig 1969; Baillie 1976; Whitmore 1984; Cranbrook and Edwards 1994). The ENSO also has an important influence. A strong event usually induces drought over much of the region (Barber and Chavez 1983; Harrison 2000b) though different areas may be affected at different times. For example, during a recent ENSO event widespread fires and droughts were reported from southern Borneo in September 1997 (Pearce 1999), but they occurred in early 1998 in northern and eastern Borneo (Toma 1999).

The ENSO is an irregular supra-annual climatic oscillation that affects the entire Pacific region. During normal years a pool of warm surface water develops around the western equatorial Pacific, built up by the constant push of the trade winds. An El Niño event occurs when this warm surface water flows back eastward like a river, eventually shutting off the cold upwelling off the Pacific coast of South America (Gill and Ramusson 1983; Webster and Palmer 1997; Guilderson and Schrag 1998). The warmer surface waters cause flooding along the eastern Pacific coast, but around Borneo the cooler conditions reduce evaporation and droughts follow. The El Niño phenomenon may be as much as 100,000 years old (Pearce 1999), but there is evidence that the intensity and frequency of El Niños have suddenly increased in recent decades (Guilderson and Schrag 1998; Huppert and Stone 1998; Salafsky 1998). Moreover, several authors have predicted from climate models that such trends may be the consequence of global warming (Meehl 1997; Timmermann et al. 1999).

Within Borneo there is considerable variation in rainfall patterns, both in terms of total amount of rain and in its seasonal distribution (Whitmore 1984; Walsh 1996; Harrison 2000b). If we compare Miri Airport, just 30 km north of LHNP, and Kuching Airport, in the south of Sarawak (see Fig. 5.1), we find

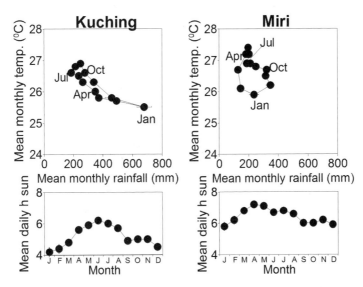

Figure 5.1. Climate at Kuching Airport (01°29'N, 110°20'E) and Miri Airport (04°02'N, 113°47'E) in southern and northern Sarawak, respectively. Annual rainfall: Kuching: 4048 ±57 mm; Miri: 2923 ±56 mm (30 km north LHNP); monthly temperatures vary by <2°, rainfall is always >100 mm. Kuching has more seasonal distribution of rainfall; montly means >400 mm during Laddas (Dec. to Feb.).

that Kuching is much more seasonal but also has a much higher total annual rainfall (Kuching: 4048 ±57 mm; Miri: 2923 ±56 mm). This means that the incidence of droughts (30 day-rolling-rainfall-total <100 mm, Brunig 1969; Whitmore 1984) is actually higher in Miri (see Fig. 5.2). In fact, in LHNP brief droughts occur in almost every 12 month period (see Fig. 5.3) and minor droughts cannot therefore be considered rare. These brief droughts occur most often in February to March, before onset of the southern monsoon (see Fig. 5.2; Seal 1957; Brunig 1969; Baillie 1976).

However, severe droughts are much rarer. Because minor droughts rarely extend to more than a month (see Fig 5.2), a parameter often used to better detect the occurrence of severe droughts in wet climates is the 3-month-shifting-average-rainfall. This method uses long-term monthly rainfall records, thus it can be seen that in both Miri and Kuching severe droughts are usually associated with ENSO events (see Fig. 5.4). In Kuching the 3-month-shifting-average-rainfall fell below 100 mm only three times in 123 years, and all were ENSO events. In Miri there were 15 such droughts in 87 years, 11 of which were associated with ENSOs. Since 1966 an increase in the severity of ENSO events has led to a corresponding increase in the severity of the droughts in Miri (see Fig. 5.4). The same pattern has been reported from Pontianak in southern Borneo (Salafsky 1998). However, it is also evident that there is no such trend in Kuching, although one of the three severe droughts was in 1997.

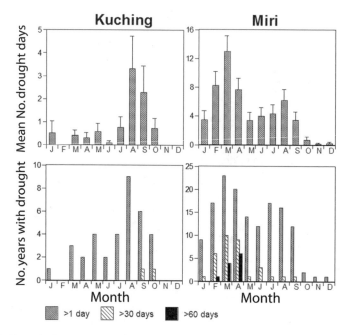

Fig 5.2. Drought frequency at Kuching and Miri airports (1969–98), defined as 30-day-rolling-rainfall-total <100 mm.

Walsh (1996) has suggested that the recent spate of droughts simply reflects long-term periodic changes in frequency, recognizing an intense drought period from 1877 to 1915 in Sandakan, northeast Borneo. However, Sandakan is much more seasonal than the rest of Borneo, so it is perhaps not the best example, and it is clear that different areas are being affected to different extents. Given the predictions of climate models (Timmermann et al. 1999), and that we are seeing worsening droughts in at least two widely separate locations, it would seem complacent to assume nothing is afoot. If the recent massive fires over much of Kalimantan and in northern Borneo are anything to go by, the impacts of droughts are certainly worsening if only because of human disturbance.

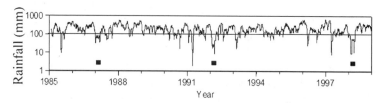

Figure 5.3. Thirty-day-rolling-rainfall-totals from the hydrological station of the Department of Irrigation and Drainage in LHNP (1985–98). ENSO events are indicated by solid squares.

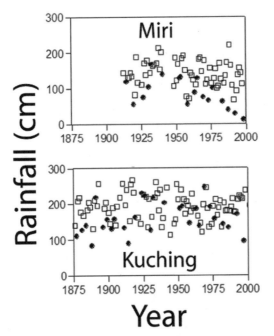

Figure 5.4. Severity of droughts in Kuching and Miri since the start of rainfall records. Rainfall is given as the yearly minimum 3-month-shifting-average, i.e., the most significant drought in each year (some years without records). ENSO events indicated by solid circles (Walsh 1996; Timmermann et al. 1999). In the case of events occurring over two years, the year with the more pronounced drought was judged as the ENSO year.

The drought in 1998 was by far the worst on record for Miri (see Fig. 5.4). Between January and March 1998 only 7% of the mean rainfall over the same three months fell in LHNP. Comparing the drought record in the 3-month-shifting-average-rainfall data from Miri against a normal distribution (from which it does not differ significantly) a drought of this magnitude would be expected to occur less than one time in 100 years. The 1997-98 drought was also very widespread and affected the entire northern region of Borneo (see Fig. 5.5; Toma 1999); given the high degree of endemism in many taxa and current fragmentation of the forest, this is an important consideration for the conservation of biodiversity.

Also interesting is that the 1997–98 ENSO did not cause serious drought in either Kuching or in Kuala Belalong, Brunei, although coastal sites in Brunei were affected (see Fig. 5.5). Both Kuching and Kuala Belalong have very high total annual rainfall, which may buffer them against most droughts. Kuala Belalong also has a large area of intact forest sufficient to maintain its own hydrological cycle. Although evapo-transpiration is not thought to contribute much to total precipitation in Borneo (Brunig 1969), it may be especially important in

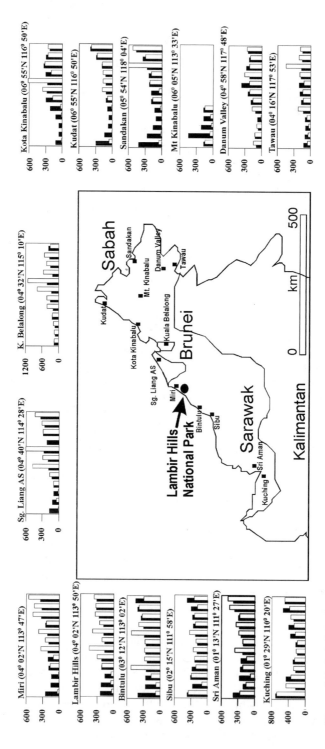

Figure 5.5. Rainfall of northern Borneo over the 1998 drought period. Solid bars indicate the long-term monthly mean rainfall and open bars the monthly rainfall during 1998 (after Harrison 2000).

periods of drought. In addition, areas that escape the most serious and wide-spread droughts will act as refuges for some species and hence will be very important for biodiversity conservation.

We can conclude that in LHNP brief droughts are not infrequent but severe droughts are rare and usually associated with ENSO events. There has been a trend of increasingly strong ENSO events, and more severe droughts as a result. The drought in 1998 was by far the most severe on record.

5.3 Impacts of the 1998 Drought

Fire. Under normal conditions the understory is always damp, so that in primary forest areas, fires are associated only with the most severe droughts. The tall, closed canopy, which permits only approximately 2% of the incident light through to the forest floor (Whitmore 1998) and imperceptible breezes, maintains a humid microclimate. In secondary forest areas, more light penetrates the increased openness of the canopy, and the soil may therefore dry out even in relatively brief droughts. As a result, secondary forests are more flammable (Woods 1989). This principle is incorporated in Sarawak agricultural methods in which the larger trees are cut down to dry out the ground about a month ahead of burning.

Forest fires are almost always started deliberately, usually to recycle nutrients and clear land, but some burn out of control (Leighton and Wirawan 1986; Woods 1989; Bertault 1991; Nykvist 1996). The recent widespread fires in Borneo (Toma 1999) made world headlines and caused not only colossal destruction of tropical forest but also global concern over the huge quantities of CO_2 and ash particles (Rosenfeld 1999) released in the air. The dry peat soil may keep fires burning for months; these fires are extremely destructive as the entire soil and root layer may be burnt away. A large area of impoverished *Imperata cylindrica* and *Baeckea frutescens* grassland known as the Sook Plain in Sabah was created during the 1915 drought, when approximately 100,000 ha of forest over peat burned (Cockburn 1974). Other large areas of grassland in central Borneo were quite probably created by fire.

After our project in LHNP started in 1992, there were not any fires—until the severe drought of 1998. The fires that burned in 1998 were slightly less destructive, being over a mineral soil, than those reported from elsewhere. During the drought a deep layer of leaf litter built up from a combination of increased leaf fall and reduced decomposition. All the fires started burning from the roadside and burned into the forest, slowly consuming the leaf litter (see Plate 5D). The leaf litter, all dead wood, and seedlings smaller than approximately 50 cm high were consumed (see Plate 5F). Slow-moving fires generate an intense heat that kills surface roots and scorches the bases of trees even if they do not burn (see Plate 5F). All saplings up to approximately 20 cm diameter were killed from the scorch damage ringing the bases of the trees. Occasionally, larger trees burnt, especially if a rotten root or split trunk provided fuel to ignite the trunk (see Plate 5F), but most large trees were not affected. Since then, self-seeding banana plants have invaded the area.

A fire in East Kalimantan during the 1982–83 ENSO event killed 90% of lianas and 90% of stems less than 5 cm dbh. For the latter this was *four times* the figure for mortality due to drought alone. Larger trees (>10 cm, dbh) were less affected and mortality was not significantly higher than in unburned plots (see Table 5.1; Leighton and Wirawan 1986). A similar pattern was observed in LHNP with almost 100% of seedlings and saplings dying, but most of the larger trees survived. Figures are limited to figs (*Ficus* spp.), but 13 out of 14 hemi-epiphytes, the only climber in the area, and 100% of freestanding trees (29 individuals, 10–30 cm dbh) died in the burned areas. This compares with 83% for hemi-epiphytes and 100% for climbers in the Kalimantan study (Leighton and Wirawan 1986).

Fires in LHNP were easily stopped by creating small fire breaks in the leaf litter. Even the path from the laboratories to the first tower, which was swept clean of leaves, acted as such a break (see Plate 5F). Given that all fires started by the road, this suggests park management here and in other national parks could easily prevent fires from entering the forest, by clearing a small break and by controlled burning along roadsides, during prolonged droughts.

Fires also exacerbate drought. In March 1998 more than 300 ha of peat forest were burning around Miri, and fires were widespread in Brunei, Sabah, and East Kalimantan (Toma 1999). Smoke formed a pall over much of northern Borneo. Smoke particles in the haze caused water to condense as tiny droplets, too small to allow cloud formation, and thus rainfall was reduced even further (Rosenfeld 1999).

Plant mortality due to drought. As might be expected, severe droughts in Borneo cause high plant mortality through increased water stress (see Plate 5A–C; Leighton and Wirawan 1986; Becker and Wong 1993; Becker et al. 1998; Kudo and Kitayama 1999). The composition and structure of a habitat may also be altered through differential mortality among tree species and size classes.

It has often been suggested that seedlings might suffer more from serious droughts because of their shallower root systems and that droughts have an important influence in determining the local distribution of species through seedling mortality (Turner 1990; Ashton 1993; Burslem et al. 1996; Newbery et al. 1996). However, a study of seedling mortality during the 1998 drought in LHNP

Table 5.1. Mortality in burned and unburned areas in Kalimantan after the 1982–83 drought (adapted from Leighton and Wirawan 1989)

	Lianas	<4cm dbh	4–10cm dbh	11–30cm dbh	>30cm dbh
Burnt	83.2% (357)	82.4% (922)	66.0% (465)	37.0% (786)	21.8% (353)
Unburnt	—	14.5%***	18%***	21%***	35%***

G-test probability, ***p <0.001

did not appear to confirm these assumptions (Delissio and Primack 2003). Overall mortality rates increased marginally (23%) compared to larger size classes and were within the range of non-drought mortality rates (see Table 5.2). Moreover, there were no significant differences in drought mortality among species. Interestingly, there was a greater proportional increase in mortality rates and height loss among taller seedlings, between non-drought and drought censuses. These results tend to suggest that seedlings are, in fact, adapted to prolonged periods of water stress (Delissio and Primack 2003; Burslem et al. 1996).

Overall tree (>10 cm dbh) mortality rates were *5 to 7 times* higher than during the previous non-drought period (see Table 5.2; Nakagawa et al. 2000), but this is probably an underestimate, because many individuals in our phenological census did not die until later in the year. The pattern for seedlings was a higher proportional increase in mortality of larger individuals between non-drought and drought periods. Plant families also fared differently with the highest mortality among dipterocarps, *15 to 30 times* the non-drought level (Nakagawa et al. 2000).

Thus, all groups of plants showed increased mortality rates during the drought. Seedlings were least affected, and the smaller trees less so than mature trees. In terms of species composition, no effect could be detected among seedlings, but among larger trees the important timber family, Dipterocarpaceae, sustained high mortality. Mortality of adult trees will clearly have repercussions for populations of pollinators, seed dispersers, and herbivores that depend on flowers, leaves, and fruit, or on their nesting sites.

Plant phenology. Because the temperature varies little throughout the year (see Fig. 5.1), we might expect rainfall and especially drought to be major determinants of plant phenology (Whitmore 1984; Reich 1995; Corlett and Lafrankie 1998). The GF phenomenon (Wood 1956; Burgess 1972; Medway 1972; Ng 1977; Appanah 1985; Ashton et al. 1988; Momose et al. 1998c; Sakai et al. 1999c) produces a strong pattern in reproductive phenology. Most researchers have not found a relationship between drought and general flowering (e.g., Burgess 1972; Ashton et al. 1988), but leaf production is thought to correlate with

Table 5.2. Plant mortality at Lambir Hills National Park during a severe drought (1998) and non-drought periods

Plant stage (dbh)	N	Non-drought Mortality (% yr^{-1})	N	Drought (1998) Mortality (% yr^{-1})	Source
Seedlings	3828	7.5	397	9.4**	Delissio and Primack
	3264	7.5			(2003)
Saplings (1–3 cm)	4278	2.8	4064	8.3***	Potts et al. (2003)
Trees (>10 cm)	1226	1.5	1234	3.7***	Potts et al. (2003)
	771	0.9	816	6.4***	Nakagawa et al. (2000)

(G-test probability, ns p>0.05, ** p<0.01, ***p<0.001)

the two annual peaks of rainfall, at least among dipterocarps (Medway 1972; Ng 1981).

At LHNP a phenology census of 413 individuals in 255 species from tree towers and a canopy walkway system clearly demonstrated the general flowering characteristic of these forests (Sakai et al. 1999c). Nevertheless, drought had a strong influence on phenology, especially in new leaf production and flowering, which increase following dry spells (Ichie et al. 2004; R.D. Harrison et al. unpublished data). The pattern was pronounced in general flowering years, but there was a tendency for the greatest flowering activity in the forest to follow periods of lower rainfall in all years. Leaf production in particular showed a biannual periodicity. Even if drought is not the trigger for general flowering, as seems likely, drought directly affects the physiology of plants during bud development and may also contribute to its cause.

In LHNP trees rarely lose a substantial part of their leaves at one time, or if they do it is only at the individual level and no pattern of population-level deciduousness can be observed. During the severe drought in 1998, however, a large number of trees lost all or part of their leaves (Plate 5B), followed by a flush of new leaves shortly after the drought (see Fig. 5.6). There was a small flowering in 1998, followed by low fruit availability. The influence of the drought on general reproductive phenology of the forest is therefore sometimes difficult to assess. It seems likely that many trees were unable to allocate resources to reproduction under leaflessness, normally a stressful condition, in which many of the individuals in our phenology census died. The fact that it was also the third year in a row in which a GF took place requires special consideration.

Herbivores, pollinators, and seed dispersers. Plant phenology affects the availability of plant resources for pollinators, seed predators, and herbivores, hence it has a major influence on the ecology of forests (Coley 1998). Given that the duration of a drought is short in relation to tree life span, the impact of the 1998 drought on phenology probably had only a transient effect on plant fitness. But the impact on herbivores, pollinators, and seed dispersers, with much shorter life spans, was much more significant.

Rates of herbivory are generally highest in tropical forests (Coley 1998), but herbivores tend to specialize more on young leaves (Aide 1993) and have less host diversity than in temperate-zone forests (Marquis 1992). This leaves their populations vulnerable to crashes if there is a break in leaf production, such as occurs during a drought (Aiello 1992). However, the most noticeable effect of drought is often an explosion of herbivore populations following the renewal of rains (Aiello 1992; Coley 1998). This has been recorded in Sarawak on at least two occasions, when unidentified lepidopteran species defoliated trees of Sapotaceae, especially *Palaquium walsurifolium* in 1953 and *Dactylocladus stenostachys* (Crypteroniaceae) in 1958 (Anderson 1961, quoted in Coley 1998). The most likely explanation for these herbivore outbreaks is that the populations of their natural enemies, especially parasitoids, were severely reduced, presumably by drought (Aiello 1992; Coley 1998).

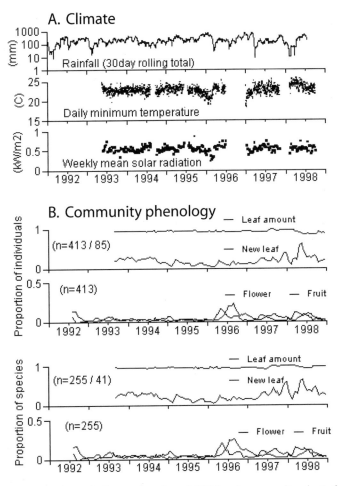

Figure 5.6. Comparison of climate data from LHNP and community plant phenology, by species and individual. A: 30-day rolling rainfall totals from the hydrological station of DID in LHNP, daily minimum temperature and weekly mean solar radiation, from the top of Tower 1 (35 m) in the Canopy Biology Plot, LHNP. B: Proportion of individuals and species with a full crown of leaves (Leaf amount: >75% of a full crown of leaves), new leaves (>10% of crown consists of new leaf), flowers and fruit by census from 1992 to 1998.

The captures of all insects, including herbivores, at the light traps on Tower 1 decreased to record low levels during the severe drought in 1998 (Itioka et al. 2003). Later, when the rain returned, the populations of some herbivores, especially Lepidoptera, exploded in coincidence with the leaf-flushing peak (Itioka and Yamauti 2004). In the roadside fig species, *Ficus fulva*, many individuals were partially, and in some cases completely, defoliated by a lepidopteran approximately one month after the drought (Harrison et al. 2003).

One extreme impact of the 1998 drought was that the pollinating and non-pollinating wasp populations of all dioecious figs under observation in LHNP became locally extinct (Harrison 2000b). There were no fig inflorescences for a period of approximately two months—roughly twice the total life span of the wasps, which prevented newly emerging wasps from finding inflorescences in which to breed. Pollinators of monoecious figs appeared less affected by the drought and recovered immediately afterward (Harrison 2000b), possibly because immigrant wasps can colonize from large distances (see Nason et al. 1996).

Fig pollinators belong to the same wasp family as many parasitoids of lepidopterans and other herbivorous insects. Because populations of their hosts crashed during the drought, and they have similar short life spans, it seems likely that such parasitoid populations also crashed or became locally extinct. As suggested earlier, a population explosion of lepidopterans after the drought may be a case of *ecological release* from natural enemies, as often occurs when herbivores are accidentally introduced to areas without natural enemies.

Unfortunately, our observations of vertebrate seed dispersers are limited and complicated by the fact that poaching occurs in the park (Shannahan and Debski 2002). For those feeding on dioecious figs, the disappearance of pollinators caused a fruiting failure and must have had a significant impact. Small fruit bats disappeared from one site (Harrison 2000b), but birds feeding on many of the monoecious figs seemed little affected.

5.4 Conclusions

Droughts as severe as that in 1998 can be described as catastrophic. The rarity of such severe droughts, which occur at intervals of several generations or more for all but the longest-lived forest trees, means that, in general, species are ill-adapted. Natural selection cannot fully prepare populations for such disasters, and local extinctions must be followed by immigration and new colonization. Fire is particularly destructive, especially over peat or in secondary forests, but even in the primary forest the understory layer is generally destroyed. Plant mortality during severe droughts is considerably higher than at normal times, and the differential mortality among species and tree size classes, with higher proportion of big trees affected, alters composition and population structure of the forest. Disruption of plant phenology, while only a temporary impact for most plant species, has greater consequences for certain herbivores, pollinators, and seed dispersers if they depend on constant food availability. The extinction of all pollinating and non-pollinating agaonid wasps on dioecious figs at LHNP is a dire case in point (Fig. 5.7). It is not unreasonable to expect that other species of short-lived insects became locally extinct at such times. The consequences for vertebrate seed dispersers, such as the hornbills and primates, may also be serious. These species require large home ranges (e.g., Bennett et al. 1997) especially at times when resources are scarce, and the size or location or connections between reserves like LHNP may be inadequate.

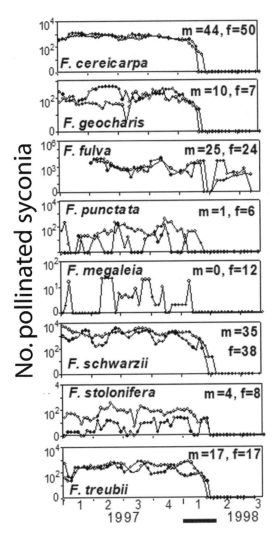

Figure 5.7. Extinction of pollinating wasps of dioecious figs during the 1998 drought at LHNP. The number of pollinated syconia on male trees indicates the population of wasp larvae developing. Pollinators live one day, hence a period of longer than one day without any pollinated infloresences on male trees indicates local extinction of the population.

The climate data presented here and by other researchers (Guilderson and Schrag 1998; Huppert and Stone 1998; Salafsky 1998) indicate that ENSO events and the droughts associated with them in Borneo have become more severe in recent decades. If, as the climate modelers are warning (Meehl 1997; Timmermann et al. 1999), this is being induced by global warming, then the implications for conservation of biodiversity are serious. It becomes all the more urgent that, as many ecologists have been saying for some time, reserves are made bigger and the matrix surrounding them managed as far as possible to provide corridors and secondary habitat (Bawa and Dayanandan 1998). This will not only prevent the extinction of some species, but will also reduce effects of drought itself by increasing rainfall through higher evapo-transpiration, and limit the spread of fires. It is also clear that new reserves are needed in places less prone to droughts and fire.

6. The Plant-Pollinator Community in a Lowland Dipterocarp Forest

Kuniyasu Momose and Abang A. Hamid Karim

6.1 Introduction

Tropical pollination biology at the community level in forests was studied first in the Neotropics (Bawa et al. 1985; Kress and Beach 1994). In a tropical rain forest in La Selva, Costa Rica, medium-sized to large bees and small diverse insects are the main pollinators in the canopy, while hummingbirds and euglossine bees are prevalent in the forest understory (Janzen 1971a; Stiles 1978; Endress 1994; Kress and Beach 1994; Rincón et al. 1999). In West Malaysia, however, plant-pollinator communities are expected to be different from those in the Neotropics, because plant reproductive phenology, fauna, and flora differ greatly between the two areas.

The phenomenon known as general flowering, or GF, occurs in West Malaysia and, as might be expected, this has consequences for the coevolutionary processes between plants and pollinators. More than 80% of the emergent and canopy tree species bloom in short periods of three to four months at irregular intervals, usually of 2 to 10 years (Ashton et al. 1988; Appanah 1993). During the remainder of the time, often for several years, both floral resources and pollinators become relatively rare. Therefore, pollinator shortages might occur unless there is a rapid response to the general flowering with population growth and adult activity (Ashton et al. 1988). According to these authors, thrips are capable of such a response. Thrips maintain a low population density using floral resources in gaps during generally flowerless seasons, and they have a short generation time of around two weeks (Appanah and Chan 1981) and high fecundity. Thus, when a general flowering starts, they can increase in numbers quickly by using the massive floral resources. However, thrip pollination of diptercarps is only known for the genus *Shorea*, section *Mutica* in the Malay Peninsula (Appanah and Chan 1981).

Are many trees that bloom in GF pollinated by thrips, or are there other types of pollinators that can quickly respond to the general flowering? This is the first question that we address. Appanah (1990) provided one clue to the answer. Carpenter bees (*Xylocopa*) shift foraging areas in GF from forest edges, much like the thrips, to the inside of closed forests. However, we question whether such shifts of foraging areas could provide sufficient pollinator populations.

Bawa (1990) stated that long-distance pollen flow is intensified in species-rich tropical rain forests, because conspecific plants are spatially isolated from each other. Hummingbirds and euglossine bees are among the most important long-distance pollinators in the Neotropics (Kress and Beach 1994), but they are absent in Southeast Asia. From La Selva, Costa Rica, 1287 species of wild flowering plants have been recorded (Hartshorn and Hammel 1994). The exact number of plant species in Lambir, Sarawak is unknown, but even when restricted to trees (>1 cm dbh) found in a 52 ha plot, over 1173 species have been recognized, and it is likely that the total number of flowering plants exceeds 2000. In and around the Canopy Biology Plot (8 ha), 999 species of flowering plants have been collected (Nagamasu and Momose 1997). Thus, species richness is very high, and conspecific plants are considered to be spatially isolated from each other. It would be expected that long-distance pollinators have important roles in the species-rich lowland dipterocarp forest. If so, what types of long-distance–specific pollinators are there? This is our second question.

Momose et al. (1998c) collected or observed flower visitors of 270 plant species of 73 families (see Appendix A). Based on that study, we describe pollination syndromes, describe the plant-pollinator community in this lowland dipterocarp forest, and try to answer the above questions. Figs (*Ficus* spp., Moraceae) are not included, and their pollination is discussed in Chapter 10.

6.2 Plant-Pollinator Interactions

Mammal pollination. Four species in three families (Leguminosae, Loganiaceae, Sapotaceae) were pollinated by bats, *Macroglossus* spp. (see Plate 7G) and one species, Sapotaceae, by squirrels and flying squirrels (Yumoto et al. 2000). Rewards were nectar in three bat-pollinated species and a berry-like sweet corolla in one bat-pollinated species and the squirrel-pollinated species. Flowers of all five species were white and emitted a strong scent, but the shapes varied widely.

Bird pollination. Nineteen species in seven families were pollinated by birds— *Nectarinia jugularis, Arachnothera longirostra,* and *A. robusta,* Nectariniidae

(Plate 7C–F). Flowers were bilabiate or tubular in shape (some that burst open); white, red, or orange in color; without scent or with a strong odor (Yumoto et al. 1997). Details of vertebrate pollination are explained in Chapter 12.

Social bee pollination. Flowers of 86 species in 42 families were predominantly visited and pollinated by the genera *Apis* (honeybees), *Trigona, Lisotrigona* and *Pariotrigona* (stingless bees), and *Braunsapis* (little carpenter bees); see Plate 6, C, D, I and Plate 9. Among them, the number of *Apis dorsata* (giant honeybees) increased greatly during the GF by colony immigration and multiplication, but they were much less abundant in non-GF (Itioka et al. 2001a). In the daytime, they were found on flowers together with other social bees and diverse insects of several families of Coleoptera, Diptera, and Hymenoptera. However, in the early morning before sunrise (0500–0600), only *A. dorsata* among social bees can forage (Dyer 1985). Two species of *Dryobalanops* (Dipterocarpaceae) and *Dillenia excelsa* (Dilleniaceae) flowered in the early morning (0500), where *A. dorsata* was an especially important pollinator for those plants.

Other social bees are not migratory and were important pollinators, especially in non-GF (Inoue et al. 1984a; Momose et al. 1996; Nagamitsu and Inoue 1997b), when *A. dorsata* was rare. They visited flowers together with diverse insects of several families of Coleoptera, Diptera and Hymenoptera. Flowers dominated by social bees were brushlike, radially symmetric, or cup-shaped, and white or yellow in color. Further themes on bees and their biology are discussed in Chapters 1, 7, 8, and 11.

Xylocopa pollination. Eight species in seven families were pollinated mainly by *Xylocopa* (carpenter bees). The flowers usually had large flowers with long pistils. The locations of anthers and stigmata fitted to the body sizes of carpenter bees, but other visitors like stingless bees were not excluded. Some *Xylocopa* flowers had porose anthers, from which carpenter bees collected pollen grains by vibrating their flight muscles, or, so-called buzz-collecting. Carpenter bees usually foraged at forest edges and open habitats but were sometimes found in the forest canopy. Although Appanah (1990) reported carpenter-bee pollination of forest trees in the Malay Peninsula, they were not among the main pollinators in the forest trees in our study site in Sarawak. However, carpenter bees sometimes visited papilionaceous flowers (usually *Megachile*-pollinated; see below) and became dominant pollinators if plants were located in gaps.

Amegilla pollination. Seventeen species in six families (Costaceae, Gesneriaceae, Marantaceae, Pentaphragmataceae, Polygalaceae, and Zingiberaceae) were pollinated only by the trap-lining long-tongued bees, *Amegilla pendleburyi* and *A. insularis* (Plate 6J). Flowers were odorless, bilabiate, colored white, yellow, purple, or orange, with nectar guides. Abundant nectar was secreted in these flowers and protected from other insects by a specialized floral shape (Kato et al. 1993a). Males of *A. pendleburyi* were observed in mating territories around flowers. Whereas *A. pendleburyi and A. insularis* forage on forest floors only, one more species of *Amegilla, A. andrewsi*, usually foraged at forest edges and open habitats. *A. andrewsi* often visited *Xylocopa*-pollinated flowers but was not dominant there.

Halictid pollination. Twenty-one species in nine families (including Zingiberaceae, Verbenaceae, Acanthaceae) were pollinated by smaller bees, *Nomia* or *Thrinchostoma* (Halictidae). Their flowers were similar to *Amegilla*-pollinated flowers in shape but smaller in size.

Megachile pollination. Megachile (Megachilidae) bees appeared twice (May–July 1993 and May–July 1996) in the 53-month census period. Plants with papilionaceous flowers (four species of two families: Leguminosae and Xanthophyllaceae) flowered in synchrony with the emergence of *Megachile* and were pollinated by them. *Megachile*-pollinated flowers seem to have shorter flowering cycles than most other plants. Nectar and pollen were protected by keel petals from other visitors. However, after visitation by *Megachile*, a small amount of pollen fell from the anthers and was deposited on the surface of petals. Stingless bees and beetles were often found collecting those pollen grains on petal surfaces, but they did not touch stigmata and therefore were not pollinators.

Butterfly pollination. There were two shapes of butterfly-pollinated flowers: one brushlike and the other tubular. Butterfly-pollinated flowers (six species in three families: Leguminosae, Rubiaceae, and Verbenaceae) were usually odorless and orange in color when fresh, but they often remained in inflorescences, turning reddish, even after pollinated. This phenomenon was common in both brushlike flowers (*Bauhinia*, Leguminosae) and tubular flowers (*Ixora*, Rubiaceae).

Moth pollination. Moth-pollinated flowers (two species in two families: Dipterocarpaceae and Lecythidaceae) were also brushlike or thinly campanulate (mostly tubular). They had an odor and were white or pale yellow in color. Plate 6B shows a unique example of moth pollination of a gymnosperm, *Gnetum gnemon* (Gnetaceae) that was found in our study site (Kato and Inoue 1994; Kato et al. 1995b).

Beetle pollination. Fifty-six species in 11 families were pollinated by beetles (Plate 6E–G). There were three types of rewards for beetles: floral tissues, stigmatic secretions, and pollen. Beetle-pollinated flowers were radially symmetrical, *urceolate*, or formed a floral chamber, and were yellow, white, or pink in color. Beetle pollination, which is described further in Chapter 9, is categorized into four types pertaining to four plant families: Annonaceae, Araceae, Myristicaceae, and Dipterocarpaceae.

Diverse insect pollination. Thirty-seven species in 22 families were visited and pollinated by several orders of insects and were not dominated by any single family. The floral characters were those common to social bee-pollinated plants. Large flower patches tended to be dominated by social bees.

Others. Flies (Culicidae, Lauxaniidae, Drosophilidae, Calliphoridae) were attracted to four species in three families (Burmanniaceae, Gnetaceae, and Triuridaceae) of forest floor plants (Kato et al. 1995b; Kato 1996) and to some trees, Moraceae (Plate 6A). Wasps (Vespidae) were attracted to *Casearia grewiaefolia* (Flacourtiaceae; Kato 1996). Cockroaches (Blattidae) were attracted to *Uvaria* aff. *elmeri* (Annonaceae; Nagamitsu and Inoue 1997a). Mechanisms of attraction of special pollinators and exclusion of other insects were uncertain in these examples. The genera *Popowia* (Annonaceae; Momose et al. 1998b) and *Horsfieldia* (Myristicaceae) attracted thrips (Thysanoptera, Thripidae) by odor and offered floral tissues and pollen as rewards. Other visitors were excluded by the small entrances of the pollination chambers. In Chapter 11, pollination by some specific insects is given a more detailed review.

6.3 Plant Habit and Pollination Syndrome

Plants pollinated by generalists (small social bees and diverse insects) were found nearly everywhere and were especially common at *intermediate strata* of closed forests (see Fig. 6.1). Plants pollinated by vertebrates were found in the *subcanopy* (including epiphytes) and *gaps*. Plants pollinated by solitary bees and lepidopterans were found among gaps and on the forest floor. Beetle-pollinated plants were found in all strata (most common at the emergent stratum) of closed forests but were not found in gaps. The significance of such patterns should be tested, while excluding phylogenetic constraints.

The pattern that may explain why trap-lining solitary bees and lepidopterans are found at the forest floor and other insect pollinators at higher strata has been

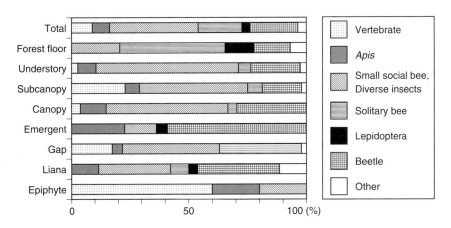

Figure 6.1. Frequency distribution of pollination systems in different plant habits: standing plants of the mature phase forest at different forest strata—(1) forest floor: <2.5 m; (2) understory: 2.5–12.5 m; (3) subcanopy: 12.5–27.5 m; (4) canopy: 27.5–42.5 m; and (5) emergent: >42.5 m—standing plants in gaps, and other life forms (liana and epiphyte).

commented upon by Heinrich and Raven (1972). Plants at higher strata can attract large numbers of pollinators by massive blooming, while plants in the forest floor cannot provide a large display. Instead, they attract specialized pollinators by offering considerable nectar per flower, protected from other visitors by a specialized floral morphology.

We found that plants in gaps also attract specialized pollinators. We believe that gap species, moreover, cannot provide large displays in part because of their need for continuous or frequent reproduction (thus production of flowers and fruit). They do not accumulate resources for periodic reproduction but reproduce continuously or frequently in an ephemeral habitat where the mortality is high and opportunity for establishment of the next generation is unpredictable (Momose et al. 1998a).

Plants pollinated by vertebrates were found in the subcanopy of mature forests or in gaps. It seems difficult to explain this pattern by some of the principles mentioned above, and behavioral characteristics of the animals should be considered.

6.4 General Flowering and Pollination

In the case of thrip pollination of *Shorea* in the Malay Peninsula, adult thrips quickly increased after GF began (Appanah and Chan 1981). In contrast, in our study site their rapid increase was not observed and they had limited roles as pollinators (Sakai et al. 1999b). In addition, carpenter bees (*Xylocopa* spp.) shift foraging areas in GF from forest edges to closed forests in the Malay Peninsula (Appanah 1990), but they were not common in closed forests during GF at our study site.

We hypothesize that some beetles can use the rapidly increasing flower resources, and several plant species reproducing in GF use those beetles as pollinators, which qualifies them as loose niche pollinators (see Chapter 1). Chrysomelids pollinating some dipterocarps fed on dipterocarp leaves in non-GF and shifted resources to floral tissues in GF, because they were collected on dipterocarp leaves in flowerless seasons.

Social bees (Meliponini, *Apis* and to a lesser extent *Braunsapis*) also had important roles as pollinators in the lowland dipterocarp forest, compared to the Neotropical forest in Costa Rica, where medium- to large-sized bees (Apidae: Centridini, Euglossini, Xylocopini, some Meliponini) are dominant (Bawa et al. 1985, Kress and Beach 1994), and the genus *Apis*, until recently (Roubik 2002) was absent. Unlike the predictable annual flowering cycles in Costa Rica (Newstorm et al. 1994a,b), the general flowering of lowland dipterocarp forests is *supra-annual*, and its intervals are not constant. Social bees can use such unpredictably fluctuating floral resources by generalizing, storing food, or migrating great distances.

Apis dorsata, as a colony in transit, can move over 100 km (Koeniger and Koeniger 1980). In Sarawak, they migrate to lowland dipterocarp forests as soon as the general flowering starts, and as the general flowering finishes they leave (Itioka et al. 2001a). In non-GF periods, their nests are apparently restricted to mountain forests (T. Inoue, personal communication) or in swamp forests (H. Samejima, personal communication).

By stabilizing the effects of temporal changes in floral resources at a colony level (Inoue et al. 1984b), *Trigona* colonies can maintain forager workers, which can quickly start foraging in response to abrupt increases of ephemeral and massive floral resources in both GF and non-GF, and then store these resources in the nest (Inoue et al. 1984b, 1990, 1993; Salmah et al. 1990; Nagamitsu and Inoue 2002). Recruitment behavior of social bees can further increase the quick exploitation of mass-flowering trees (Roubik 1989, Roubik et al. 1995).

6.5 Long-Distance Pollinators

Hummingbirds (long-billed nectar-feeding birds) and euglossine bees (long-tongued bees) are important long-distance–specific pollinators in the Neotropics (Kress and Beach 1994; Roubik and Hanson 2004). In these areas, hummingbirds and euglossine bees are diverse and coexist by visiting different floral resources (Janzen 1971a; Stiles 1978; Roubik and Hanson 2004). Long-billed nectar-feeding birds and long-tongued bees also are found in Southeast Asia; the former are spiderhunters and sunbirds, and the latter are *Amegilla*. However, the species richness of these long-distance–specific pollinators is much lower in Southeast Asia. Only three species of long-billed nectar-feeding birds (*Arachnothera longirostra*, *A. robusta* and *Nectarinia jugularis*) and two species of long-tongued bees (*Amegilla pendleburyi* and *A. insularis*) were pollinators in our study site. The proportion of plant species pollinated by these long-distance–specific pollinators is smaller in Lambir than in La Selva (bird: 7.0 vs. 14.9%; long-tongued bee: 6.3 vs. 8.7%), as shown in Fig. 6.2).

Plant species pollinated by mammals and lepidopterans (other types of long-distance pollinators) are also less frequent in Lambir than in La Selva (mammal: 1.5 vs. 3.6%; lepidopteran: 3.3 vs. 12.3%). *Xylocopa*, Halictidae, and *Megachile* are also long-distance–specific pollinators in Lambir (not specified in the data set of Kress and Beach 1994). However, plant species pollinated by them in Lambir represent only 2.9%, 7.7%, and 1.5% of the whole, respectively (Momose et al. 1998c). Some beetles may be capable of moving long distances (Young 1988). However, if dipterocarps, which are pollinated by beetles feeding on floral tissues in GF, are excluded, the frequency of beetle pollination is similar between Lambir (10.7%) and La Selva (12.7%).

Long-distance pollinators have less important roles in the species-rich lowland dipterocarp forest of Lambir than in the Neotropical forest. They require a continuous supply of rich resources, because their costs for body maintenance and foraging are high (Heinrich and Raven 1972), and irregular and ephemeral floral resources in lowland dipterocarp forests are inadequate for their survival.

Highly eusocial bees (*Apis* and Meliponini) are not specific pollinators but

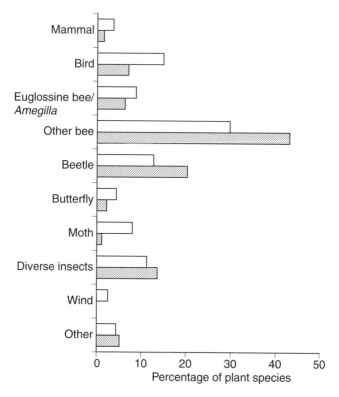

Figure 6.2. Comparison of the frequency distributions of pollination systems between two tropical regions: (*open bars*) a Neotropical lowland rain forest in La Selva, Costa Rica (from Kress and Beach, 1994) and (*shaded bars*) a Southeast Asian lowland dipterocarp forest in Lambir, Sarawak, Malaysia (Momose et al. 1998c). The frequency of different pollination syndromes was significantly different between two sites ($G = 18.0$, $P<0.05$; wind pollination was excluded because of insufficient information in Lambir).

generalists in the sense that they use a wide range of floral resources, but these are flexible, based on competition with other foragers. Such bees communicate with colony members and can harvest floral resources effectively (Seeley 1985; Roubik 1989). According to pollen analyses at bee nests by Nagamitsu and Inoue (2002), at any one time, they often major on one or a few plant species that offer the richest floral resources (see also Seeley 1985). In this case, conspecific plant individuals can be selectively visited by social bees. Honeybees, in particular, are considered to have a wide foraging area (5 km or more), which permits long-distance pollen transfer (Seeley 1985). To attract them, plants must have a reproductive phenology of the mass-flowering type. This might be another way of achieving effective long-distance pollen transfer, and it is a more favored strategy in lowland dipterocarp forests.

7. Floral Resource Utilization by Stingless Bees (Apidae, Meliponini)

Teruyoshi Nagamitsu and Tamiji Inoue

In this chapter, we examine patterns of floral resource utilization, mechanisms of floral resource partitioning, and foraging responses to general flowering (GF) according to studies on coexisting stingless bee species in the Lambir Hills National Park, or, LHNP (Nagamitsu and Inoue 1997b, 1998, 2002; Nagamitsu et al. 1999b). First, floral resource utilization is described, based on flower visitation and pollen diets. Second, mechanisms of resource partitioning are determined by field experiments and morphological analysis. Third, responses in colony number and foraging activity to GF are demonstrated at the local community scale by censuses of non-arboreal, ground-level colonies.

7.1 Introduction

Stingless bees (Hymenoptera, Apidae, Meliponini) are distributed in tropical and subtropical areas throughout the world (Michener 2000). They are social insects and live in perennial colonies with queens, workers (sterile females), and males (drones). Workers collect materials to construct and defend their nests and gather foods to maintain metabolism and reproduce. Foraging bees may collect nectar and occasionally honeydew from bugs and scale insects, while only female bees collect pollen, and even flesh from dead animals or inorganic salts from various sources (Roubik 1989). Among resources, nectar and pollen of flowers are the food sources of most species. Stingless bees are abundant flower visitors in tropical rain forests. In La Selva, Costa Rica, medium-sized bees that include meliponine, halictid, and megachilid bees are believed to pollinate 14% of plant species (Kress and Beach 1994). In Lambir, social bees that include honeybees, stingless bees, and the subsocial allodapine bees (small carpenter bees that make nests in stems) apparently pollinate the largest number (32%) of plant species (Momose et al. 1998c). However, interaction between plants and stingless bees is not only mutualistic but also antagonistic. Nectar and pollen robbing by stingless bees reduces reproductive success of many plants (Roubik 1982, 1989).

Stingless bees are generalist foragers and may visit flowers of more plant species than coexisting solitary bees and wasps (Heithaus 1979). Plant taxa of nectar and pollen sources are shared among stingless bee species as well as between stingless bees and honeybees (Koeniger and Vorwohl 1979; Roubik et al. 1986; Wilms and Wiechers 1997; Eltz et al. 2001). Such generalized utilization of common resources results in interference and exploitative competition that reduces not only foraging efficiency at feeding patches (Johnson and Hubbell 1974; Roubik 1980) but also pollen and nectar harvest of colonies (Roubik et al. 1986; Wilms and Wiechers 1997). Availability of either foods or nest sites is likely to limit population density of stingless bees (Hubbell and Johnson 1977; Inoue et al. 1993; Eltz et al. 2002). Effects of competition on population density, however, have not been well confirmed (Roubik 1983; Roubik and Wolda 2001).

Different foraging strategies also allow stingless bee species to share the same type of resources by partitioning them in different times and places (Johnson 1982). These foraging strategies differ according to variation in foraging traits, such as body size, energetic cost of foraging, aggressiveness, communication, and recruitment. With respect to variation in aggressiveness, two mechanisms have been proposed for resource partitioning. First, the more aggressive species monopolize clumped and rich resources, whereas the less aggressive species are excluded from the resources and forage on scattered or poor resources (Johnson and Hubbell 1975; Johnson 1981). Second, early-arriving, less aggressive species are temporally replaced with late-arriving, more aggressive species, because more aggressive species require more time to discover new resources than do the less aggressive ones (Hubbell and Johnson 1978).

Aggressive foraging behavior of stingless bees has been rarely investigated in Asia. Behavior of stingless bees and honeybees at artificial feeders was observed in Sri Lanka (Koeniger and Vorwohl 1979) and Peninsular Malaysia (Khoo 1992). Koeniger and Vorwohl (1979) suggested that aggression of stingless bees compensated for disadvantage due to smaller foraging area of stingless bees than that of honeybees. Khoo (1992) observed that more aggressive species that arrived later at feeders excluded less aggressive species that had already visited the feeders. These studies, however, were conducted using both artificial and highly concentrated resources and were not designed to examine how aggressiveness affects the partitioning of floral resources.

Frequent partitioning of common resources by foraging in different times and places may be a unique feature of social insects, which often evaluate changing resource availability as they communicate the locations of optimal feeding sites

(Seeley 1995, Davidson 1998). However, there are also other factors that may enable stingless bee species to partition floral resources among plant taxa having different floral traits. A lowland mixed-dipterocarp forest in LHNP has extremely high tree species richness (LaFrankie et al. 1995), which is suitable to study partitioning of a wide taxonomic range of floral resources. Because the architecture of lowland mixed dipterocarp forests is complex (Ashton and Hall 1992), stingless bees may partition flowers in different locations in the forest. Furthermore, flowers of different taxa are morphologically diverse, and thus stingless bee species may specialize on flowers of a particular morphology. Previous studies of flower visits and pollen diets of stingless bees have not been well designed to detect partitioning in relation to such variation in floral features.

Another characteristic of lowland mixed dipterocarp forests is general flowering (Ashton et al. 1988). General flowering (GF) may have great importance to the population and behavior of pollinators, caused by changing availability of floral resources. In Panama, the El Niño-southern oscillation produced a floral resource flush in the dry season, followed by an increase in bee populations (Roubik 2001). In Malaysia, specific responses of some pollinators, such as thrips, chrysomelid beetles, and giant honeybees, have been proposed. Thrip populations rapidly increase due to their short generation time and high fecundity (Appanah and Chan 1981). Giant honeybees immigrate to lowland mixed dipterocarp forests as soon as general flowering starts, and when it finishes they leave these particular forests (Itioka et al. 2001a). Stingless bees, however, show none of these responses, i.e., rapid population growth (Inoue et al. 1993), changes in food types (Nagamitsu and Inoue 2002), or migration over a long distance (Inoue et al. 1984a).

7.2 Stingless Bees in Lambir Hills National Park

The taxonomy of stingless bees of Southeast Asia has advanced due to an early foundation (Schwarz 1937, 1939), excellent revisions (Sakagami 1975, 1978), and the relatively small number of species. In Asia and the Sunda islands, there are three genera, *Trigona, Pariotrigona,* and *Lisotrigona*, and three subgenera, *Lepidotrigona, Homotrigona,* and *Heterotrigona* in the genus *Trigona*—as shown in Table 7.1 and Plate 9 (Michener 2000). There have been different systems for classifying stingless bees (Moure 1961; Wille 1979). Although these systems are still modified, an intermediate system of special interest for studies in Southeast Asia was presented (Sakagami 1982), as follows: In the subgeneric system of Sakagami (1982), the subgenus *Heterotrigona* in Michener (2000) was further divided into eight subgenera and five species groups. A key to workers of the Sumatran species, which includes most of the regional species, is available (Sakagami et al. 1990).

The distribution of Indo-Malayan stingless bees is described based on records from 10 localities and collections of the Canopy Biology Program in Sarawak, or, CBPS (see Table 7.1). CBPS collected 25 species in six locations throughout

Table 7.1. Distribution and abundance of stingless bees in the Indomalayan region

Genus (Subgenus)[a] (Subgenus)[b]	SL	ID	BU	TH	IC	MA	BO	BO'	SM	JA	TA	LM	SB
Pariotrigona													
1. *pendleburyi**						+	+	+	+				1
2. sp. nov. aff. *pendleburyi*								x					
Lisotrigona													
3. *cassiae*	x	+											
4. *scintillans*				+	+	+	+	+	+			8	1
Trigona (Lepidotrigona)[a]													
5. *nitidiventris**				+		+	+	+	+	+		45	23
6. *trochanterica*						+			+				11
7. *terminata**				+	+	+	+	+	+	+		46	69
8. *ventralis ventralis**		+		+	+	+	+	+	+	+		278	2
9. *v. hoozana*											+		
Trigona (Homotrigona)[a]													
10. *fimbriata**				+	+	+	+	+	+			310	52
Trigona (Heterotrigona)[a]													
11. (Heterotrigona) *itama*				+		+	+	+	+	+		1016	455
12. (H.) *erythrogastra*						+	+	+				113	
13. (Lophotrigona) *canifrons*				+		+	+	+	+			405	153
14. (Geniotrigona) *thoracica*				+		+	+	+	+			28	34
15. (Odontotrigona) *haematoptera**							+	+	+			46	
16. (Platytrigona) *hobbyi*							+	+					
17. (Tetrigona) *apicalis*			x	+	+	+	+	+	+	+		151	174
18. (T.) *peninsularis*				+	+	+		+					
19. (T.) *melanoleuca*				+		+	+	+	+			34	
20. (T.) *binghami*			+					+	+				
21. (Sundatrigona) *moorei*						+	+	+	+			29	64
22. (S.) *lieftincki*									+				25
23. (Tetragonula) *atripes*				+	+	+	+	+	+				36
24. (T.) *collina*			+	+	+	+	+	+	+			186	132

Table 7.1. *Continued*

Genus (Subgenus)[a] (Subgenus)[b]	Locality[c]											Abundance[d]	
	SL	ID	BU	TH	IC	MA	BO	BO'	SM	JA	TA	LM	SB
25. (T.) fuscibasis				+		+	+		+				22
26. (T.) rufibasalis						+	+	+				29	
27. (T.) reepeni			+		+		x	+	+			11	56
28. (T.) iridipennis	+												
29. (T.) bengalensis	+												
30. (T.) pagdeni			+	+	+								
31. (T.) fuscobalteata*			+	+	+	+	+	+	+			380	112
32. (T.) geissleri			+		+	+	+					2	
33. (T.) gressitti				+									
34. (T.) melina			+		+	+	+	+				30	5
35. (T.) melanocephala						+	+					58	
36. (T.) melanocephala small type							+						
37. (T.) drescheri					+		+	+	+				86
38. (T.) minangkabau							+	+					102
39. (T.) minangkabau f. darek								+					142
40. (T.) hirashimai			+		+								
41. (T.) pagdeniformis			+										
42. (T.) laeviceps	+	x	+	+		+	+	+	+			74	369
43. (T.) laeviceps small type							+	+					7
44. (T.) sarawakensis													
45. (T.) zucchii					+								
46. (T.) minor					+								
47. (T.) forma B	+												
Total	2	5	4	22	11	29	29	27	25	8	1	3279	2133

[a] Michener (2000)

[b] Sakagami (1982)

[c] Sri Lanka (SL), India (ID), Burma (BU), Thailand (TH), Indochina (IC), Peninsular Malaya (MA), Borneo (BO), collections of Canopy Biology Program in Sarawak (BO'), Sumatra (SU), Java (JA) and Taiwan (TA)

[d] Lambir Hills National Park (LM) and central Sumatra (SB)

x: unpublished data by Sakagami, *some varieties were distinguished by Schwarz (1939)

Sarawak (Inoue et al. 1994). The distribution shows clearly the central and peripheral regions with respect to the number of species. Evidently, Borneo and Malaya represent the center of species richness, followed by Sumatra and Thailand. CBPS found one new species of *Pariotrigona* in LHNP (Inoue et al. 1994). The range of *Trigona minangkabau* in Borneo was found to include LHNP and Mulu National Park. Distinct forms in *T. melanocephala* and *T. laeviceps*, likely to be different species, were recorded from Borneo. The CBPS collection lacks six of the species recorded by Schwarz (1939).

Assemblages of stingless bees collected on flowers of various plant species are similar between central Sumatra and LHNP, as shown in Table 7.1 (Inoue et al. 1990; Nagamitsu et al. 1999b). The relative abundance of some subgenera in the system of Sakagami (1982), however, differs between the two sites. Compared with central Sumatra, subgenera *Lepidotrigona*, *Homotrigona*, *Heterotrigona*, and *Lophotrigona* were abundant in LHNP, whereas *Sundatrigona* and *Tetragonula* were rare in LHNP. This difference is likely due to variation in habitat types between the two sites. The study sites in central Sumatra included secondary forests and disturbed areas, whereas primary forests were mainly surveyed in LHNP.

7.3 Patterns of Floral Resource Utilization

Stingless bees have been regarded as generalists in floral resource utilization, which rarely specialize in particular plant taxa with unique floral traits. However, some preferences in foraging stratum were suggested by a light-trap study (Roubik 1993). We examined whether stingless bee species prefer flowers in specific locations in forest architecture based on flower visits of 11 abundant species of the genus *Trigona*—species 8, 10, 11, 12, 13, 17, 24, 31, 34, 35, 42 in Table 7.1 (Nagamitsu et al. 1999b). Flower visits were determined by the presence of species in collections of stingless bees that visited flowers of a plant within a flowering period. We obtained a total of 100 collections from canopy, gap, and understory flowers in LHNP during 1993 and 1996. Two of the 11 species, *T. fuscobalteata* and *T. melanocephala*, showed non-random patterns in their flower visitation. That relationship was indicated by the proportion of collections that contained the two species, in relation to the flower locations (see Fig. 7.1A). *Trigona fuscobalteata* frequently visited canopy and gap flowers, while *T. melanocephala* most often visited understory flowers. Based on the same data set, we examined preference in floral shapes and found that *T. erythrogastra* more frequently visited deep flowers with complex shapes and closed petals rather than shallow flowers in cup, whorl, or brush shapes (see Fig. 7.1B). Thus, some preferences in flower locations and floral shapes were observed in three species; such preferences promote floral resource partitioning, in particular between *T. fuscobalteata* and *T. melanocephala*.

To examine partitioning of nectar sources between the two species, we mea-

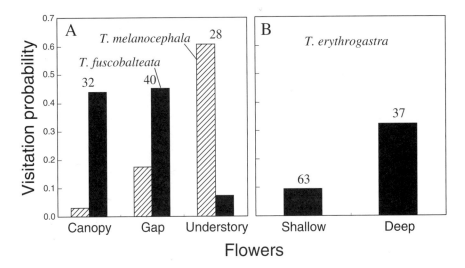

Figure 7.1. Visitation probability and number of recorded bees. (**A**) *Trigona fuscobalteata* and *T. melanocephala* at canopy, gap, and understory flowers; and (**B**) *T. erythrogastra* at shallow or deep flowers (after Nagamitsu et al. 1999b).

sured sugar concentration of nectar loads brought to nests (Nagamitsu and Inoue 1998). The sugar concentration of nectar may become higher in canopy and gap flowers than in understory flowers, because sunny, dry, and windy conditions in the canopy and gaps cause water in the nectar of flowers to evaporate. Thus *T. fuscobalteata* is expected to have nectar loads of higher sugar concentration than *T. melanocephala*. The sugar concentration of nectar loads differed among stingless bee species and among times of day (see Fig. 7.2). Sugar concentration increased from the morning to the afternoon, as expected, by nectar evaporation. *Trigona fuscobalteata* had nectar of the highest sugar concentration of six examined species. Sugar concentration of nectar loads of *T. melanocephala*, however, was not the lowest of the six species.

To examine partitioning of pollen sources between *T. fuscobalteata* and *T. melanocephala*, we collected pollen loads brought to nests, identified morphological types of pollen grains in the loads, and calculated interspecific similarity of pollen diets (Nagamitsu et al. 1999b). The mean rank of pollen diet similarity was lowest between *T. fuscobalteata* and *T. melanocephala* of six pairs of four examined species (see Fig. 7.3). However, similarity varied temporally and was extremely high at a period in 1996, when GF occurred. Except for this period, pollen diet similarity between *T. fuscobalteata* and *T. melanocephala* was very low, indicating almost complete partitioning of pollen sources—and a response to competition.

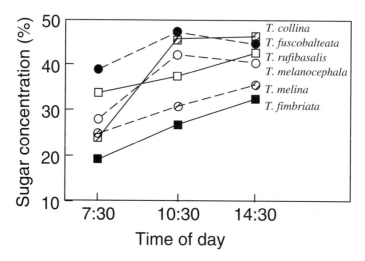

Figure 7.2. Temporal changes in sugar concentration (% total dissolved solids, per volume) of nectar loads for six *Trigona* (after Nagamitsu and Inoue 1998, © Blackwell Publishing).

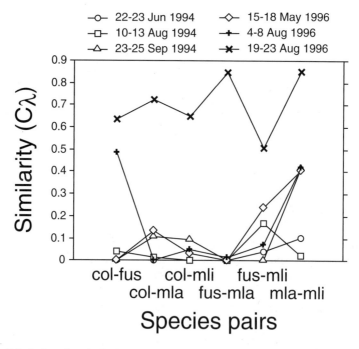

Figure 7.3. Pollen diet similarity (Morishita similarity index) at six periods in 1994 and 1996 in pairs of four *Trigona* species, *T. collina* (col), *T. fuscobalteata* (fus), *T. melanocephala* (mla), and *T. melina* (mli); after Nagamitsu et al. (1999b).

7.4 Mechanisms of Floral Resource Partitioning

Floral resource utilization of *Trigona* species in LHNP showed preference in flower locations and floral shapes; it indicated partitioning of nectar and pollen between species that preferred different flower locations. Preferences in flower visitation may result from specific foraging traits, and matching between the foraging trait variation and floral resource heterogeneity may result in floral resource partitioning.

Trigona melanocephala that preferred understory flowers was expected to forage the lowest stratum in the forest, whereas *T. fuscobalteata* was predicted to have the highest foraging stratum. To examine this prediction, we conducted field experiments three times using honey-water feeders that were vertically arranged at towers in LHNP (Nagamitsu and Inoue 1997b). In all three experiments, *T. fuscobalteata* did not visit the feeders frequently. *Trigona melanocephala*, however, frequently appeared in all the experiments, and always preferred the lowest feeders more than other feeders (see Fig. 7.4).

Other *Trigona* and honeybees (*Apis koschevnikovi*) did not show stable patterns in foraging heights (Roubik et al. 1995, 1999). Possible factors for the variation in foraging heights are different physiological traits that fit sunny (canopy and gap) and shaded (understory) conditions. In La Selva, Costa Rica, *Trigona* species that visited *Justicia aurea* flowers differed between sunny and shaded places (Willmer and Corbet 1981). *Trigona ferricauda* and *T. angustula* (called *jaty* in that publication) preferred flowers in sunny places, whereas flowers in shaded places were visited by *T. fulviventris*, which was also more often caught in the understory than in the canopy in another light-trap study (Roubik 1993). These findings in La Selva and LHNP suggest that different foraging strata in relation to sunny and shaded conditions cause floral resource partitioning.

Floral morphology determines accessibility to floral resources. Only flower visitors with long tongues obtain nectar secreted in the bottom of long, narrow corollas. Such morphological matching between floral parts and feeding organs is a major mechanism of floral resource partitioning. In general, in LHNP, *T. erythrogastra* preferred deep flowers and thus was expected to have longer tongues than the other species. To compare morphology among *Trigona* species in LHNP, we measured lengths of seven body parts in 17 species, and we performed discriminant analysis to separate these species based on the morphological characters (Nagamitsu and Inoue 1998). Seven characters were summarized in two variables and were plotted on a coordinate of the two variables (see Fig. 7.5). *Trigona erythrogastra* and *T. thoracica* were located far from the other species on vectors of tongue lengths (PL and GL), which indicates that the two species have the longest tongues relative to their body sizes. Although flower visits of *T. thoracica* were not recorded, a morphological match between *T. erythrogastra* and its visited flowers was found.

In addition to the matching between foraging trait variation and floral resources, we found that aggressive interference caused temporal resource parti-

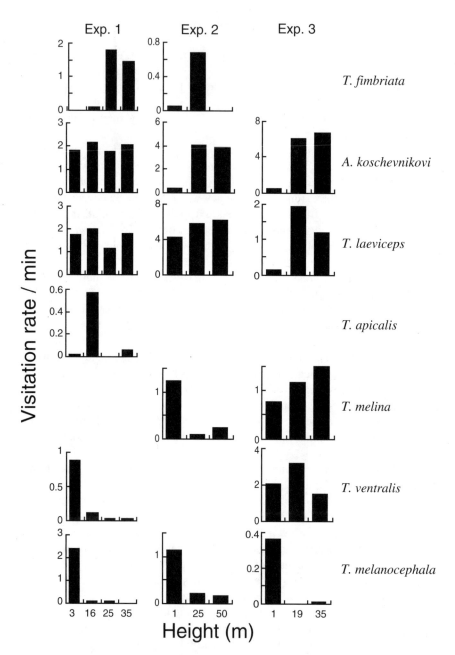

Figure 7.4. Rates of visitation to feeders at different heights by six *Trigona* species and a honeybee, *Apis koschevnikovi*. Results of three experiments are shown.

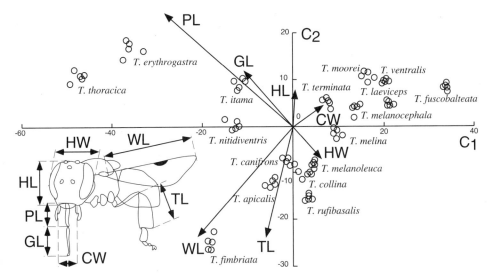

Figure 7.5. Morphological measurements and canonical discriminant analysis by seven morphological characters, head width (HW), head length (HL), prementum length (PL), glossa length (GL), clypeal width (CW), wing length (WL), and tibial length (TL). Circles indicate bees on first two canonical variables (C1 and C2); arrows indicate vectors for standardized coefficients for all seven characters (after Nagamitsu and Inoue 1998, © Blackwell Publishing).

tioning within a patch. There are two mechanisms for the temporal partitioning between more aggressive and less aggressive species, which was proposed from studies on Meliponini in Costa Rica (Johnson and Hubbell 1975; Hubbell and Johnson 1978; Johnson 1981).

One mechanism is that more aggressive species monopolize rich resources, while less aggressive species that are excluded from the rich resources use relatively poor ones. We observed such partitioning within a flower patch, in which the nectar production rate changed during a day (Nagamitsu and Inoue 1997b). Flowers of a *Santiria laevigata* opened early in the morning, and the nectar production rate peaked around noon (see Fig. 7.6).

Thirteen *Trigona* species visited the flowers of this tree, and seven of them were abundant. One species, *T. canifrons*, foraged aggressively and visited most frequently when nectar production peaked. *Trigona canifrons* excluded three *Heterotrigona* species from the flowers at midday but did not greatly affect the *Lepidotrigona*. Although it is not clear why *Lepidotrigona* were more tolerant to aggression than *Heterotrigona*, the foraging method observed in the *Lepidotrigona* species was the same as 'insinuation' (Johnson 1982).

The other mechanism for temporal resource partitioning results from a trade-off between finding and defending a resource. This trade-off results in temporal partitioning, in that early-coming, less aggressive species are replaced with late-coming, more aggressive species. To examine this trade-off, we observed ag-

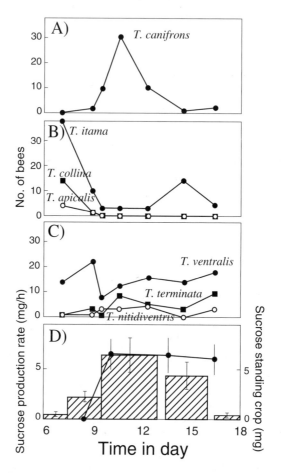

Figure 7.6. Diurnal changes in flower visits by seven *Trigona* and nectar secretion in a canopy tree, *Santiria laevigata*. Flower visits (N bees collected in 10 min at flowers) of (**A**) aggressive *T. canifrons*, (**B**) non-aggressive *Heterotrigona* species, and (**C**) non-aggressive *Lepidotrigona* species (after Nagamitsu and Inoue 1997b).

gressive behavior, interference competition, and time of the first arrival at honey-water feeders located on the towers in LHNP (Nagamitsu and Inoue 1997b). Six *Trigona* and one honeybee, *Apis koschevnikovi*, frequently visited the feeders. *Trigona fimbriata*, *T. apicalis*, and *T. melina* were aggressive toward others. *Trigona fimbriata* dominated *T. apicalis* and *T. melina*, because it won the physical battles, and its interference more effectively reduced honeybee visits than did that of the other two species. *Trigona ventralis*, *T. laeviceps* and *T. melanocephala* were not aggressive. Visits by *Trigona ventralis* were reduced less after encounters with the aggressive species than *T. laeviceps* and *T. melanocephala*. Rank of the aggressive dominance based on these findings tended to be negatively correlated with rank of the first arrival at the feeders (see Fig. 7.7).

This negative correlation suggests a trade-off between searching and defensive abilities of *Trigona* in LHNP. Sympatric ant species also have such a trade-off (Fellers 1987; Holway 1999). Both ant and stingless bee workers forage from their nest by communicating food locations to nest mates and bring food to their nest. If the number of experienced foragers is limited, investment of the foragers in the defense of resources that have been found already reduces the number of foragers assigned to search for new resources. Such allocation of workers may cause the trade-off between search and defense in foraging tasks.

7.5 Responses to General Flowering

General flowering increases the number of flowering species in a forest, the number of flowering plants of each species, and the amount of floral resources of each plant (Sakai et al. 1999c). In LHNP, general flowering occurred in 1992, 1996, 1997, and 1998. These events provided an opportunity to investigate responses of the populations and behavior of stingless bees to large fluctuation of floral resources.

Floral resources are likely to limit nest density of the aggressive *Trigona* in Costa Rica and Sabah (Hubbell and Johnson 1977; Eltz et al. 2002). If this were

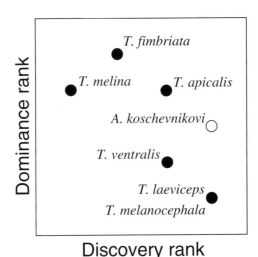

Figure 7.7. Trade-off between search and defense for *Trigona* (closed circles) and a honeybee *Apis koschevnikovi* (open circle). Searching ability is ranked by time of the first arrival at feeders. Defending ability is ranked based on aggressive behavior and interference competition at feeders.

true in LHNP, nest density would track the fluctuation of floral resources. To examine change in nest density, we surveyed subterranean *Trigona* nests (which are located reliably) within the 8 ha Canopy Biology Plot. We found 14 nests of five *Trigona* species, which were naturally aggregated at the bases of large trees on ridges (see Fig. 7.8).

This distribution agreed with observations in Brunei and Sabah (Roubik 1996a; Eltz et al. 2001) but contrasted to uniform patterns of nest dispersion observed in Costa Rica (Hubbell and Johnson 1977). Our observed nest density ranged from 1.1 to 1.6 ha^{-1} during 1992 and 1996. This nest density is similar to those in continuous forest habitats in Borneo, Danum Valley, and Deramakot in Sabah (0.0 to 2.1 ha^{-1}) (Eltz et al. 2001) and Belalong in Brunei (1.2 ha^{-1}

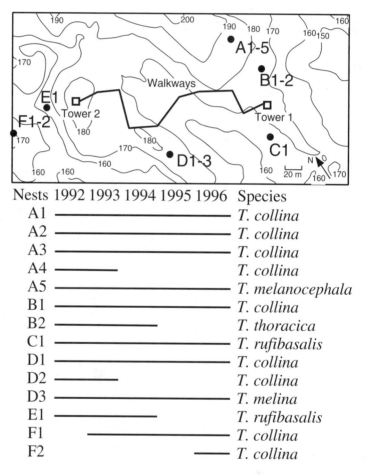

Figure 7.8. Spatial distribution of subterranean *Trigona* nests in Canopy Biology plot and their active periods (1992–96).

Trigona nests at ground level) (Roubik 1996a). The nest density increased immediately after GF. The increase, however, was slight compared to the fluctuation of floral resource harvest in GF and non-GF intervals.

Although nest density increased only slightly, forager returns to nests of four *Trigona* species were 1.4 to 4.8 times more frequent in 1996 than in 1994 (see Fig. 7.9; Nagamitsu and Inoue 2002). The increase in forager returns was probably due to worker population growth of each colony during GF in 1996. The proportion of nectar and pollen foragers did not differ statistically between 1994 and 1996, because of large variation in these variables within a year. This result suggests that allocation of foragers to nectar and pollen collection responds to resource fluctuation in temporal scales much shorter than GF to non-GF cycles.

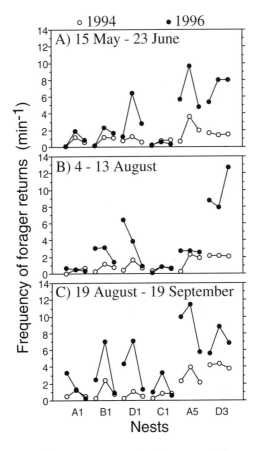

Figure 7.9. Frequency of forager returns to six subterranean *Trigona* nests. Nests codes are as in Figure 7.8. Observations were conducted in three periods (**A**, **B**, and **C**) in 1994 (*open circles*) and 1996 (*closed circles*). Lines connect observations at 07:30 (*left*), 10: 30 (*center*) and 14:30 (*right*); after Nagamitsu and Inoue (2002).

Furthermore, pollen diet similarity was stable in 1994 and 1996. Pollen type richness, diversity, and evenness of pollen diets in *T. collina*, *T. melina,* and *T. melanocephala* were temporally stable, which was in agreement with Eltz et al. (2001, 2002). Pollen type richness was highest in *T. melina*, followed by *T. melanocephala* and *T. collina*. Eltz et al. (2001) obtained similar results. The relationship of pollen diet similarity among the three species also agrees between Eltz et al. (2001) and our study. This consistency suggests that the observed proportion in pollen diets of each species is stable through temporal and spatial variation in floral resources.

8. Honeybees in Borneo

David W. Roubik

My intention here is to re-create the steps in honeybee evolution, many of which were defined during studies in Borneo and Malaysia. Methods include biogeographic analysis, molecular studies, and pollen identification or *palynology*. I then discuss how honeybee ecology is connected to the great tree stature, canopy and flowering characteristics, and rarity in space and time of floral resources in the rain forest of Borneo.

8.1 Introduction

Many aspects of honeybee ecology and evolution directly illuminate tropical biology, all the way from paleobiology to community patterns. Mutualisms, pollination ecology, interaction with predators and natural enemies, floral parasitism, canopy biology, traditional human use of honey and wax, and community structure can be profitably examined from a perspective of honeybees. Borneo, situated in a geographic center of honeybee distribution and diversity, opens the full range of discussion about the biology of tropical, advanced social insects—a dominant and persistent feature of terrestrial biomes at all latitudes.

Discoveries involving the honeybees are tantalizing because they place familiar organisms in a new light. For instance, Tanaka et al. (2001a, 2004) performed DNA analysis of three honeybee species on Borneo, finding relatively deep geographic differentiation in one species yet very little in the others. Nat-

ural history and historical biogeography held the key to understanding this pattern. One of those common honeybees, however, was not even recognized as a
species endemic to Southeast Asia until recently (Engel 1999; Otis 1996). Perhaps because *Apis koschevnikovi* is found only in wet, primary forests, it remained unknown to researchers for such a long time.

In addition to *Apis koschevnikovi,* the most common forest honeybee in Borneo, the three best known honeybee species of Indo Malaya are *A. dorsata,
A. cerana,* and *A. florea.* Distributed as far as Nepal and Sri Lanka, with the
first (and larger) two represented as far east as the Philippines and *A. cerana* as
far west as Oman, their general biology has been reviewed recently by Seeley
(1985); Ruttner (1988), Dyer and Seeley (1991a,b); Punchihewa (1994); Kevan
(1995); Kiew (1997); Smith et al. (2000) and Dyer (2002). Although one of the
two dwarf honeybees *A. florea* is apparently absent in Borneo, its sister species
A. andreniformis lives there.

8.2 Out of Borneo?

Not all honeybees are ancient inhabitants of the forests or other habitats they
now occupy. The giant honeybee, like the dipterocarp trees, is not among the
oldest members of lowland forest in Southeast Asia, which has been in existence
perhaps twice as long as the giant honeybee (see Fig. 8.1; Morley 2000). This
large, migratory honeybee has had remarkable success and impact by exploiting
loose pollination niches, although it remains enigmatic just how far it may have
coevolved with its host flowers. With the fossil pollen record of the rain forests,
Morley (1998, 2000) documents the arrival of dipterocarps in Southeast Asia
20 million years ago. The age of the *A. dorsata lineage* is around 35 million
years (see Table 8.1), and it may be older than the node or branch indicated in
Fig. 8.1. This bee may be exceptional. Its seasonal long-distance migrations
allow re-colonization and promote low genetic diversity (Oldroyd et al. 2000;
Tanaka et al. 2001a,b, 2004). The strong flight allowed dispersal to islands that
later become completely isolated, thus bees adaptively radiated to produce new
species on Sulawesi and the Philippines. However, Engel (1999) has reservations
about calling the bees different species, so that a future consensus will need to
be formed.

Although some may be relatively new arrivals to Borneo, I suggest that *Apis
koschevnikoi, Apis andreniformis,* the ancestor of the widespread Asian hive bee,
Apis cerana, and perhaps the common ancestor of *A. dorsata* and the so-called
western hive bee *A. mellifera,* are recent colonists of Asia that came from Borneo
or the block of rain forest in Sundaland. The species that we recognize today
were derived from the fragmentation, differentiation and isolation of a previously
huge ancestral population of honeybees, which I suggest may have gotten its
start in Borneo and neighboring land masses. The Asian hive bee *Apis cerana*
likely returned to its ancestral habitats on the margins of Borneo (Hepburn et

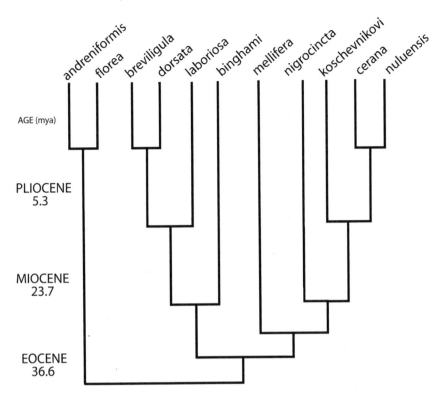

Figure 8.1. Hypothetical scheme of honeybee evolution (after Engel 1999; Tanaka et al. 2001).

al. 2001) as a species adapted to open and more seasonal, or cooler and drier habitats, in contrast to the core Bornean wet forest region. During this transphyletic migration, ancestral *A. cerana* lost whatever place it had in mature Bornean wet forests, where today it is rare (but see Osawa and Tsubaki 2003; Tanaka et al. 2004).

Where did the first Bornean honeybees come from? The answer begins with an outline of geological evolution in Southeast Asia (Hall and Holloway 1998). The Eocene through the Pleistocene for Borneo involves three distinctive periods (see Fig. 8.1), each with extraordinary consequences for evolution due to intervals of the island's separation and re-unification with the mainland. Originally, the tropical forests of eastern Asia extended through Borneo and southwestern Sulawesi. The movement of tectonic plates then tore Sulawesi from Borneo 34 million years ago. Uplands of the Kinabalu mountain region were formed from 20 million to 4 million years ago until the most recent glacial period, which ended approximately 12,000 years ago. Extensive glaciers formed on Kinabalu,

Table 8.1. Honeybee distribution and biological characteristics

Species	origin (mya)	bee size (mm)	colony size*	Range
Apis (fossil spp)	37–57	16	?	Eurasia
A. florea group (2 spp)	>37?	8	2	S. Asia, Sunda
A. dorsata group (4 spp)	>37?	21	40	Asia, IndoPacific, Himalayas
A. mellifera	10 to 37	12	60	Africa, Europe
A. koschevnikovi	5–10?	11	2	SE Asia
A. nigrocincta	4–10?	11	2	Sulawesi
A. cerana	2 to 4	11	30	Asia, IndoPacific
A. nuluensis	2 to 3	11	2?	Mt. Kinabalu

* While fossils seem to be of workers (a helper caste) because they have the wax mirror and other characteristics found in modern worker *Apis*, the fossil record contains no large nest, or hexagonal wax comb with stored food and immature bees. Only such concrete evidence would help establish that fossil bees had colonies with workers and queens, like modern *Apis*. A worker is in most respects like a normal female, solitary bee (at least more so than is a queen bee) except it does not mate to produce diploid female offspring (Michener 1974; Starr 1987).
mya = million years ago.

thereby dividing populations that could not adapt to the cold conditions within the barrier mountain ranges (Hadisoesilo et al. 1999; Tanaka et al. 2004).

The Miocene was a time of maximum extension of tropical humid forests in Asia. Shortly after its end and the beginning of the Pliocene, approximately 4.5 million years ago, Borneo first became an island. That crucial period in Bornean biogeography followed the world climate changes at the end of the Miocene. But in the next million years the first of the glaciations led to large fluctuation in sea level. That was repeated many times. The glacial periods were cooler, drier, and extensive, on the order of 100,000 years, while the interglacials (such as that we are living in today) lasted around 10,000 years. Not only were separations and connections between continental Southeast Asia, Borneo, Sumatra, Java, and Sulawesi made and broken several times by those cycles, both climate and vegetation underwent considerable modification.

During the mid-Tertiary, as hypothesized in Fig 8.1, honeybee lineages were separated into two groups. One of the large ancestral populations of bees extended from the Asian mainland to its extremes in the Borneo-Sulawesi peninsula, and potentially all the way to Africa. Later, as substantial cooling of climate and a great reduction in forest cover occurred in the Miocene, major lineages divided again at least two times, which fostered the evolution of the two common migratory honeybees of Africa and Asia—*Apis dorsata* and *Apis mellifera*. Later, the populations of cavity-nesting honeybees produced a highly adaptable invading bee, which would in 2 million years colonize areas as separate as Oman and Iran, the high Himalayas, Russia, Japan, and all of tropical Asia. This is *Apis cerana*, presumably out-competed by its forest siblings on much of the island of Borneo.

8.3 Honeybee Evolution

Engel (1998) suggests that honeybees, genus *Apis*, may have existed since the Eocene began 57.8 million years ago. The first honeybees ostensibly came from Asia and are now best represented by the smallest and largest living *Apis*, or by fossils of extinct species. The giant and dwarf *Apis* do not build nests in cavities but instead form their nest of a wax comb under a branch or rock ledge.

How and when did honeybees arrive in Borneo and Sundaland? Early *Apis*, as mentioned above, may actually have evolved in Borneo, or they may have arrived from elsewhere in what is currently continental Asia, but the ancestral populations were not exactly like any modern *Apis* (see below). Giant honeybees and *A. mellifera* are excellent dispersers. If any single biological trait is responsible for the fast pace of honeybee evolution since the Miocene, it is probably mobility of such colonies, and their ability to construct a nest of wax from provisions in the form of honey, carried with them on their dispersal flights. They are thus somewhat independent of their environment and can emigrate readily, unlike the Meliponini, the other eusocial bees. This remarkable dispersal ability also fostered the speciation and local adaptation to diverse habitats from the mountains to the lowlands on Borneo, and within a relatively short time.

The giant honeybees fly 10s to 100s of kilometers between seasonal nesting sites (Koeniger and Koeniger 1980; Parr et al. 2000; Neumann et al. 2000). Modern giant honeybees—Himalayan *A. laboriosa*, the widespread *A. dorsata*, and endemic Indonesian and Philippine *A. binghami* and *A. breviligula*, respectively—all measure at least 20 mm in body length and 15 mm in forewing length (see Table 8.1). That is decidedly large for bees, considering the world bee fauna (see Michener 2000). The most diverse honeybee fossils include *A. vetustus*, *A. armbrusteri* and *A. henshawi*, of Oligocene age. Those were less than 16 mm in length and 10 mm in forewing length (Zeuner and Manning 1976; Engel 1998). Even the oldest fossil *Apis* is only slightly larger than *Apis mellifera* (the common Western hive bee introduced throughout the world) but such fossils are classified by paleontological study as most like the dwarf honeybees (Engel 1998, 1999). Dwarf honeybees today are remarkably small, not exceeding 8 mm body size nor forewing length of 6 mm, the same as a large Southeast Asian species of *Trigona* (Meliponini). Thus, the first *Apis* to occur in the region of Southeast Asia, now extinct, was probably like *Apis florea* in its biology but larger than the dwarf honeybees of today.

Evolution of a completely new species and its eventual separation from relatives is a slow process. When studied with an array of techniques, evidence for animals shows that species formation may require some 2 million years (Avise 2000) and the species may endure 2 million years before final extinction. During a period of coalescence, or, lineage sorting, members of a population steadily reduce the amount of breeding with an antecedent population to eventually breed only among themselves. Natural selection in the particular environment pushes their features away from those of ancestors. Changes occur in outward characteristics, such as behavior, ecology, physiological response, and appearance, while the genes organizing this transformation increase in relative frequency and representation in breeding individuals. Another major type of evolutionary change involves genes that are neutral or silent. These obey the laws of genetic mutation, which produces one change in a gene each 10,000 years, or one genetic change per year among 10,000 genes (organsims have on the order of 50,000–100,000 genes). Such genes are especially valuable to the study of change over time.

The first cavity-nesting honeybee may have lived in small colonies making a few combs within small tree cavities in the rain forest—as *A. koschevnikovi* does today. However, in view of the relatively recent appearance and more derived evolutionary position of this species (see Fig. 8.1), compared to the common western honeybee, *Apis mellifera*, colony size and habitat may have matched other criteria. The common ancestor of the cavity nesters was potentially a very widespread population with both cavity-nesting and open-nesting colonies, which in fact is well known for *Apis mellifera* in the tropics today.

Apis florea is the mainland Asian dwarf honeybee and does not exist in Borneo or eastern Java, although slight ecological divergence and some notable morphological change has occurred in the other dwarf bee, *A. andreniformis*, which occupies not only Sundaland but also Southeast Asia, in sympatry with *A. florea*. Molecular data have shown, however, only slight divergence between species in the silent (non-coding) regions of COI, which is a portion of mitochondrial DNA that serves as a molecular clock (Tanaka et al. 2001b). Thus speciation was recent, and perhaps the degree of divergence still varies widely, depending on timing of separation and interbreeding. In any event, the biogeographic model for these species is one of divergence on Borneo when populations were separated after the Miocene and then during glacial periods. Colonization of mainland Southeast Asia from Borneo or Sumatra then occurred when dry land again connected them to the mainland. The dwarf honeybees are relatively poor dispersers, being the only widespread Asian *Apis* never to have reached the oceanic island portions of the Philippines.

If the giant honeybees consist of four species, their diversity may, like that of *Apis cerana* and its allies (*nigrocincta, nuluensis,* and *koschevnikovi*), derive largely from evolution during the glacial periods, when many tropical climates were more like the temperate grasslands that favored grazing mammals rather than forest browsers. The forest bees perhaps became anachronisms, while the major geographic radiation occurred with the dispersal of species adapted to drier, more open habitats and non-forest resources, such as bamboos and other grasses (Roubik 1988, 1989; Kiew 1997).

8.4 Biogeography

Of the known Sundaland *Apis*, only *A. cerana* colonized the temperate zone from the tropics, extending now to eastern Russia, Japan, and Korea, as well as to the Himalayas (Ruttner 1988; Hepburn et al. 2001). For Borneo and the Indopacific area, dispersal across the sea is a point of interest. Oversea dispersal can occur in honeybees by rafting, that is, drifting with trees or branches in the flotsam of rivers and carried to the sea. Both the open-nesting species and those that use cavities potentially disperse by this means. Present-day *Apis dorsata* in Southeast Asia nest upon the tallest forest trees, like *Koompassia excelsa* (Leguminosae) which often grows along rivers, and upon, for example, large Dipterocarpaceae, Tiliaceae, and Sterculiaceae (Seeley et al. 1982). By nesting high in trees they may escape predation from the sun bear. The giant bees also are common in coastal mangrove forest (Burkill 1919). A floating tree swept by a storm into a raging river, and then to the sea could harbor colonies in its branches. Cavity-nesting honeybees could also have traveled within floating trees but evidently because of ocean currents never arrived at the Moluccas, New Guinea, or Australia. Depending on the wind speed, the flying swarms easily reach more than 20 or 30 kilometers from their origin (Dyer and Seeley 1991a,b; Dyer 2002). Gaps of several dozen kilometers separating islands from the mainland, particularly between Borneo and continental Southeast Asia, would only have allowed colonization by entire dispersing bee colonies when the sea gaps were minimal, such as during the glacial periods. At this time, dry-land connections existed as far as eastern Borneo, Palawan and Sumatra, and portions of islands never connected to the mainland, for example Mindanao and Sulawesi, were joined together. These two portions of Sulawesi both contain the endemic cavity-nesting honeybee *Apis nigrocincta* (Otis 1996). Sea level recedes to as much as 200 meters below its current depth during glaciation, and the lowest sea level, allowing the greatest possibility for dispersal to remote islands, occurred in one of the first Pliocene glaciations, about 3 million years ago. At this time, the volcanic Sunda islands from Java eastward had not yet emerged from the sea.

Waif dispersal by fertilized females and very small reproductive swarms of dwarf honeybees might have allowed some colonization of islands. This could occur with a much greater frequency than the transportation of entire nesting colonies, thus sizable founding populations would be propagated across gaps over the water. One argument against this, however, is the absence of dwarf honeybees from the Philippines, despite the presence there of other primitively eusocial bees, such as bumblebees (Starr 1989). Furthermore, no bumblebees exist on Borneo, and the origin of the Philippine bumblebees was, contrary to most groups of plants and animals that arrived in the oceanic Philippines Proper, Taiwan, rather than Borneo (Starr 1989).

The island biogeography and speciation models given here are similar to those for the other highly eusocial bee group, stingless bees: Meliponini. Meliponine biogeography for Asia was summarized by Sakagami et al. (1990). Between Borneo and Peninsular Malaysia, 5 of 30 local species in each area are endemic, while 25 are shared between them. At least one stingless bee species is endemic to Sulawesi (S.F. Sakagami, 1984 personal communication). Thus one-sixth of the local fauna is endemic on Borneo and Peninsular Malaysia. The pattern is almost the same for *Apis*, with one-sixth and one-fifth endemic species, respectively. This analysis suggests that while elements of each area have combined via exchanges during the glacial periods, the relatively brief interglacials resulted in production of new species at the same rate for Meliponini and Apini. However, the widespread populations of Meliponini, those found through India to Borneo, include only two species of *Trigona* (*iridipennis* and *ventralis*), while three *Apis* span this distance. The contrast implies that the stingless bees have more restricted requirements for food and shelter. Major portions of biology of these two groups differ (Michener 1974; Roubik 1989; Peters et al. 1999). *Apis* can disperse as a colony with little need to prepare beforehand, but stingless bees must prepare a new nest site first. Stingless bees forage using relatively small areas or odor trail, while *Apis* can forage 15 kilometers from its nest (Seeley 1985; Dyer 2002). Additionally, queen *Apis* always mate with several to many drones, while for stingless bees, normally only one drone fertilizes the queen. The genetic basis for adaptive radiation is therefore more limited for stingless bees, and they virtually never emigrate as colonies (Roubik 1989).

8.5 Honeybee Ecology in the Borneo Forest

The honeybees used about 18% of the 89 flower species that appeared only during a period of general flowering. The giant *Apis dorsata* and the resident *A. koschevnikovi*, and the more disturbed habitat species (*A. andreniformis* and *A. cerana*) rapidly harvest forest resources from some 20% of local flowering plants to propagate their swarms. The single permanent forest *Apis* at the Lambir Hills National Park (LHNP) is *A. koschevnikovi*, which has a very small colony size of a thousand or so bees (see Table 8.1 and Plate 9E). Perhaps this bee is able to persist in the forest because it maintains such small colonies, which nevertheless recruit very rapidly to harvest resources (see Plate 9D) and reproduce at a fast pace during GF.

The giant honeybee has become a specialist in long-distance migration and may track the regional blooms caused by GF, as it moves between the mature forest and areas that have more annual or seasonal flowering. This pattern suggests that a pollinator deficit exists in the forest of LHNP during GF, although this species competes successfully with the resident eusocial bees and presumably could persist in the rain forest if resources were adequate (see Salmah et al. 1990).

Kiew (1997) points out interesting differences found in the Malay language with reference to the number of giant honeybee colonies in single trees. A *lebah camok* refers to less than seven colonies of *Apis dorsata* on a tree, while *lebah tualang* is a much larger aggregation found only in extensive forest areas on the huge tree *Koompassia*, usually a few dozen to nearly 200 colonies in Peninsular Malaysia (see Fig. 8.2C; Itioka et al. 2001a).

Due to palynological studies that identified major and minor pollen types in

Figure 8.2A. Migratory swarm of *Apis dorsata* resting on a low tree branch.

nests of *A. dorsata*, we have a good idea how this bee uses lowland mixed dipterocarp forest in northern Malaya (see Table 8.2). The pollen data, taken both from nectar and pollen-only sources like grasses or the leguminous shrub *Mimosa pudica*, are complemented by the data taken at Lambir Hills from hundreds of flowering tree species. Some implications of the data are examined below.

The closed–nest-cavity *Apis* endures cold by having a larger colony size and honey stores that permit food to be converted to heat as needed. In addition, they were likely adapted to disturbed or marginal habitats and escaped some predators by living in tree cavities, while adapting to some new predators, such as *Vespa* wasps (Ono et al. 1995) and the ever-present *Helarctos* (sun bears) and sloth bears (Roubik et al. 1985; Roubik 1996a). Where the ferocious sun bear was absent, such as in the Philippines, giant honeybees completely shifted their nesting biology and used substrates near ground level (Starr et al.1987). In Cambodia, where nests also are common near ground level, the sun bear is

Figure 8.2B. Individual nest of *Apis dorsata* on a tree branch in the upper forest canopy.

Figure 8.2C. Aggregation of honeybee nests, *Apis dorsata*, in the high branches of a forest tree, *Koompassia excelsa*. Photograph by T. Inoue.

still present, evident from tree trunks that have been ripped apart by its depredations, but large forest trees are becoming rare (D. Roubik, personal observation).

8.6 Pollination Ecology

In contrast to bee-flower coevolution or implied mutualism (Barth 1985), migratory colonies of *Apis dorsata* can be megaparasites of the floral community, but evidently not so often in the old forest at LHNP. When giant honeybees take floral food from male flowers but do not visit female flowers, as they seem to do in bamboos, some palms, and rattans (*Korthalsia, Calamus, Arenga*)—which was revealed by pollen analysis from multiple bees nests—as major pollen sources (see Table 8.2; Kiew 1993, 1997), then floral *visitation* is not the same as *pollination*. Visitation of nectarless flowers for pollen, when the female flowers occur on separate plants or are separate from male flowers in space or time, qualifies many flower visitors as non-pollinators or thieves (Roubik 1989). Tropical *Apis*, in general, use large quantities of pollen from grasses and wind-pollinated plants, making them opportunists that effectively use recently disturbed habitats and the extensive stands of pioneer plant species occurring there (Roubik 1989; Villanueva and Roubik 2004). At the same time, giant honeybees often collect their nectar from large, open flowers giving easy access: for example, *Durio, Eugenia, Elaeocarpus,* and other trees including Dipterocarpaceae and Sapotaceae (see Plates 6C, 7G). Because such flowers are not dioecious and present rewards to the bees where both stigma and anther occur, the giant honeybees are probably active and important pollinators.

In contrast, both *Apis dorsata* and *A. cerana* often visit the male flowers of large palm inflorescences, while shunning the female flowers, and this automatically makes them floral parasites or thieves (Kiew 1993). The same observation was made both for African *A. mellifera* and for Meliponini in Neotropical forests (Roubik and Moreno 1990). A general predilection of *Apis* for large

Table 8.2. Recorded visitation by *Apis* to flowering trees in LHNP by the Canopy Biology Program in Sarawak

Flowering tree family and genus		Honeybee present
Alanginaceae	*Alangium*	*Apis dorsata*
Asteraceae	*Vernonia*	*Apis dorsata*
Bombacaceae	*Durio*	*Apis dorsata*
Burseraceae	*Dacroides*	*Apis dorsata, A. koschevnikovi*
Clusiaceae	*Mesua* 2 spp.)	*Apis dorsata, A. koschevnikovi*
Dilleniaceae	*Dillenia*	*Apis dorsata*
Dilleniaceae	*Tetracera*	*Apis koschevnkovi*
Dipterocarpaceae	*Dipterocarpus* (3 spp.)	*Apis dorsata*
Dipterocarpaceae	*Dryobalanops* (2 spp.)	*Apis dorsata*
Elaeocarpaceae	*Elaeocarpus*	*Apis koschevnikovi*
Euphorbiaceae	*Cleistanthus* (2 spp.)	*Apis dorsata*
Euphorbiaceae	*Endospermum*	*Apis koschevnikovi*
Euphorbiaceae	*Trigonopleura*	*Apis dorsata*
Ixonthaceae	*Allantospermum*	*Apis dorsata*
Leguminosae	*Spatholobus* (2 spp.)	*Apis dorsata, A. koschevnikovi*
Meliaceae	*Walsura*	*Apis koschevnikovi*
Myrtaceae	*Eugenia* (2 spp.)	*Apis dorsata*
Orchidaceae	*Coelogyne*	*Apis dorsata*
Rosaceae	*Parastemon*	*Apis dorsata, A. koschevnikovi*
Sapotaceae	*Payena*	*Apis dorsata*
Sterculiaceae	*Pterocymbium*	*Apis dorsata, A. koschevnikovi*
Sterculiaceae	*Scaphium*	*Apis dorsata*

bunches of male flowers, including wind-pollinated Fagaceae, Euphorbiaceae (*Macaranga*), and many grasses, for example wind-pollinated *Zea mays*, generally signifies that pollen is being removed but often little or no pollination service is provided. In Neotropical forests, now-naturalized African *Apis* often takes pollen in early morning from the flowers of nocturnally flowering species, pollinated by bats or moths (Roubik 1989). However, there may be no fitness cost to the plant, provided that sufficient pollen is available for wind pollination to still occur, and then the bees are merely commensals: that is, neither useful nor harmful. Like the animals that forage on the fruits produced during the mass flowering and mast fruiting episodes of tropical Asia, honeybees are literally living off the fat of the land.

In Bornean primary forests, *Apis dorsata* and *A. koschevnikovi* (see Plate 9A,D) are the only honeybees that appeared frequently at flowering canopy trees or at baits (Roubik 1996a; Roubik et al. 1995, 1999; Momose et al. 1997), although this was not found at Pasoh Forest Reserve in Peninsular Malaysia (Osawa and Tsubaki 2003). The observations in LHNP forest canopy indicate 29 flower species were visited by these two honeybees, thereby making the *Apis* visitors of 10% of the species and 22% of the plant families (see Table 8.2). No other honeybee species were noted, although *Apis andreniformis* was seen on *Mimosa* at the roadside. Similarly, in a slightly higher-elevation forest in Brunei at 500 to 600 meters, Roubik (1996) recorded *A. andreniformis* and *A. cerana* only a few times, among the many *A. koschevnikovi* and *A. dorsata* at sugar or salt baits near ground level. In primary forest in Japan (Inoue et al. 1990; Kato et al. 1990) the honeybee *A. cerana* visits 12% to 15% of local flower species, while in Sumatran localities, *Apis cerana* was reported very rarely in primary rain forest (Salmah et al. 1990), but often in the canopy in Peninsular Malaysia (Osawa and Tsubaki 2003).

In the more seasonal monsoon forest of Thailand, but in a lowland mixed diptercarp forest (like LHNP), Seeley et al. (1982) reported that colonies of *Apis cerana* were very unlikely to abandon a nest site during a year, while dwarf honeybees did so often (30%), and giant honeybees completely left the forest during the dry season. They returned four or five months later, in the wet season.

A much longer study was carried out at LHNP, using light traps that capture noctural insects. Light traps were operated within the canopy on one of the canopy walkway tree towers. *Apis dorsata* flies during nights that are illuminated by a moon in half-phase or greater. However, the results of light trapping, as known to any entomologist, are far better when there is no moon. Light traps were used only on those nights, but *A. dorsata* was still trapped in early morning (Kato et al. 1995a). There were many individual bees during the years of El Niño-Southern Oscillation (ENSO) and the GF in 1996 (a non-ENSO year) and immediately thereafter in 1992 to 1993 and in 1997, but almost none in other years (Itioka et al. 2001a). The presence of *A. dorsata* in LHNP, representing the aseasonal lowlands, occurred primarily during GF in February through July, although the bees also were present during April in 1993, where no general flowering occurred (Roubik et al. 1995; Itioka et al. 2001a). Giant honeybees were in the montane region during October and November 1991, when rainfall there was highest (Roubik 1996a; Cranbrook and Edwards 1994). They apparently inhabit the wet upland forests in low numbers, but return to the lowlands and presumably reproduce there (drones were found in the light trap catches) at odd intervals determined by ENSO and GF events.

The giant honeybees are probably canopy specialists, although easily found at *Mimosa pudica* on the ground, where they and many other bees gather pollen in the morning. They are, however, as ephemeral as the flowers in many regards, being all but absent in this forest during most times when the six-month mass flowering is not taking place (Nagamitsu 1998; Itioka et al. 2001a, but see Roubik et al. 1995). *Apis koschevnikovi* is a steady forest component, visiting flowers during the entire year. However, its presence at flowering trees was detected less often than that of *Apis dorsata*. The latter visited a total of 23 of the 28 species recorded with honeybees, while *A. koschevnikovi* visited only 9 of the 270 plant species with recorded flower visitors.

8.7 The Forest and the Bees

By what mechanisms do the rain forest honeybee, *Apis koschevnikovi*, and the giant honeybee coexist? They differ in size and tongue length, which helps to separate some Meliponini in resource use. *Apis koschevnikovi* evidently does not use most of the resources used by *A. dorsata*. About half of the former's resource species are shared with the giant honeybee (see Table 8.2). *Apis koschevnikovi* does overlap spatially with giant honeybees in using trees at all heights in the canopy, where both recruit intensely in a short time (see Plate 9D, Roubik et al. 1995, 1999; Roubik 1996a,c).

Borneo forest is among the most heterogeneous habitats on Earth in terms of flower species availability. It therefore makes sense that the fragmentation of resource abundance over time has reduced consumer species and driven up the proportion of those able to wait out periods of resource scarcity—namely, the bee colonies that either store food or migrate (Roubik 1979, 1990). If one takes the ratio of angiosperm to bee species, not counting wind-pollinated plants un-attractive to bees, there is a large deficit in bee species in the tropics (Roubik 1992, 1996b). Griswold et al. (2000) partly confirm this pattern by comparing several places in Costa Rica and North America, finding that, when the highly eusocial bees like Meliponini are not considered, bee diversity is about the same. Better tests of the general proposal that highly eusocial bees dominate the bee community structure are forthcoming, but the correlative evidence is compelling.

The ENSO in Southeast Asia has created an ecological gauntlet through which any sustained population of bees must pass: that is, the supra-anual flowering of the majority of forest tree species. While the most plausible means for a bee to avoid local extinction is to enter diapause, to generalize among flowers of a given size, or to migrate between habitats where flowering is adequate, the net outcome should be a general reduction in the local richness of bee species. Other flower visitors would thereby assume a more important role in the pollination of forest trees. In some deserts, bees can delay up to 14 years before adult emergence, a bizarre timing to coincide with a flowering event of the floral host (Danforth 1999; Minckley et al. 1999). This kind of lengthy development period is still unknown in tropical forests and is unlikely. With a local flora consisting of thousands of species, and many loose pollination niches, bet hedging on rare mutualists is common, and generalist pollinators and plants are numerous.

Honeybees in extensive forest have been noted for wide breadth of pollen species used, with 245 species estimated from a year's colony samples of *Apis mellifera* in one Panamanian forest (Roubik 1988, 1992, 1996d; Villanueva and Roubik 2004). However, there is a surprising degree in diet specialization, also noted for *A. dorsata* and *A. cerana* (Kiew 1997). The latter have few actual studies in rain forest, but Kiew reports that dipteorpcarps and male flowers of palms are used extensively. This last point makes them pollen thieves, in the same way that pollen from palms implies that many of the visits by honeybees and stingless bees to palms, grasses, and other plants do not occur for pollina-tion, but rather for feeding (Roubik 1988, 1989).

Foraging behavior of the two forest honeybees was investigated in Borneo (Roubik et al. 1995, 1999). There are immediately discernible interactions of bees at artificial feeders, as noted by Koeniger and Vorwhol (1979) and Roubik (1980). At feeding stations, in marked contrast to normal behavior at flowers, bees fight. The grappling and attempted stinging behavior that I witnessed of

Apis koschevnikovi and *A. dorsata* at sugar solution feeders was extraordinary. At one point the *A. dorsata* began to attack and to sting me. Was this displaced nest-defense behavior, as suggested for fighting by honeybees at feeders (Roubik 1989)? It seems likely. However, the general result makes one aware that honeybees, when recruiting to a rich resource, do not mix well with competing foragers. They may, by their competitive recruitment ability, tend to dominate floral patches as individual colonies.

Trees that must be outcrossed to produce seeds and fruit are common in tropical forests (Bawa 1990; Sakai et al. 1998b; Momose et al. 1997). A honeybee model has recently been proposed to account for outcrossing over relatively long distances in dense forest (Roubik 1999). Bees forage where they have been directed, which usually requires some trial flight or brief periods of searching without finding a reward (Seeley 1995). After a foraging site is known, multiple forage flights, going to and from the nest, are initiated. When no more food is available at the site, a forager either returns to the nest or goes to another site that she remembers from recent foraging experience. African honeybees in the primary forest of Gabon rapidly traveled up to 1.6 kilometers between feeding stations (Roubik 1999). While I could not determine whether individual bees had returned to their nest before shifting between feeding sites, the elapsed time between observations I made at the different sites would have allowed them to do so. The important event was the bee's appearance among multiple sites, culminating in a maximum observed separation distance of 1.6 kilometers. An even more rapid movement between tree towers in Lambir, and along the canopy walkway, and between canopy and ground level, was seen for *A. koschevnikovi* and *A. dorsata* (Roubik et al. 1999). The distances that were emcompassed were 220 meters to 640 meters, but this study was not designed to examine maximum foraging ranges.

Let us now return to *lebah tualang* and the multiple colonies of *Apis dorsata* that occupy their special forest niche as a dense aggregation of large bees upon tall forest trees during GF. The competitive interactions among many colonies within patches of flowers, comprised of either part of a tree canopy or individual small canopies, is a force driving tree reproduction. A simple computation illustrates the point. Given a colony with 15,000 bees, 5,000 of which are foragers (Dyer and Seeley 1994), a tree with 50 nests therefore contains 250,000 foraging bees. They have, conservatively, a maximum flight range of 5 kilometers, and each will undoubtedly visit an average of something like 50 flowers before returning to her nest. Foragers make perhaps five flights each day. My computation from these figures suggests that a single bee tree in primary forest can carry the genetic material of some 63 million flowers from place to place during a day, over an area of 80 km^2.

Honeybees in the Borneo forest are responsible, during a GF, for dispersing the genes of approximately one million flowers per square kilometer per day. What if their nest tree was occupied by only five colonies? The first prediction is that competition would certainly not force them to forage as widely. Average foraging area would likely be less than a kilometer from the nest, as recently found in India for *A. dorsata*, in a forest habitat that does not support many colonies of bees (P. Batra, personal communication). The five colonies would, in fact, disperse the genes of somewhat more than a million flowers per square km per day, but the area over which they would travel would be on the order of only 4% that of the *lebah tualang* in mature forest. Outcrossing, and the genetic diversity of tree populations pollinated by the bees, would decline.

By decreasing honeybee competition with other colonies, and with other bees, the unique pollination ecology in the Borneo forests would undergo drastic change. With an uninterrupted presence of migratory honeybee colonies since the Eocene, and the presence of permanent colonies nesting in tree hollows within the forest since the Miocene, honeybees have left their imprint on the forest, other bees, and other pollinators.

9. Beetle Pollination in Tropical Rain Forests

Kuniyasu Momose

9.1 Introduction

Lists provided by Irvine and Armstrong (1990) and the Canopy Biology Program in Sarawak, or, CBPS (Momose et al. 1998c) show that 42 families of Coleoptera include pollinating beetles (see Table 9.1). Generalist-pollinated plants are almost always visited by beetles that feed on nectar or pollen. However, a close association with beetles as pollinators is revealed in mechanisms for excluding other kinds of flower visitors. Furthermore, the morphology of flowers and floral rewards are the relevant and conspicuous mechanisms.

9.2 Annonaceae

Flower visitors other than beetles are excluded both by floral morphology and rewards. Gottsberger (1970, 1989a,b) reported the details of beetle pollination of Annonaceae in Brazil, while pollination of Annonaceae in the Asian tropics was reported by Rogstad (1994) in *Polyalthia*. At Lambir Hills, 20 of 66 Annonaceae were pollinated by beetles (Momose et al. 1998c). In most species the beetle pollination syndrome can be predicted from the shape of flowers, but a few exceptions have been found: for example, cockroach pollination in *Uvaria* aff. *elmeri* (Nagamitsu and Inoue 1997a) and thrip pollination in

Table 9.1. Families of Coleoptera known to include pollinating beetles

Beetle families		
Alleculidae	Coccinellidae*	Mordelidae*
Allocorhynidae	Corylophidae**	Nitidulidae*
Anthicidae*	Curculionidae*	Oedemeridae*
Anthribidae**	Cryptophagidae*	Phalacridae
Biphyllidae**	Dermestidae*	Ptiliidae**
Brentidae**	Elateridae*	Pythidae
Bruchidae	Helodidae	Rhipiphoridae
Buprestidae*	Lagriidae*	Rhynchophoridae**
Byturidae	Lampyridae**	Scarabaeidae*
Cantharidae*	Languriidae	Scolytidae**
(Carabidae**) probably as predators	Lycidae*	Scraptiidae
Cerambycidae*	Melandryidae**	Staphylinidae*
Chrysomelidae*	Meloidae	Tenebrionidae**
Cleridae*	Melyridae*	Trixagidae

* Listed in Irvine and Armstrong (1990) and found in Lambir
** Not listed in I and A but found in Lambir
No mark: listed in I and A but not found in Lambir

Popowia pisocarpa (Momose et al. 1998c). The pollination systems of beetle-pollinated Annonaceae in Lambir (Momose et al. 1998c) are introduced below.

Flowers are protogynous and the petals form a chamber. In the female phase, beetles are attracted by odor and enter the floral chamber. One to several beetle species per plant species were collected. The genera *Carpophyllus* (Nitidulidae), *Endaenidius*, *Endaeus* (Curculionidae), *Proagopertha* (Scarabaeidae), and unidentified genera of Chrysomelidae were attracted (see Table 9.2). Beetles fed on a stigmatic secretion or petals. Oviposition was not observed. When flowers turned into male phase, anthers dehisced and petals forming floral chambers fell down, whereupon beetles flew away with pollen on their bodies.

Some of the most primitive families of Magnoliales, Magnoliaceae (Thien 1974), Degeneriaceae, and Winteraceae (Thien 1980) have similar mechanisms of pollination but are visited by diverse insects. Flowers are protogynous. Petals form a floral chamber, but they do not tightly enclose sexual parts, as do Annonaceae. Diverse flower visitors feed on pollen or stigmatic secretions. This is considered to be a prototype of the beetle pollination found in Annonaceae, which has more specialized pollination systems than primitive families. The same systems are also found in other families of Magnoliales, Calycanthaceae (Grant 1950) and in Eupomatiaceae (Irvine and Armstrong 1990).

Table 9.2. Beetle-pollinated plants found in Lambir Hills National Park, Sarawak

Species	Color	Flower shape	Sexual separation	Main pollinators*	Minor visitors
Annonaceae Type					
Annonaceae					
Cathostemma aff. *hookerii*	purple	cup	protogyny	en3	
Enicosanthum coriaceum	white	chamber	protogyny	*Proagopertha* sp. (Scarabaeidae), G1, G2, en1, en5, en6	
Enicosanthum macranthum	white	chamber	protogyny	e4	
Fissistigma paniculatum	yellow	chamber	protogyny	en1, en2, en3, en4, e3	
Friesodieisia glauca	yellow	chamber	protogyny	c4, en2, en5	
Friesodieisia filipes	yellow	chamber	protogyny	c, en	
Goniotharamus sp. nov.	yellow	chamber	protogyny	c3, en4	
Goniotharamus uvarioides	white	chamber	protogyny	c	
Goniotharamus velutinus	yellow	chamber	protogyny	c3, c4, c5, c7, en4	
Meiogyne cylindrostigma	pink	chamber	protogyny	en, *Brachypeplus* sp. (Nitidulidae)	Biphyllidae
Monocarpia euneura	yellow	chamber	protogyny	c4, en3, en5	
Polyalthia cauliflora	yellow	chamber	protogyny	c1, c2, c3, c4, c6, c7, e1, en2, en3, en5	
Polyalthia hypogaea	white	chamber	protogyny	en1, en5, G1.	
Polyalthia motoleyana	white	chamber	protogyny	A1	
Polyalthia rumphii	yellow	chamber	protogyny	en6	
Polyalthia sp. nov.	yellow	chamber	protogyny	c4, en1, A1	
Polyalthia sp. nov. 2	purple	chamber	protogyny	en7	
Pyramidanthe prismatica	yellow	chamber	protogyny	A1, G3	
Sphaerotharamus insignis	red	chamber	protogyny	en7	
Anomianthus sp. nov.	yellow	chamber	protogyny	e1	
Araceae Type					
Araceae					
Homalomena propinqua	white	spathe chamber	monoecy	*Parastasia* sp. (Scarabaeidae), *Dercetina* sp. (Chrysomelidae)	

Table 9.2. *Continued*

Species	Color	Flower shape	Sexual separation	Main pollinators*	Minor visitors
Myristicaceae Type					
Ebenaceae					
Diospyros dicotyoneura	white	urceolate	dioecy	Staphylinidae, Nitidulidae	Tenebrionidae
Myristicaceae					
Gymnacranthea contracta	white	campanulate	dioecy	Curculionidae, Chrysomelidae	Ptiliidae
Kema tridactyla	yellow	chamber with slits	dioecy	en5	
Knema cineria var. *sumatrana*	yellow	chamber with slits	dioecy	Staphylinidae	
Knema latifolia	yellow	chamber with slits	dioecy	Curculionidae	
Sterculiaceace					
Heritiera borneensis	white	campanulate	monoecy	Curculionidae, Chrysomelidae	
Heritiera sumatrana	red	campanulate	monoecy	Chrysomelidae	Curculionidae
Sterculia laevis	red	urceolate	monoecy	Chrysomelidae	Mordellidae
Sterculia stipulata	red	urceolate	monoecy	Chrysomelidae	Staphylinidae
Dipterocarpaceae Type					
Dipterocarpaceae					
Hopea pterigota	yellow	spiral, cup	none	Chrysomelidae, Curculionidae	Elateridae, Lagriidae, Nitidulidae, Staphylinidae, Tenebrionidae
Shorea agami	yellow	spiral, cup	none	Chrysomelidae, Curculionidae	
Shorea beccariana	yellow	spiral, cup	none	Chrysomelidae, Curculionidae	Elateridae, Staphylinidae
Shorea bullata	yellow	spiral, cup	none	Chrysomelidae, Curculionidae	Scarabaeidae, Staphylinidae
Shorea confusa	purple	spiral, cup	none	Chrysomelidae, Curculionidae	
Shorea balanocarpoides	yellow	spiral, cup	none	Chrysomelidae, Curculionidae	

Table 9.2. Continued

Species	Color	Flower shape	Sexual separation	Main pollinators*	Minor visitors
Shorea falcifeloides	white	spiral, cup	none	Chrysomelidae, Curculionidae	Cerambycidae, Cleridae
Shorea ferruginea	yellow	spiral, cup	none	Chrysomelidae, Curculionidae, Nitidulidae	
Shorea havilandii	white	spiral, cup	none	Chrysomelidae, Curculionidae	Elateridae, Lagriidae, Staphylinidae, Tenebrionidae
Shorea macrophylla	pink	spiral, cup	none	Chrysomelidae, Curculionidae, Cleridae, Nitidulidae	
Shorea macroptera	pink	spiral, cup	none	Chrysomelidae, Curculionidae	Scarabaeidae, Nitidulidae, Coccinellidae, Cryptophagidae, Corylophidae,
Shorea ochraceae	yellow	spiral, cup	none	Chrysomelidae, Curculionidae	
Shorea parvifolia	yellow	spiral, cup	none	Chrysomelidae, Curculionidae	
Shorea patoiensis	white	spiral, cup	none	Chrysomelidae, Curculionidae	Anthicidae, Staphylinidae, Tenebrionidae
Shorea pilosa	yellow	spiral, cup	none	Chrysomelidae, Curculionidae, Nitidulidae	
Shorea smithiana	yellow	spiral, cup	none	Chrysomelidae, Curculionidae, Nitidulidae	
Shorea superba	white	spiral, cup	none	Chrysomelidae, Curculionidae	Scarabaeidae
Shorea xanthophylla	yellow	spiral, cup	none	Chrysomelidae, Curculionidae, Cleridae	
Vatica micrantha	yellow	spiral, cup	none	Chrysomelidae, Curculionidae	Cleridae, Elateridae, Lagriidae, Scarabaeidae
Vatica parvifolia	white	spiral, cup	none	Chrysomelidae, Curculionidae	

*: e: *Endaeus*, en: *Endaenidius* (Curculionidae); c: *Carpophyllus* (Nitidulidae); A: subfamily Alticinae, G: subfamily Galerucinae (Chrysomelidae) Numbers after alphabetical codes indicate morphological species.

9.3 Araceae

Here the entire inflorescence acts just like a single flower of Annonaceae and the pollination mechanism is similar. This type is known in Neotropical Araceae (Young 1986), Cyclanthaceae (Beach 1982), and Balanophoraceae (Borchsenius and Olesen 1990). Irvine and Armstrong (1990) also reported the pollination of Balanophoraceae from Australia. *Homalomena propinqua* (Araceae) in LHNP (Kato 1996) has the pollination system discussed here (see Plate 6F).

The inflorescence is monoecious. Female flowers are located in the lower part and covered with the bract (spathe) which together form a chamber. Male flowers are in the upper part of an inflorescence, protruding from the bract chamber. First, the female phase lasts for two days. Two species of beetles, *Parastasia* (Scarabaeidae) and *Dercetina* (Chrysomelidae), visited flowers in the female phase (see Table 9.2), where they fed on staminodes. Male *Dercetina* excluded other males from the inflorescence as its mating territory. When the inflorescence converted to male phase, the bract closed tightly to protect female flowers, and beetles were excluded from the chamber. Male phase continued for one day, then beetles moved to upper parts of the inflorescence, fed on pollen, and flew away.

9.4 Myristicaceae

Flower visitors other than beetles are excluded by floral morphology. Armstrong and Durmmund (1986) reported pollination of *Myristica* from India, and Armstrong and Irvine (1989) from Australia. The same system is also found in *Diospyros dicotyoneura* (Ebenaceae), *Sterculia*, *Helitiera* (Sterculiaceae), and *Knema* (Myristicaceae) in Lambir Hills National Park (Momose et al. 1998). The flowers are unisexual, dioecious (Myristicaceae and most of Ebenaceae), or monoecious (*Sterculia* and *Helitiera*), as seen in Plate 6G. Rewards are pollen in male flowers, but female flowers mimic the males and offer no food for pollinators. Petals (Myristicaceae and Ebenaceae) or sepals (Sterculiaceae) are urceolate and form a chamber that excludes bees and flies. The floral chamber entrance is located at the bottom, and sexual parts at the top. Beetles of Chrysomelidae, Curculionidae, Nitidulidae, and Staphylinidae visited the flowers (see Table 9.2). One to many beetle species per plant species were collected. Armstrong and Irvine (1989) reported that beetles stay longer in male flowers than in female flowers of *Myristica*.

9.5 Dipterocarpaceae

Flower visitors other than beetles are excluded by the type of reward offered. Some *Shorea* (Dipterocarpaceae) were reported to be thrips-pollinated in Pasoh, Malay Peninsula (Appanah and Chan 1981). However, in Lambir Hills National

Park, Sakai et al. (1999b) show experimentally that *Shorea parvifolia* is beetle-pollinated. Not only this tree, but 21 additional species of *Shorea*, *Hopea,* and *Vatica* in Lambir are pollinated by beetles (Momose et al. 1998c).

The flowers are hermaphroditic and cup-shaped. Flowers open in the evening and last for one or two days. Nectar is not secreted, but two species of *Shorea* section *Anthoshorea* have nectaries. A number of beetle species of several families including Chrysomelidae, Cleridae, Curculionidae, and Nitidulidae (see Table 9.2 and Plate 6E) visit flowers and feed mainly on petals, but nectar (section *Anthoshorea*) and perhaps pollen are also rewards. Oviposition by the beetles was not observed. Even curculionid beetles visit flowers to feed on petals but do not oviposit. Some thrips are also found, but pollen transfer by thrips is not as effective as beetles according to experiments by Sakai et al. (1999b).

9.6 Characteristics of Beetle Pollination

As described above, the first three types of beetle pollination reflect the behavior of beetles that tend to stay in enclosed spaces. Petals, sepals, or bracts form a chamber in which beetles remain. Because beetles may remain for several days, the floral sexual functions must be separated spatially (unisexual flowers) or temporally (protogyny) if outcrossing is to occur.

Because tropical rain forests contain high plant species diversity and low plant population density, or local dominance, a specialized relationship between plants and pollinators is needed for effective pollen transfer. The order Magnoliales provides good examples in evolution of beetle pollination in tropical forests. The most primitive families of this order are found in temperate regions: Magnoliaceae in the Northern Hemisphere, and Winteraceae in Southern Hemisphere. They are pollinated by diverse insects (Thien 1974, 1980; Frame 2003). Specialization for beetle pollination occurs independently in two major tropical families, Annonaceae and Myristicaceae. Still, the greatest number of such species is recognized within Magnoliales.

Finally, the Dipterocarpaceae possess a pollination system that is different from other types of beetle pollination. Plants do not have chambers, diverse and unspecialized beetles visit flowers, and they move vigorously among them. Plants of this type flower only during GF (Momose et al. 1998c; Sakai et al. 1999c). During GF such a large number of plants, including emergent trees, bloom successively within short periods that pollinator shortage might occur unless pollinators can quickly respond to the general flowering (Ashton et al. 1988). Some beetles are considered able to use such suddenly increasing flower resources. Chrysomelids pollinating some dipterocarps feed on leaves of dipterocarps in non-GF and shift resources to floral tissues in GF, because they were collected on dipterocarp leaves in flowerless seasons (Nagamitsu, personal observation; Yamauchi, personal observation). The association between such beetles and flowering dipterocarps may be maintained steadily and thus permit rapid availability of beetles as pollinators during a general flowering episode.

10. Seventy-Seven Ways to Be a Fig: Overview of a Diverse Plant Assemblage

Rhett D. Harrison and Mike Shanahan

The Fici of Borneo show quite a series of adaptations, both in their shape and size, to varied biological conditions, and well deserve special investigation.
—Odoardo Beccari (1904)
Wanderings in the Great Forests of Borneo

10.1 Introduction

Figs (*Ficus*: Moraceae) have been described as the "most distinctive of the wide-spread genera of tropical plants" (Janzen 1979). They are renowned, at least among biologists, for their intricate relationship with obligate species-specific pollinators, the fig wasps (Agaonidae: Chalcidoidea) (Herre 1989; Weiblen 2000; Kjellberg et al. 2001; Machado et al. 2001), and as keystone resources for fru-givorous mammals and birds (Terborgh 1986; Shanahan et al. 2001a). Numerous indigenous cultures in the tropics venerate figs for their fecundity and vitality (Corner 1985; Xu et al. 1996; Simoons 1998). With approximately 750 species in a roughly pan-tropical distribution (see Table 10.1), *Ficus* is a species-rich genus (Berg 1989). However, especially remarkable is the diversity of fig species that coexist in local assemblages. Whether in the Neotropics, Africa, or the Indo-Australian region, *Ficus* is invariably one of the most speciose genera in any lowland tropical forest. At Lambir Hills National Park (LHNP), Sarawak there are an extraordinary 77 species and 6 varieties.

High species diversity, a unique pollination system within the genus, and oft-cited importance to vertebrate seed dispersers make figs an ideal topic for com-parative study (Bronstein and McKey 1989), and figs have frequently been used as a model system for examining evolutionary theory (Herre 1987; Kjellberg et al. 2001; West et al. 2001; Weiblen 2002; Molbo et al. 2003). Less attention has been given, however, to comparative ecological study of figs in the same

Table 10.1. Taxonomic diversity of *Ficus* in Borneo,
Papua New Guinea (PNG), and globally (Corner 1965;
Berg 1989)

	Subgenera	Sections	Species
Asia-Australia	4	15	>500
Borneo	4	10	>160
PNG	4	11	>140
Africa	4	7	105
Neotropics	2	2	140
Global	4	18	>750

assemblage (but see Herre 1989; Herre 1996). Indeed the title of Janzen's seminal review of fig biology, "How to Be a Fig" (Janzen 1979), reveals a misperception prevalent at the time—that figs are largely ecologically uniform. Subsequent studies have gradually changed this view (Herre 1989; Compton 1993; Herre 1996) but perhaps nowhere is its fallacy more apparent than in Southeast Asia, where fig diversity reaches its zenith (see Table 10.1). In this chapter, we give an overview of the rich assemblage of figs at Lambir Hills to illustrate some of the many ways there are of being a fig.

10.2 Natural History of Figs

The Chinese character for fig is 無花果, meaning fruit without flowers. It reflects an ancient confusion concerning fig pollination, and one that persists in folklore throughout the tropics. The fig's tiny flowers line the inside of the inflorescence (see Plate 11A,B), and the pollinating fig wasp (see Plate 11A) enters through a narrow bract-lined ostiole. Inside the inflorescence, fig wasps scatter pollen and lay eggs on flowers by inserting their ovipositors down the styles. Hence there is an obligate relationship between the fig and its fig wasp, as neither is able to reproduce without the other.

There are two modes of pollination in figs: monoecy and dioecy. In monoecious figs, wasp larvae and fig seeds mature in the same inflorescence. The style length is unimodal and the wasps' ovipositors can reach most ovules (Verkerke 1988; Ganeshaiah et al. 1995; Nefdt and Compton 1996). Ovules that receive an egg develop into a gall, and the wasp larva feeds on the gall tissue. Non-pollinating wasps also exploit fig ovules (Kerdelhue and Rasplus 1996; West et al. 1996), but most species oviposit through the wall of the inflorescence (see Plate 11C,D,F) (Boucek 1988). Seeds *only* develop in pollinated ovules not utilized by the wasps.

In dioecious figs there are two types of inflorescence borne on different trees (Beck and Lord 1988). Female trees have inflorescences with long, thin-styled female flowers (see Plate 11B). Pollinating wasps enter but fail to lay eggs because their ovipositors cannot reach the ovules. They therefore die without reproducing, but usually successfully pollinate female inflorescences (Patel et al. 1995; Anstett et al. 1998). Male trees bear inflorescences with short-styled female flowers (gall flowers) and male flowers (see Plate 11A). The pollinating wasps are able to oviposit in any ovule, thus very few or no seeds are produced. Female trees produce only seeds, while male trees produce pollen and pollinators (Galil 1973; Beck and Lord 1988).

The emergence of adult wasps, a few weeks later, is coincident with the maturation of the fig's male flowers (see Plate 11G). *Male wasps emerge and mate with females still in their galls.* Females then chew their way out and collect pollen, either passively or by loading pockets on the mesothorax, as shown in Plate 11G (Kjellberg et al. 2001). Meanwhile, the wingless males cut an exit tunnel and in some cases scatter across the surface of the inflorescence, thereby distracting predatory ants while the females escape, as shown in Plate 11H (Harrison 1996). In a brief adult life span, which is 1 to 3 days (Kjellberg et al. 1988), the female wasps must locate a fig with receptive inflorescences to reproduce (see Plate 11H,I). The asynchronous flowering among fig trees ensures receptive inflorescences are almost continuously available, at the population level.

The inflorescences ripen after emergence of the wasps and are eaten by vertebrate frugivores, the most important groups being birds, bats, and primates (Shanahan et al. 2001a). Moreover, production of large quantities of fleshy fruit, particularly at times of the year when other fruits may be scarce, makes figs an important resource for wildlife (Terborgh 1986; Lambert and Marshall 1991; Kinnaird et al. 1999; Shanahan et al. 2001a).

Figs have an ancient pedigree and are believed to have originated during the Cretaceous, around 90 million years ago (Murray 1985; Machado et al. 2001). A rough congruency between fig and fig wasp phylogenies, in particular the correspondence between of specific clades of pollinators and figs, is suggestive of co-speciation between fig wasps and their hosts (Weiblen 2002). However, strong evidence is lacking, and host switching has also occurred. Moreover, a breakdown in species-specificity, albeit usually allopatric, is not rare (Rasplus 1994; Kerdelhue et al. 1997; Lopez-Vaamonde et al. 2002), and recently, pairs of pollinator species coexisting on the same host were discovered in Panama (Molbo et al. 2003).

10.3 The Fig Assemblage at Lambir Hills

Lambir Hills is a relatively small island of primary lowland dipterocarp forest surrounded by secondary forest, oil palm plantations, and shifting cultivation. A small area of kerangas heath forest occurs along the ridge that forms the summit of Bukit Lambir. The fig assemblage described here was collected from all habitats within the park, including a 52 ha long-term ecological dynamics plot (Lee et al. 2002) and surrounding areas.

The figs of Lambir Hills are remarkably diverse and represent slightly less than half the Bornean fig flora (see Table 10.2; Plate 10; Appendix B). A total of 77 species and 6 varieties have been recorded, which is about 40% more species than reported for Gunung Palung in southwest Borneo and Kutai in east Borneo (Laman and Weiblen 1999). However, Mount Kinabalu, which has high β-diversity through its altitudinal gradient, and has rare soil types (Wong and Phillipps 1996), has 82 species (Corner 1964; J. H. Beaman unpublished data). Gunung Palung has notably fewer dioecious figs compared to Lambir Hills and Mt. Kinabalu, and most are common and widespread, indicated by the high degree of overlap with the flora of Lambir Hills. However, the focus of research at Gunung Palung was on the large monoecious hemi-epiphytes (Laman 1995; Laman 1996b), so it is possible some of the more inconspicuous dioecious fig species were missed. Nevertheless, less than 60% of the species at Lambir Hills are shared with either Mt. Kinabalu or Gunung Palung, although many species have broad geographic distributions (Corner 1965). Laman and Weiblen (1999) suggested that differences between local assemblages may be due to large-scale habitat associations. Certainly, casual observations in limestone forest at Niah and peat swamp forest at Loagan Bunut, 30 km to 50 km from Lambir Hills, suggest that the *Ficus* flora is different at each site.

   The degree of endemism varies strikingly among monoecious and dioecious figs, as shown in Table 10.2 (Corner 1967). Many monoecious figs have distributions extending from the Asian mainland to Papua New Guinea and sometimes Australia (Corner 1965, 1967). There are just four endemic monoecious species: two have an unusual growth form, being small understory vinelike hemi-

Table 10.2. Comparison of the *Ficus* flora*: Lambir Hills, Mt. Kinabalu in north Borneo (Corner 1964; John Beaman, unpublished data), Gunung Palung, Southwest Borneo (Laman and Weiblen 1999), and for Borneo as a whole (Corner 1965). The number of species by section (Corner 1965) are given and numbers in parentheses are the numbers of species shared with Lambir Hills.

Ficus	Lambir Hills	Mt Kinabalu	Gunung Palung	Borneo	Borneo endemics
Subgenus *Urostigma*					
Section *Urostigma*	2	0	1 (1)	4	0
Section *Conosycea*	27	17 (8)	27 (18)	36	4
Subgenus *Pharmacosycea*					
Section *Oreosycea*	1	2 (1)	0 (0)	5	0
Subgenus *Sycomorus*	0	1 (0)	0 (0)	1	0
Subgenus *Ficus*					
Section *Ficus*	10	11 (6)	4 (4)	20	11
Section *Rhizocladus*	9	11 (7)	8 (6)	17	5
Section *Kalosyce*	4	6 (1)	2 (2)	12	8
Section *Sycidium*	10	18 (10)	6 (6)	23	8
Section *Neomorphe*	1	1 (1)	1 (1)	1	0
Section *Sycocarpus*	13	15 (11)	6 (5)	24	14
Total	77	82 (45)	56 (43)	143	50

* Includes recent additions, nomemclature changes and undescribed spp from each site

epiphytes, and two are known only from Mt. Kinabalu (Corner 1965; Kochummen 1998; Laman and Weiblen 1999). In contrast, 46 out of 97 dioecious fig species are endemic to Borneo. That pattern, the higher endemism among dioecious figs relative to monoecious figs, is consistent across the Indo-Australian region (Corner 1967).

10.4 Life Histories of *Ficus* at Lambir Hills

Some rare figs at Lambir Hills. Brief mention should be made here of the figs in sections *Urostigma, Oreosycea,* and *Neomorphe,* whose species at Lambir Hills are known from very few individuals (see Table 10.3). A single individual each of *F. virens* and *F. caulocarpa* (section *Urostigma*) was recorded. Both are widespread monoecious hemi-epiphytes. *Ficus virens* is more common in parts of Papua New Guinea and Australia, where it is an extremely large strangler (Shanahan et al. 2001b; R.D. Harrison personal observation). Superficially at least, both species appear to have ecologies similar to the other monoecious hemi-epiphytes in section *Conosycea*. A single adult individual and a few samplings of *F. vasculosa* (*Oreosycea*), a large monoecious canopy tree, were recorded in the 52 ha plot. It is possibly infrequent throughout Borneo, as it rarely is recorded. Section *Oreosycea* is distributed from Madagascar to New Caledonia and several of its species are widespread but rare (Corner 1985). Only in New Caledonia does it attain higher densities and, significantly, higher levels of endemism (Corner 1970). Corner speculated that *Oreosycea* might be the remnant of an ancient fig-dominated forest, but it seems likely that extremely

Table 10.3. Comparison of ecological charactersitics among sections of *Ficus* at Lambir Hills

Ficus	#spp	Sex[1]	Growth-form[2]	Density[3] (ha^{-1})
Subgenus *Urostigma*				
Section *Urostigma*	2	m	he	0.06
Section *Conosycea*	25	m	he	0.02–0.46
Subgenus *Pharmacosycea*				
Section *Oreosycea*	1	m	lt	0.25
Subgenus *Ficus*				
Section *Ficus*	10	d	st/e	0.04–5.38
Section *Rhizocladus*	9	d	c	0.04–0.10
Section *Kalosyce*	4	d	c	0.04–0.29
Section *Sycidium*	10	d	he/s	0.02–0.25
Section *Neomorphe*	1	d	lt	0.06
Section *Sycocarpus*	13	d	st	0.02–15.00

[1] m=monoeclous; d=dioeclous
[2] he=heml-epiphyte; lt=large tree; st=small tree; e=epiphyte; c=climber; s=shrub
[3] Range across species occuring in the 52 ha plot at Lambir Hills

low density is characteristic of the section. *F. variegata* (*Neomorphe*) is a large dioecious tree with cauliflorous figs that are dispersed by bats. This species also has a wide distribution but within a north-south orientation: Japan to Australia. Usually described as common, only a few juveniles were recorded in the 52 ha plot at Lambir Hills, and *F. variegata* appears infrequent in Borneo. However, at higher latitudes it is often a dominant tree, especially in disturbed forests (Walker 1976; Spencer et al. 1996).

Growth habits. Figs in Lambir Hills exhibit an extraordinary variety of growth habits (see Plate 10). Monoecious figs are either hemi-epiphytes (*Urostigma* and *Conosycea*) or large trees (*Oreosycea*), while dioecious figs include root-climbers (*Rhizocladus* and *Kalosyce*), small understory hemi-epiphytes (*Sycidum*), shrubs or small trees (*Ficus* and *Sycocarpus*), and large canopy trees (*Neomorphe*) (see Table 10.3).

Such general terms, however, do not adequately represent the diversity of fig life forms. For example, the monoecious hemi-epiphytes include species with vinelike stems, which have sometimes mistakenly been called climbers (Corner 1985). Although most species remain dependent on their hosts for support, two species have been recorded as stranglers at Lambir Hills: *F. kerkhovenii* and *F. stricta* (Harrison et al. 2003). There are also the banyans—species that have secondarily lost the hemi-epiphytic habit and root directly in the ground, such as *F. microcarpa* and *F. superba*, common in coastal vegetation. This growth form, however, has not been observed at Lambir Hills. Among the dioecious hemi-epiphytes (section *Sycidium*) most are small and shrublike but can be quite catholic in surfaces they colonize. In Lambir Hills, *F. heteropleura*, for example, has been recorded on dead stumps, cliffs above streams, and as shrubs on sandy riverbanks, in addition to the normal hemi-epiphytic habitat on buttresses or smaller understory trees. Dioecious hemi-epiphytes also include species with vinelike stems, such as *F. sinuata*. Finally, in the section *Ficus* (normally, small trees) there is an epiphyte, *F. deltoidea* var. *borneensis*. This is only known as an epiphyte and is probably a separate species. *F. deltoidea* consists of wispy shrubs in nutrient-poor environments, including kerangas forest on the crest of Bukit Lambir.

Rarity and abundance. Unsurprisingly, growth habit appears to be a determining factor in the distribution and abundance of figs. In the 52 ha plot at Lambir Hills, *F. delosyce* (*Conosycea*) was the most common monoecious hemi-epiphyte with 24 individuals, and *F. heteropleura* (*Sycidium*) was the most common dioecious hemi-epiphyte with 13 individuals (see Appendix B). Both are extremely rare compared to common tree species. Moreover, 6 out of 28 species of monoecious hemi-epiphyte and 2 of 10 dioecious hemi-epiphytes found in Lambir Hills were not recorded in the 52 ha plot. Similar low densities in natural forest have been reported for monoecious hemi-epiphytes from Gunung Palung, India, West Africa, and Panama (Todzia 1986; Michaloud and Michaloud 1987; Laman 1996b; Patel 1996a). Dioecious climbers (*Rhizocladus* and *Kalosyce*) were also rare. The most common, *F. aurantiacea*, had just 13 individuals in the 52 ha plot (see Appendix B).

In contrast, 9 species of dioecious trees and shrubs (*Ficus* and *Sycocarpus*) had more than 50 individuals in the 52 ha plot; *F. stolonifera* had 730. Seven out of 23 species, however, were not recorded in the plot. Abundance, therefore, varies considerably among species with different habits, and the hemi-epiphytes and climbers are rare. The small dioecious trees are common. In secondary forests, they may dominate the vegetation (Corner 1967,1988; Harrison 2000a; Harrison et al. 2000).

Distributions. Within the dipterocarp forest at Lambir Hills, clay-dominated *udult* soils are found predominantly at lower elevations and sandier *humults* on the ridges. Superimposed on this is a complex topography (Lee et al. 2002). Overall, figs appear more common on richer soils and become progressively more difficult to find as one climbs up the hill, the exception being the afore-mentioned *F. deltoidea* var. *deltoidea*, restricted to sandstone outcrops on Bukit Lambir. However, only in the 52 ha plot have both the soils and the figs been surveyed in detail. There, among monoecious hemi-epiphytes (*Conosycea*), only 2 out of the 11 most common species showed a significant association with soil or topography. *F. xylophylla* was found on sandier soils and *F. kerkhovenii* on steeper slopes (see Fig. 10.1) (Harrison et al. 2003). Dioecious hemi-epiphytes (*Sycidium*) had similarly scattered distributions, while the dioecious climbers (*Rhizocladus and Kalosyce*) were scattered but almost entirely restricted to lower, clay-rich areas.

Among the dioecious trees in section *Ficus*, two distribution patterns can be recognized. The first is associated with steep slopes. Pioneers such as *F. aurata* and *F. fulva* colonize large landslide gaps and are common along roadsides (Harrison et al. 2000). Other species are scattered on the sandier areas of the plot, and as exemplified by *F. setiflora,* are small understory trees (Metcalfe et al. 1998).

The dioecious trees and shrubs in section *Sycocarpus* are almost entirely con-fined to the clay-rich soils at the lower end of the plot. Those figs colonize a variety of gaps including tree-fall gaps, smaller landslides, and stream sides. They have strongly overlapping distributions, and it is possible to encounter several species close together (see Fig. 10.1). Most are geocarpic, with inflores-cences borne on specialized stolons that run along the ground. By rooting from the stolons, new stems can grow several meters away from the main trunk, presumably an advantage for species colonizing unstable sites. A closer exam-ination of the geocarpic figs reveals a spectrum of pioneer ecologies, with spe-cies segregating according to colonization microsite, light environment, and maximum diameter (R.D. Harrison, unpublished data).

Fig phenology. In contrast to the supra-annual GF of many species in the forests in Southeast Asia (Corlett and Lafrankie 1998; Sakai et al. 1999c), figs flower continuously. The classic reproductive phenology of tropical figs involves the

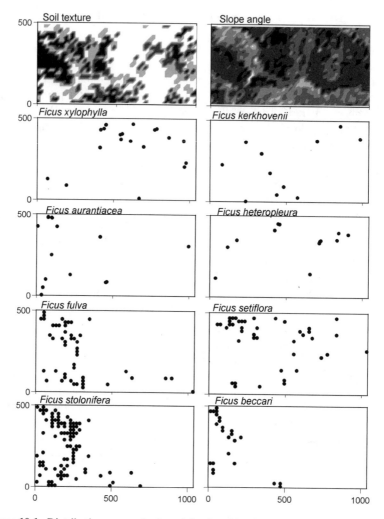

Figure 10.1. Distribution maps of selected figs Lambir Hills, 52 ha plot (Lee et al. 2002; Harrison et al. 2003; R.D. Harrison, unpublished data). For soil texture lighter shades indicate a higher proportion of sand. For slope angle lighter shades indicate steeper slopes.

production of large, highly synchronous fruit crops (Janzen 1979; Milton et al. 1982; Bronstein et al. 1990; De Figueiredo and Sazima 1997). However, broader studies have revealed a diversity of phenological types (Corlett 1987, 1993; Damstra et al. 1996; Patel 1996b; Spencer et al. 1996). At Lambir Hills there was considerable variation in phenology, evident among species from different sections, in flowering frequency, fruit crop sizes, and crop synchrony (see Table 10.4, 10.5). Monoecious figs had infrequent and large synchronous crops, while dioecious figs tended to have smaller, more frequent crops, which were highly asynchronous in some species (Harrison 1996, 2000a; Harrison et al. 2000).

Fig pollinators reproduce in receptive inflorescences and thus their populations can be modeled using data on the phenology of fig trees. In contrast to the large populations of figs required to maintain pollinator populations in monoecious species (Bronstein et al. 1990; Anstett et al. 1997), Kameyama et al. (1999) found that frequent flowering of dioecious *F. schwarzii* (*Sycocarpus*) at Lambir Hills enabled pollinators to persist in small clumps of trees. Moreover, a single large male individual of the asynchronous flowering species *F. cereicarpa* (*Sycocarpus*) maintained continuous production of fig wasps over an 18-month period (Harrison 2000a). The pollinator populations of dioecious figs, however, are vulnerable to catastrophic disturbances such as droughts (see Chapter 5).

Seed-dispersal syndromes. Figs have often been referred to as keystone resources because of the importance their fruits have to vertebrates, particularly at times when little other fruit is available (Terborgh 1986; Lambert and Marshall 1991; Borges 1993; Kinnaird et al. 1999). However, figs present their fruit in many different ways, which in turn means fig species vary in their importance to particular frugivores (Kalko et al. 1996; Shanahan and Compton 2001; Shanahan et al. 2001a). In Lambir Hills, among 34 species of figs there were five recognizable seed dispersal syndromes, each predominantly composed by different frugivores, as shown in Fig. 10.2 (Shanahan 2000).

The size and color of fruit, its height above ground, and crop size appeared to determine which animals ate the fruit, while a frugivore's size, sensory and locomotory physiology, and foraging height determined guild membership. Im-

Table 10.4. Comparison of the phenology of figs from six sections of *Ficus* at Lambir Hills. The number of species and individuals under observation, the percentage of individuals that fruited, and the mean frequency of crops produced is given.

Section	No. spp.	No. individuals	% fruiting	Frequency (yr^{-1})
Conosycea[1]	15	52	73	0.75
Ficus[2]	1	52	98	3.51
Rhizocladus[1]	6	11	81	1.13
Kalosyce[1]	3	12	83	2.55
Sycidium[1]	3	3	100	2.8
Sycocarpus[2]*	1	63	100	4.06

[1] Biweekly censuses from October 1994 to September 1998 (Harrison, unpublished data)

[2] Biweekly censuses from June 1996 to September 1998 (Harrison et al., 2000; Harrison 2000)

* Other species under observation fruited asynchronously, hence crops could not be easily descerned.

Table 10.5. Seed dispersal guilds among six sections of *Ficus* (34 species) at Lambir Hills (adapted from Shanhan 2001)

Section	Fig spp.	Guild	Position[1]	Ripe color[2]	Fruit size[3] (mm)	Crop size[4]	Crop synchrony[5]	Frugivore spp.[6]
Conosycea	15	Canopy	a	R-B	10–35	10^{3-4}	High	22 (10–32)
	1	Understorey	a	R-B	15–20	10^{3}	High	10
	1	Bat	a	G	30–35	10^{3}	High	5
Ficus	2	Understorey	a	R	10–20	10^{1-3}	High	13 (8–18)
Rhizocladus	1	Understorey	c	R	5–10	10^{3-4}	Medium	4
Kalosyce	2	Arboreal mammal	c	R/O	50–70	10^{1-2}	Low	3 (2–4)
	1	Bat?	c	O	20–25	10^{2}	Medium	1
Sycidium	4	Understorey	a/c	R	6–12	10^{2-3}	High	8 (6–10)
Sycocarpus	7	Terrestrial Mammal	c/g	R/W/Br	12–40	10^{0-2}	Low	2 (1–3)
	2	Bat	c	G	20–25	10^{1-2}	High	2

[1] a=axial; c=cauliflorous; g=geocarpic
[2] R-B=red turning black; G=green; R=red; O=orange; W=white; Br=brown
[3] Range across species
[4] Range of means across species
[5] Synchrony at seed dispersal: High=1–3 weeks; Medium=3–6 weeks; Low=6+ weeks, crops often indistinguishable
[6] Mean number of species across fig species, range in parenthesis

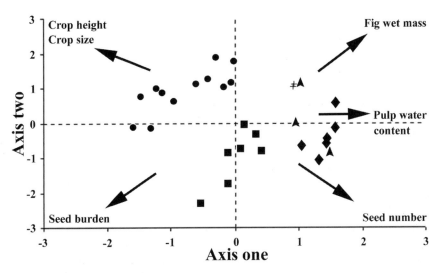

Figure 10.2. Principal components scatter-plot of 34 *Ficus* species at Lambir Hills based on fruit mass, seed number, seed burden (seed mass/total mass), pulp water content, crop height, and crop size (Shanahan 2000; Shanahan and Compton 2001). Arrows represent the direction of a trait-influenced species: 1st PC axis explained 48% and the 2nd 28% total variance, respectively. Dispersal guilds: ● canopy birds and mammals, ♦ understory birds and mammals, + arboreal mammals, ▲ fruit bats, and △ terrestrial mammals.

portantly, frugivore diversity differed substantially among guilds. Canopy figs attracted a total of 53 species of birds and mammals, while only 4 species of terrestrial mammals were observed eating cauliflorous and geocarpic figs, as listed in Table 10.5 (Shanahan 2000). Canopy figs are probably also an important resource for terrestrial frugivores, because large quantities of fruit fall to the forest floor (Heydon and Bulloh 1997).

Guilds correspond roughly to different sections of the fig genus (see Table 10.5). The monoecious hemi-epiphytes (section *Conosycea*) have dark red to black fruit dispersed by canopy birds and mammals. Small dioecious trees (section *Ficus*) with small red fruit were dispersed predominantly by small birds of the understory or open areas, such as bulbuls. Likewise, dioecious climbers with small red fruit (section *Rhizocladus*) were dispersed by understory birds, while those with very large fruit (section *Kalosyce*) were mainly eaten by arboreal primates. Finally, the dioecious pioneers with cauliflorous and geocarpic figs (section *Sycocarpus*) were dispersed by terrestrial mammals, such as mouse deer, rodents, and tree shrews. Bat-dispersed figs were of two types. Two species were dioecious pioneers (section *Sycocarpus*) with cauliflorous fruit, and one was a monoecious hemi-epiphyte (section *Conosycea*) that bore fruit among its leaves. Nevertheless, fruit of each species is green and odorous. Thus, while a phylogenetic pattern may be strong, bat-dispersed figs illustrate variation in seed dispersal within sections.

Further segregation of species within these broad guilds is evidently possible. For example, dispersal by canopy birds and mammals may be divided into species with smaller or larger fruit. Species with larger fruit were found higher in the canopy and eaten to a greater extent by a sub-guild of large canopy frugivores, particularly hornbills (Shanahan and Compton 2001). Geocarpic figs also display considerable variation in the presentation of fruit. Some species (e.g., *F. treubii* and *F. stolonifera*) have small figs (10–15 mm diameter) that are soft-walled and eaten whole, while others (e.g., *F. geocharis, F. megalia,* and *F. uncinata*) are large (20–30 mm diameter) with a thick rubbery wall concealing a sweet sticky mass having embedded seeds. The very different handling required to eat such distinct fruits suggests they are targeted for different dispersers.

Clearly, the figs at Lambir Hills demonstrate an extraordinary variety of seed dispersal syndromes and must also vary among frugivores. Monoecious canopy species support very diverse frugivore assemblages, but they normally have low densities of fruiting individuals. The higher densities and fruiting frequencies of dioecious figs mean they may provide a more dependable resource for their few species of frugivores. Such figs may be particularly valuable in the understory habitat where fruit is relatively scarce (Loiselle and Blake 1999; Shanahan and Compton 2001). Thus, while the keystone paradigm may still be appropriate, the concept takes on different meanings according to the figs in question.

10.5 Niche Specialization Among Monoecious Hemi-epiphytes

The foregoing discussion illustrates large differences in the ecologies of figs from various sections of the genus. Here, the diversity of figs at a finer taxonomic scale is revealed by considering niche specialization within the section *Conosycea*: the monoecious hemi-epiphytic figs. Harrison et al. (2003) conducted a survey of 226 hemi-epiphytic figs in approximately 120 ha of Lambir Hills National Park (LHNP). Those figs had colonized a tremendous diversity of host taxa (35 families, 73 genera, and 107 species among 181 individuals) and showed no evidence of host preference. The fig species segregated into five guilds according to host size (diameter at breast height, or, dbh), height of colonization, and light environment, which corresponded to host canopy strata, as listed in Table 10.6 (Harrison et al. 2003). Species on subcanopy hosts were more abundant and colonized trunks, small and large branch crotches, and branch limbs, whereas species on emergent and canopy hosts were restricted to large branch crotches at the base of the host's crown. This suggests a trade-off inherent to the hemi-epiphytic habit. Hosts with higher canopy positions, and hence better light environments, have lower densities and stricter micro-site requirements (Harrison et al. 2003).

As mentioned earlier, hemi-epiphytic figs are rare. Only 1.77% of trees over 30 cm dbh were occupied by such figs. Evidently, either microsites are extremely limiting or seeds fail to reach them, or a combination of both factors. In a study

Table 10.6. Niche-differentiation among eleven monoecious hemi-epiphytic figs (section *Conosycea*) at Lambir Hills (Harrison et al. 2003)

Type[1]	*Ficus* species	n	Growth form[2]	Crown area (m²) maximum	Host-dbh[3] (cm) mean (s.d.)	Height of attachment[4] (m) mean (s.d.)	CI[5] index mean	Fruit size class[6]
Emergent								
	F. cucurbinita	12	radial	2200	97 (30.9)	29 (9.9)	4.2	3
	F. dubia	17	radial	1370	96 (27.0)	30 (7.7)	4.4	3
	F. subcordata	9	radial	2900	110 (50.3)	34 (7.6)	4.4	3
Canopy								
	F. kerkhovenii	22	strangler*	2990	74 (30.6)	26 (6.0)	4.4	2
	F. stupenda	14	radial	2670	89 (58.1)	26 (7.2)	4.4	3
	F. xylophylla	26	radial	790	74 (26.2)	26 (6.2)	4.1	3
Subcanopy								
	F. binnendykii	13	emergent	910	48 (24.3)	21 (9.4)	3.5	1
	F. delosyce	31	radial	750	42 (23.4)	19 (8.4)	3.7	1
	F. subgelderi	30	radial	1010	53 (28.9)	21 (8.2)	3.8	2
Other								
	F. pisocarpa	13	radial	800	67 (29.6)	26 (8.3)	3.3	2
	F. palungensis	22	vine-like	180	32 (17.3)	16 (7.3)	3.1	2

[1] Types were significantly different from one another ($p < 0.05$; Mahlanobis distances, canconical discrimination analysis)

[2] Radial=short or no trunk, branches radiate out beneath host crown; Strangler=similar to radial initially, with greater penetration of host crown and eventual strangling of host; emergent=long trunk with narrow crown emergent above host crown; Vine-like=hemi-epiphytic with several cable-like stems climbing along branches and sometimes on to other hosts

[3] Measured at 1.3 m or above buttresses

[4] Measured at the point from which aerial-roots descend with a handheld clinometer

[5] CI index=Crown illumination index, an observer assessed index of the amount of light reaching the hemi-epiphyte crown. Index ranges from 0 (no light) to 5 (full overhead and lateral light) (Harrison 2003).

[6] Classes: 1<10mm diameter; 2=10–20mm diameter; 3>20mm diameter

* Six individuals were free-standing having strangled their hosts.

of seedling establishment in Gunung Palung, Laman (1995) found that the water-retention capacity of microsites with some humus was critical but suggested that microsites rarely received seeds (Laman 1996a). Moreover, when the seed rain is augmented, a single host may support several hemi-epiphytic figs (Patel 1996a; R.D. Harrison, personal observation). Hence, it seems likely that low seed rain limits the abundance of hemi-epiphytic figs in forest habitats.

Dispersal of fig wasps. All figs share a similar pollination system. Female fig wasps, carrying pollen from their natal fig, must disperse during a brief adult life to a fig tree of the appropriate species in order to reproduce. Asynchronous flowering among fig trees ensures receptive figs are available year-round, but the low densities and infrequent flowering of some fig species require their pollinators to disperse substantial distances. From paternity analyses in Panama, the pollinators of several monoecious figs were estimated to disperse 5 km to 14 km (Nason et al. 1996; Nason et al. 1998). However, the extraordinary ecological diversity of figs, in particular the higher densities and frequent flowering of some dioecious species, suggested that pollinator dispersal varied among species. Therefore, a study using sticky-traps suspended from a canopy crane at Lambir Hills was initiated to investigate fig wasp dispersal (Harrison 2003).

Fig wasps constituted the majority of insects caught above the canopy (≥ 45 m height) and flew significantly higher than other insects (Harrison 2003), which indicates they use wind-assisted dispersal (Ware and Compton 1994a,b). Nevertheless, their abundance was remarkable and demonstrates the very high fig wasp production in tropical rain forests. The large crop sizes (10^4–10^6 inflorescences) of some figs and the emergence of several 10s to 100s of fig wasps per inflorescence could account for such abundance. Captures among fig wasp genera, however, were not uniform, as listed in Table 10.7 (Harrison 2003). The species richness of monoecious fig pollinators was greater than the number of host fig species at Lambir Hills, implying that some species must have arrived from forests with different assemblages of figs. The nearest source areas lie 30 kilometers distant, but the diversity of the fauna suggests some may have come from much farther. In contrast, fewer than 25% of the expected dioecious fig pollinator fauna was collected. Smaller crop sizes among their hosts may explain the low abundance of such species. However, the high densities and frequent flowering of many dioecious figs reduces any need for long-distance pollinator dispersal. Short-range active dispersal is likely to be more important and would explain why the traps intercepted relatively few pollinators of the dioecious figs. Reduced pollinator dispersal would also help to explain high rates of fig endemism (see Table 10.2) and the vulnerability of pollinators to local extinction (Harrison 2000b).

Fig wasps also differed in their temporal patterns of activity and flight altitude, suggesting adjustments between fig ecology and pollinator dispersal. Wind speeds are higher during the day and increase with distance above the canopy. *Waterstoniella*, a wasp that flew at night and closer to the canopy than other such genera, pollinated 9 out of 11 common monoecious figs in Lambir Hills.

Table 10.7. Number of fig pollinator species caught using sticky-traps over 10 days at Lambir Hills, the number of host fig species at Lambir Hills and in Borneo, and the mean height of captures (Harrison 2003*)

Fig pollinator genera	No. of fig pollinator species	No. of host fig species Lambir Hills	Borneo	Mean height of captures (m)
Monoecious fig pollinators				
Dellagaon	2	1	4	38 ± 15.2
Dolichoris	2	1	5	62 ± 15.0
Eupristina	17	8	11	52 ± 14.3
Platyscapa	2	2	3	61 ± 10.1
Waterstoniella	26	16	21	39 ± 09.9
All genera	49	30	44	
Dioecious fig pollinators				
Blastophaga	1	11	19	15 ± 00.0
Ceratosolen	2	13	28	75 ± 00.0
Kradibia	1	1	6	59 ± 13.4
Lipporhopalum	3	10	14	40 ± 18.1
Wiebesia	4	14	29	44 ± 06.9
All genera	11	49	96	

* Figures have been updated.

10.6 Discussion

It is the extraordinary diversity of co-existing species that makes tropical rain forests so special, and two patterns are often mentioned. Whitmore (1998) calls attention both to the presence of large numbers of rare species and to large genera with suites of closely related species. Figs are found in all tropical lowland rain forests and are invariably diverse. Indeed, *Ficus* is often the most species-rich genus in any particular forest, and many of its species are rare. Put simply, "Figs define tropical forests" (Corner 1967). *Ficus* is the richest genus at Lambir Hills, which is the most diverse forest, in terms of tree species, in the Old World. By considering the figs at Lambir Hills we may hope to gain some insights into why tropical forests have so many species.

Borneo inherited a taxonomically diverse fig assemblage (see Table 10.1). With 77 species in the fig community at Lambir Hills, where 9 sections of the genus *Ficus* co-occur, the study assemblage is diverse but comparable to other sites in Borneo (see Table 10.2). According to molecular studies, different sections of the genus appear to correspond to monophyletic groups, although a complete phylogeny still eludes us (Weiblen 2000). Hence, by comparing the ecology of figs from different sections, and then the more closely related species within a section, we can obtain a hierarchal perspective on diversity within the genus.

Above, we provided a broad-brush overview of various ecological traits, including life forms, abundance, distribution, phenology, and seed-dispersal guilds, and found that the life-histories of figs from different sections are divergent. For example, most monoecious hemi-epiphytic figs (section *Conosycea*) have scattered, low-density distributions and produce large, synchronous crops of red-black figs at infrequent intervals. The fruit is eaten by many canopy birds and mammals. In contrast, the dioecious geocarpic figs (section *Sycocarpus*) are gap pioneers found at high densities on richer soil. Their frequent (often asynchronous) small crops of figs are eaten by one or two species of terrestrial mammals.

Unfortunately, phylogenetic constraint prevents us from understanding the relative roles of common ancestry versus selective pressures in determining many of these traits. The occurrence of two or more seed dispersal syndromes in the same section, however, and shared syndromes among different sections suggest seed dispersal traits, at least, are quite malleable. In comparison to dioecious figs, the low flowering frequency of monoecious figs is remarkable, considering their extraordinarily low densities. The pollinators must therefore disperse long distances to encounter a receptive tree (Nason et al. 1998; Harrison 2003). However, the slight pollinator shortage may also reduce the exploitation of ovules by pollinator larvae (Herre 1989), or lead to other advantages tied to rarity (see Chapter 1). Large crops also attract wide-ranging dispersers, such as hornbills (Shanahan and Compton 2001). Thus, despite many apparent advantages of the dioecious condition in figs, low densities and large infrequent crops may stabilize monoecy (Harrison and Yamamura 2003).

At a finer scale, ecological diversity was also evident when more closely related species within a section were considered. Despite a superficial similarity, monoecious hemi-epiphytic figs correspond to 5 guilds among 11 common species at Lambir Hills, based on the canopy strata they occupy (see Table 10.6) (Harrison et al. 2003). Other characteristics, such as position of colonization, habit, and fruit size also varied among species and correlated to canopy strata. Moreover, niche differentiation is clearly not restricted to the monoecious hemi-epiphytic figs. Pioneer figs in sections *Ficus* and *Sycocarpus* evidently vary in their microhabitat preferences, phenology, and seed dispersal syndromes. Thus, even within a section there is considerable ecological diversity among fig species.

High species richness in the fig assemblage at Lambir Hills would, therefore, appear to stem largely from their ecological diversity. *Ficus* has many more species than related genera. Globally, Moraceae consists of 37 genera with approximately 1100 species of which 750 are figs, while *Artocarpus*, the most diverse genus in Borneo after *Ficus*, has just 12 species in the 52 ha plot (Lee et al. 2002), 24 species in Sarawak and Sabah, and 55 species throughout its range (Kochummen and Go 2000).

Another aspect, so fundamental to fig biology it should hardly need stating, is their unique pollination system. Fig pollination is best understood as wind-pollination with a motor. Wind-pollination becomes inefficient when individual densities are low and hence is rare in tropical forests (Bawa et al. 1985; Momose et al. 1998). Figs, however, circumvent the problem because the short-range active attraction of fig wasps to volatile cues released by a receptive fig greatly increases the efficiency, and figs normally enjoy very high pollination success— despite the low densities of many species. The relative importance of passive wind-dispersal and active flight also appears to vary among species, enabling the system to accommodate the ecological diversity of their hosts (Harrison 2003). Finally, the extreme specificity of the system permits large assemblages of figs to coexist.

The differences in ecology among different sections suggest evolutionary independence. Competition from figs in other sections may have prevented ecological transgression, or, perhaps more likely, the maintenance of certain traits has restricted adaptive possibilities. The diversity of figs in Southeast Asia remains unparalleled. Who would believe that a feeble shrub with a few fruit buried in the leaf-litter of a tree-fall gap and a huge strangler, its massive crown filling the canopy with hundreds of thousands of fruit, were from the same family, let alone the same genus (Corner 1988)?

11. Ecology of Traplining Bees and Understory Pollinators

Makoto Kato

11.1 Introduction

In the tropical rain forests of Southeast Asia, the dominant pollination agents of canopy trees are eusocial apid bees and beetles, as discussed in other chapters. The prevalence of native honeybees is unique to the Paleotropics, while the prevalence of meliponine bees and beetles is common in the Neotropics (see Basset et al. 2003). In addition to those observed at canopy trees, there are many other pollination systems, especially in the understory of the forest at Lambir Hills National Park (LHNP). In this chapter, I review pollination systems in which various insect groups other than *Apis* and beetles take part.

Long-tongued traplining bees. In the understory of the lowland dipterocarp forest at LHNP, there are many plant species with long-tubed flowers in families Zingiberaceae, Costaceae, Marantaceae, Acanthaceae, Gesneriaceae, Polygalaceae, and Loganiaceae. The species richness of Bornean gingers, in particular, is remarkably high (Sakai and Nagamasu 1998). Most ginger species bloom more than once a year, or continuously with short interruptions, without joining mass-flowering events (Sakai 2000). They produce abundant but relatively dilute nectar, less than 30% sugar by weight.

The long flowers are mainly visited and pollinated by long-billed spiderhunter birds, *Arachnothera* spp. (Nectariniidae), or long-tongued bees, including those of a normally short-tongued family, belonging to four genera: *Amegilla* (Apidae),

Thrinchostoma, Nomia and *Sphecodes* (Halictidae) (see Plate 6J). The *Amegilla* bees found on these understory flowers belong to two species of subgenus *Glossamegilla*, which are shade-loving, swiftly-flying, long-tongued *trapliners* (Kato 1996). Their two species, *A. pendleburyi* and *A. insularis*, are middle-sized bees with similar tongue length (13.2–22.0 mm). *Thrinchostoma, Nomia,* and *Sphecodes* are small-sized but long-tongued halictid bees, while their tongue lengths (8.2–9.0 mm) are shorter than those of *Amegilla* species. *Nomia* sp. 1 in Kato (1996) should be called *Sphecodes*, which is thought to be a cleptoparasite of *Thrinchostoma* or *Nomia*. Most of these long-tongued halictid bees are also shade-loving, swift-flying trapliners, but some *Nomia* species visited long-tubed flowers of small or subcanopy trees, such as *Vitex, Sphenodesme* (Verbenaceae) and *Trigonia* (Trigoniaceae).

The understory bees fly very swiftly near the ground, and it is difficult to collect them in flight. Their foraging behavior differs slightly between females and males. Females usually enter almost all flowers that they encounter, to forage nectar and sometimes collect pollen, but males often skip available flowers in their search for females. These bees apparently never visited flowers at the canopy or subcanopy, or in sunny habitats, but visited only understory flowers. These bees are active throughout a year, and some bee-pollinated gingers may serve as keystone species for survival of the traplining bees using nectar and pollen of understory flowers (Sakai 2000).

Traplining bee species do not always specialize on single flower species, but each individual bee seemed to specialize during a certain period. In Borneo, there are only two *Glossamegilla* species, and they contribute to pollination of many species of gingers and other long-tubed understory plants. Thus, the local or perhaps temporary flower specialization by these bees seems likely to have fostered diversification and speciation among plants.

Intensive observation of bee visits to ginger flowers showed that gingers belong to three pollination *guilds,* i.e., plants pollinated by spiderhunters, medium-sized *Amegilla,* or small halictid bees (Sakai et al. 1999a). Among species in each guild, convergence of tongue and bill lengths is found. This pattern contrasts with the community-level *character displacement* of tongue length in bumblebees found in temperate regions (Inouye 1977). While there are only two *Glossamegilla* bees in Borneo, there are nine *Glossamegilla* species in Sumatra (Lieftinck 1956), including a large-sized *Glossamegilla* bee, *Amegilla elephas,* whose tongue is among the longest of bees in the world. This bee visits exceptionally long-tubed red ginger flowers, which are sometimes also visited by spiderhunters (Kato et al. 1993a). The difference in species richness of *Glossamegilla* between Sumatra and Borneo may originate from paleogeographic history; that forest is richer in plant species than those of Borneo (Holden 2002).

The counterparts of *Amegilla* bees in the Neotropics are euglossine bees (Apidae, Euglossini). Both *Amegilla* and euglossine bees have remarkably long tongues and trapline relatively rare long-tubed flowers. In contrast with *Amegilla* bees, euglossine bees also visit epiphytic orchid flowers in canopy layers to collect floral oils or scents and long-tubed flowers at various layers in forests (Janzen 1971a; Roubik and Hanson 2004).

Leafcutter and carpenter bees. The flowers of papilionaceous legumes (Fabaceae) and Xanthophyllaceae have structures that prevent unspecialized foragers from entering but are visited and pollinated by bees of the genera *Xylocopa* (Apidae), *Megachile* and *Chalicodoma* (Megachilidae), which can open tightly closed petals and enter the flower. Most of these plant species joined mass-flowering events, and population levels of these bees and their cleptoparasitic meloid beetles greatly increased during the mass-flowering year, as shown in Fig. 11.1 (Kato et al. 2000).

The only nocturnal bee at Lambir Hills is *Xylocopa myops*, while *Apis dorsata* sometimes forages just before sunrise and just after sunset. The nocturnal carpenter bee was observed at flowers of *Duabanga grandiflora* (Sonneratiaceae) at Matang in Kuching. Since this bee is attracted to light, we can analyze its long-term population fluctuation by using light-trap data. The bee population was roughly stable for six years, irrespective of drastic changes of floral resources.

Small solitary bees. Generalist flowers of some subcanopy trees, such as *Aporusa, Cephalomappa, Cleistanthus, Drypetes, Homalanthus, Tapoides* (Euphorbiaceae), *Grewia* (Tiliaceae), and *Vernonia* (Asteraceae), attracted a number of insects including small solitary bees, such as *Hylaeus* (Colletidae, Hylaeinae), *Lasioglossum* (Halictidae), *Braunsapis* and *Ceratina* (Apidae, Xylocopinae). These flowers are minute, but produce much nectar and pollen mainly due to their abundance.

Vespid and eumenid wasps. Vespid and eumenid wasps hunt various insects and visit flowers to feed on nectar. Since wasps having alternative food items, that is, insect prey, which continuously sustain their population levels, they may be reliable pollinators for some unpredictably-flowering understory plants. This should be especially important when the population level of bees declines, following a decrease in flowers or general flowering. While generalist flowers often attract various wasps, small flowers of an understory shrub species, *Casearia grewiaefolia* (Flacourtiaceae) attracted only eumenid wasps. In a subtropical forest on Amami Islands in Japan, male and female flowers of a dioecious understory shrub species—*Psychotria sepens* (Rubiaceae)—were visited only by vespid wasps of *Vespa* and *Vespula* (Kato 2000).

Dipterans. Flies play a less important role in pollination than do hymenopterans and coleopterans at Lambir Hills. This contrasts with various temperate forests (Kato et al. 1990), subalpine meadows (Kato et al. 1993b), and wetlands (Kato and Miura 1996), where syrphid flies are important visitors and pollinators of various plant species. However, some understory plant species were visited not by bees, but by dipterans.

Burmannia lutescens (Burmanniaceae) and *Sciaphila secundiflora* (Triurida-

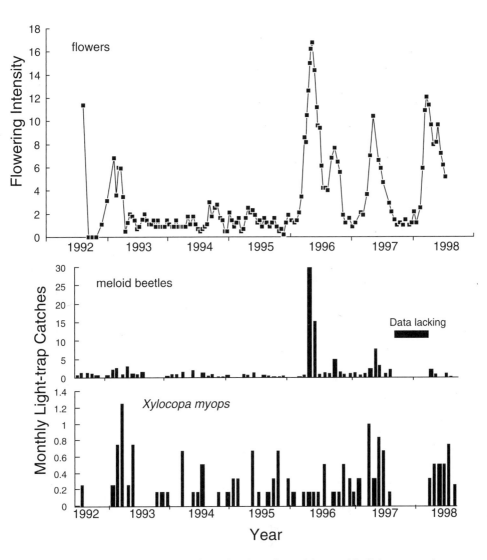

Figure 11.1. Temporal changes in flowering intensity and in monthly light-trap catches of meloid beetles and a nocturnal carpenter bee, *Xylocopa myops*, for six years at Lambir Hills National Park, Sarawak. Because meloid beetles are cleptoparasites of megachilid and xylocopine bees, fluctuation of the meloids is thought to reflect that of host bees.

ceae) are small non-green, mycotrophic herbs, and their flowers, which provide small amount of rewards (nectar and pollen, respectively), were rarely visited by culicid mosquitoes and calliphorid flies. Mosquitoes have long proboscides, which are used to suck blood from terrestrial vertebrates, the juice of fruits, or floral nectar. Both sexes of mosquitoes use fruit juice and floral nectar, whereas only females suck blood. The high density of mosquitoes in some tropical rain forests may explain the reason why mosquito pollination has evolved there.

Diploclisia kunstleri (Menispermaceae) has a cauliflorous inflorescence with minute monosexual flowers, which were visited by syrphid and calliphorid flies. Gymnosperm lianas, *Gnetum cuspidatum* and *G. leptostachyum*, had *strobili* producing pollination droplets on ovules and nectar on bracts, and the strobili were visited by various flies (Kato et al. 1995b). Since aphids are not so common in the tropics, unlike the temperate regions, aphid-eating syrphid flies are not abundant in the tropics. Insect-parasitic tachinid flies, important flower visitors in temperate forests, are also scarce in the tropics. The flies contributing to pollination in the tropics belong to Calliphoridae, Lauxanidae, Drosophilidae, and other families, which mostly breed in decaying plants, animals, fungi or vertebrate feces.

A unique pollination system in which cecidomyiid gall midges play an important role in association with parasitic fungi is found in *Artocarpus integer* (Moraceae) (Sakai et al. 2000). The fungus *Choanephora* (Choanephoraceae, Mucorales, Zygomycetes) infects male inflorescences, and gall midges *Contarinia* oviposit on the male inflorescence, where larvae feed on the mycelia. The gall midges are also attracted to female inflorescences, and pollinate them.

Butterflies. A large number of lepidopteran species live in the forest, and most of them visit various flowers to sip nectar with their long tongues. However, they play a much less important role in pollination than do bees.

Butterflies, which are diurnal, sometimes pollinate tubular flowers where the stigma and anthers protrude from the corolla. Butterfly-pollinated flowers include *Ixora* (Rubiaceae), *Bauhinia* (Leguminosae: Caesalpinioideae), and probably some epiphytic plants of Loranthaceae. Red or yellow tubular flowers of epiphytic plants, *Amylotheca* and *Tritecanthera* (Loranthaceae), are pollinated by spiderhunters, but are sometimes visited by papillionid butterflies such as *Trogonoptera brookiana* (see Plate 2E).

Moths. Most moths are nocturnal, and some of them pollinate nocturnal flowers. For example, *Barringtonia sarcostachys* (Lecythidaceae) is visited by nocturnal sphingid moths (Momose et al. 1998c). Other nocturnal tubular flowers such as *Tabernaemontana* (Apocynaceae) would also be visited by sphingid moths. Some of these nocturnal flowers may be visited also by fruit bats.

The small dioecious gymnosperm shrub, *Gnetum gnemon tenerum*, has slender strobili, which emit a unpleasant odor in the evening (Kato and Inoue 1994). Ovules of female strobili and trace ovules of male strobili secrete pollination droplets containing sugar. Pyralid and geometrid moths are attracted to the strobili by the odor and intake the droplets. These nocturnal moths fly between strobili with their proboscides covered by pollen.

Thrips. Thrips are known as important pollinators of dipterocarp canopy trees of section *Mutica*, genus *Shorea* (Dipterocarpaceae) in Peninsular Malaysia (Appanah and Chan 1981), but not in Borneo (Sakai et al. 1999b). However, *Horsfieldia grandis* (Myristicaceae) and *Popowia pisocarpa* (Annonaceae) were observed pollinated by thrips in LHNP (Momose et al. 1998a).

Cockroaches. Orthopteroid insects are an archaic group that appeared in the Paleozoic, but among them, mutualism with flowering plants is very rare. Cockroaches of Blattellidae are nocturnal, active fliers and often found on various flowers during the night. Cockroach pollination is reported in *Uvaria elmeri* (Annonaceae) at Lambir Hills (Nagamitsu and Inoue 1997).

11.2 Pollination Systems in the Understory

Pollination systems in the forest understory of LHNP were diverse. By comparing pollination systems in the forest understory in Sarawak with qualitative data on systems in the forest canopy, three characteristic features of pollination systems in the forest understory were found: (1) the prevalence of species-specific interactions, (2) the prevalence of pollination by long-tongued, traplining solitary bees, (3) a lack of contribution to pollination by mass-recruiting eusocial bees (honeybees and large stingless bees), large *Xylocopa* bees, thrips, bats, and also of wind pollination, all of which were observed in the forest canopy and subcanopy. Since eusocial bees can communicate the foraging site to nest members and recruit them (Roubik 1989), they can harvest vast quantities of floral reward as soon as flowers appear. In turn, forest understory flowers are scattered sparsely in the forest understory; thus they are not used by these mass-recruiting eusocial bees.

The differences in pollination systems among vertical levels would originate from differences in the pollinator's preference for specific vertical levels and specific light intensities and differences in distribution patterns of floral benefits to which some pollinators responded. The long-tongued solitary bees had a strong preference for flying at ground level and in shaded habitats. They may be habituated to finding resources there, but the preference may also reflect specific traits, such as sensitivity to normally unattractive red flowers (Chittka et al. 1994). Study of insect vision in the understory would further elucidate understory pollination at LHNP. Honeybees and large stingless bees seemed to prefer high, sunny habitats and responded by aggregating on honey-baits set at ground level (Salmah et al. 1990; Roubik et al. 1999). Vertical foraging habits and possible preferences by flower-visiting insects will be revealed by analyses on the collections of Malaise interception traps and honey-bait traps at various vertical levels.

12. Vertebrate-Pollinated Plants

Takakazu Yumoto

12.1 Introduction

Vertebrate pollination occurs exclusively among birds and mammals, with a few exceptions of fish and reptiles (Faegri and Pijl 1979). Pollination by birds is much more commonly found in the tropics, where several groups of birds are specialized to feed on floral nectar. The hummingbirds (Trochilidae) in the Neotropics have more than 374 species, sunbirds and spiderhunters (Nectarinidae) in the Paleotropics have 127 species, honeycreepers (Drepanididae) in Hawaii have 23 species (6 are now extinct), and honeyeaters (Meliphagidae) in Australia and New Guinea have 136 species. Members of these four families include 660 of a total of approximately 9000 bird species on Earth. Thus, bird-pollination prevails in the tropics.

Pollinating mammals mainly include bats (Hopkins 1984; Cunningham 1995), but a few cases of non-flying mammals are known (Janson et al. 1981; Cathew and Goldingay 1997). The kinds of bats visiting flowers for nectar and pollen are limited in the tropics: long-tongued bats (Glossophaginae) in the Neotropics and fruit bats (Pteropodidae) in the Paleotropics. Other animals believed to act as pollinators are the marsupials of Australia (Carthew 1994), the primates in Madagascar and South America (Prance 1980; Garber 1988; Ferrari and Strier 1992; Overdorff 1992; Kress 1994), and rodents in Central America (Lumer 1980).

Many flowers display special pollination syndromes—ornithophily and chiropterophily—mainly in the tropics (Faegri and Pijl 1979). For example, 100 species out of approximately 600 angiosperms are estimated to be pollinated by hummingbirds in a montane forest in Costa Rica (Feinsinger 1983). In a lowland forest in Costa Rica, 39 subcanopy trees and shrubs out of 220 were pollinated by hummingbirds, 2 canopy tree species out of 52 were pollinated by hummingbirds, and 8 were pollinated by bats (Bawa 1990). Even in a warm temperate forest in Yakushima Island, Japan, 3 species (*Camellia japonica, Camellia sasanqua, Taxillus yadoriki*) out of 36 woody plants were pollinated by the Japanese white eye (*Zosterops japonica*) (Yumoto 1987). But in the Asian tropics, very few studies have been carried out.

12.2 Vertebrate Pollination: Before and During General Flowering in 1996

In our Canopy Biology Program in Sarawak (CBPS), flower visitors of 270 plant species were observed or collected before and during the general flowering (GF) in 1996 (Chapter 4). Plants likely pollinated by social bees comprised 32%, followed by beetle-pollinated plants with 20% (Momose et al. 1998). Only 19 species in seven families were pollinated by birds (7%); 4 species in three families were pollinated by bats (1.5%); and 1 species were pollinated by squirrels (see Plate 7; Momose et al. 1998c). In Lambir Hills National Park (LHNP) the proportion of plants identified as vertebrate-pollinated was relatively small compared to La Selva, Costa Rica, where 14.9% of plants were pollinated by hummingbirds and 10% by bats (Kress and Beach 1994).

Before GF in 1996, only three species of mistletoes, *Amylotheca duthieana, Trithecanthera sparsa,* and *T. xiphostachys* (see Plate 7; Yumoto et al. 1997) and eight of gingers—*Amomum roseisquamosum, Etlingera inundata, E. punicea, Hornstedtia reticulata, H. leonurus, H. minor, E. velutina,* and *Plagiostachys strobilifera* (Sakai et al. 1999)—were pollinated by spiderhunters. The pollinating birds were the long-billed spiderhunters *Arachnothera robusta,* found mainly in the canopy, and the little spiderhunters *A. longirostra,* found mainly on the forest floor (see Plate 7). The little spiderhunters also visited canopy mistletoes *A. duthieana* and *T. sparsa* with less frequency, while the long-billed spiderhunters visited the terrestrial ginger *E. inundata* on the forest floor. The copperthroated sunbirds *Nectarinia calcostetha* sometimes visited flowers with a shorter floral tube, the mistletoe *A. duthieana* in the canopy, and the ginger *P. strobilifera* on the forest floor. No bird-pollinated canopy or subcanopy trees were observed before GF in 1996.

During GF from March to September 1996, 8 species of subcanopy trees (out of 49 spp.); *Durio oblangus, D. kutejuensis, D. grandiflolus* (Bombacaceae), *Ardisia macrophylla* (Myrsinaceae), *Tarenna* (Rubiaceae), *Madhuca, Palaquium beccarii, Palaquium* sp. (Sapotaceae), 1 species of understory shrub (out of 38 spp.), *Praravinia* sp., and 1 species of mistletoes (unidentified) were observed to be pollinated by birds. The pollinating birds were the crimson sunbird (*Ae-*

thopyga siparaja), the plain sunbird (*Anthreptes simplex*), the long-billed spiderhunter, the little spiderhunter, the yellow-eared spiderhunter (*A. chysogenys*), the spectacled spiderhunter (*A. flavigaster*), the scarlet-breasted flowerpecker (*Prionochilus thoracius*), the yellow-vented flowerpecker (*Diacaeum chrysorreheum*), the orange-bellied flowerpecker (*D. trigonostigma*), and the greater green leafbird (*Chloropsis sonnerati*). Three species of subcanopy trees (out of 49 spp.) *Fagraea racemosa* (Loganiaceae), *Parkia singularis*, and *P. speciosa* (Leguminosae) were observed to be pollinated by bats. In addition, the flowers of bird-pollinated mistletoes and gingers were observed in the GF. All the above vertebrate-pollinated plants, with the exception of mistletoes and gingers, only bloomed during the GF.

The number of bird species foraging on flowers increased as GF progressed. While this occurred, the fruit/flower or fruit set ratio of *Trithecanthera sparsa* (less specialized bird-pollination) increased, but that of *T. xiphostachys* (more specialized bird-pollination) decreased significantly (T. Yumoto, unpublished data), as explained later in this chapter.

12.3 Bird Pollination of Loranthaceae

Flowers of mistletoes were mainly observed before GF, in August 1992 and March 1994; additional observations were made during GF in 1996. All of the three species—*Amylotheca duthieana, Trithecanthera sparsa,* and *T. xiphostachys*—were found to be pollinated by spiderhunters. Flower characteristics, flower visitors, nectar secretion, and timing of foraging were studied in detail (Yumoto et al. 1997). All plants were parasites, growing upon Dipterocarpaceae, from 12 m to 27 m aboveground.

Flower characteristics. Flowers of *A. duthieana, T. sparsa,* and *T. xiphostachys* were reddish orange, yellow, and pink in color, respectively. All flowers were cylindrical with corollas of different lengths. The longest corolla was that of *T. xiphostachys* (16.8 cm) and the second longest that of *T. sparsa* (7.7 cm), with *A. duthieana* shortest (4.6 cm) (see Fig. 12.1). The lobes of *A. duthieana* were reflexed, *T. sparsa* suberect, and *T. xiphostachys* patent. The length of lobes was 4 mm, 18 mm, and 23 mm for each species, respectively.

Flowers of the three above-mentioned species lacked odor or nectar guides. They were sessile but reflexed upward. The inflorescence of *T. xiphostachys* is a spike of flowers, with an axis 33 cm long with a 20 cm sterile tip. These flowers lacked lips and a margin, therefore having no perch or landing place for pollinators. Flowers showed diurnal anthesis. Each flower of *T. xiphostachys* opened before 04:00 and dropped its corolla at 16:00. They had thick corolla petals.

Flowers of all three species were bisexual and homogamous. Anthers were situated as high as the tip of lobes, and the stigma was slightly higher than the top of the stamen. The separation of anthers and stigma in space, herkogamy, was apparent for those three species.

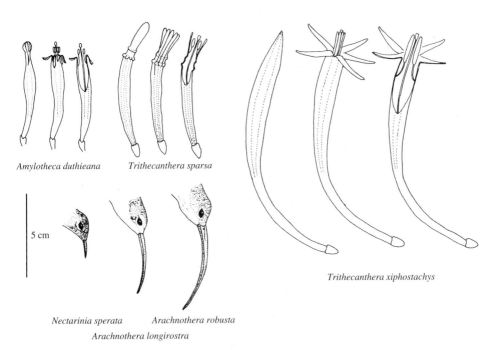

Amylotheca duthieana   *Trithecanthera sparsa*

5 cm

Trithecanthera xiphostachys

Nectarinia sperata   *Arachnothera robusta*
Arachnothera longirostra

Figure 12.1. Morphological features of bird flowers and bird bills in Lambir Hills National Park (Yumoto et al. 1997).

Explosive flower opening was found in *A. duthieana*, but not in two *Trithecanthera* species. In *A. duthieana*, expansion of stamens caused the tubular corolla to split along the petal junctions.

Nectar secretion. The mean nectar volume of *A. duthieana* was 2.8μl, and the mean sugar concentration was 8% at the morning sampling (measurements made with a handheld refractometer, weight sugar/weight solution). The nectar level in the corollas was from one-fourth to one-third. At the noon sampling, the volume and the sugar concentration was much less than in the morning. As for *T. sparsa*, the mean nectar volume in the morning sampling was 4.4μl and the mean sugar concentration was 16%. The volume increased up to 15.8μl at noon with nearly the same sugar concentration. The highest nectar level was one-fourth at the morning sample; at the noon sampling it was four-fifths. The nectar volume of *T. xiphostachys* at the morning sampling was from 130.0μl to 155.6μl, the nectar level was from one-third to two-fifths of the corolla. At the noon sampling the volume of nectar was from 184.8μl to 267.7μl, and the nectar level was from one-half to two-thirds of corolla. The sugar concentration was fairly constant at 17% to 18%.

Flower visitors. *A. duthieana* was visited mainly by the long-billed spiderhunter. The little spiderhunter and the copper-throated sunbird were observed to visit in much lower frequency. Both species visited only once during the observation period, and nectar foraging was observed only in a little spiderhunter. The little spiderhunter stayed on *A. duthieana* only for 33 seconds but visited 18 flowers. The copper-throated sunbird stayed for 15 seconds and visited no flower. Some flowers of *A. duthieana* had holes made by robbing birds visiting flowers, but no bird was observed feeding through them.

Flower visitors of *T. sparsa* also included the long-billed spiderhunter. A little spiderhunter was observed to visit flowers once during two days of observation. It stayed on *T. sparsa* for 3 minutes 30 seconds and visited 34 flowers. Both the long-billed spiderhunters and the little spiderhunter were observed taking nectar. One long-billed spiderhunter made a slit in the corolla of *T. sparsa*. As for *T. xiphostachys*, only the long-billed spiderhunters were observed to visit flowers. No other birds were seen near the plant.

Long-billed spiderhunters are the largest among nectarivorous birds in the study site (7.5 cm in body length). Little spiderhunters (6 cm) and copper-throated sunbirds (5 cm) were very cautious when they approached flowering plants, and stayed for just a short time. In particular, the little spiderhunter looked around carefully with a warning note "jwe, jwe." A long-billed spiderhunter was observed to chase a little spiderhunter that was flying near the flowering mistletoes.

The little spiderhunter has a bill long enough to remove nectar from flowers of *A. duthieana*, *T. sparsa*, and *T. xiphostachys* and to pollinate them. The morphological characteristics of long-billed spiderhunters seem the best suited for a long corolla, among nectarivorous birds in the study site. Aggressive behavior by the long-billed spiderhunters has possibly limited the frequency of visits of other nectarivorous species to Loranthaceae.

When a long-billed spiderhunter visited flowers of *A. duthieana* and *T. sparsa*, it perched on the branch near flowers (see Plate 7E). But when it visited *T. xiphostachys*, it always perched on a sterile tip of the inflorescence. Spiderhunters seldom hovered while feeding at flowers and only two cases were observed, when long-billed spiderhunters visited *T. sparsa*.

As for insects, an unidentified skipper butterfly (Hesperiidae) was observed to visit flowers of *A. duthieana* and *T. sparsa*, where it apparently foraged nectar. Many stingless bees flew around the flowers of *A. duthieana* and *T. sparsa* but did not touch the anther or stigma. During two days of observation on *T. sparsa*, a Rajah Brooke's birdwing butterfly (*Trogonoptera brookiana*), as shown in Plate 2E, tried to visit to the flowers once. No insect was observed visiting flowers of *T. xiphostachys*.

From all three Loranthaceae, pollen was observed on feathers at the base of the bill and on the forehead and crown of both long-billed spiderhunters and little spiderhunters. Less pollen was attached to the feathers of the chin. As for copper-throated sunbirds, no nectar feeding or pollen on the bird body was observed.

Visits of long-billed spiderhunters to *A. duthieana* were limited to a short period from 06:26 to 08:06, and after that no bird visited. Long-billed spiderhunters stayed on *A. duthieana* for an average 2 min (range: 1 min 5 sec to 4 min 30 sec, N=12) and visited a mean of 27 flowers (range: 12-46 flowers, N = 12) when about 250 flowers were in bloom. Long-billed spiderhunters probed the corolla tube with a regular motion that lasted 0.5 to 1.0 sec.

The long-billed spiderhunter visited *T. sparsa* almost constantly during 07:26 to 13:42. It stayed an average of 50 sec (range: 30 sec to 3 min, N = 26) and visited 9.3 flowers (range: 5–28 flowers, N = 26) with 65 flowers in bloom. On each flower, bill insertions lasted about 1 sec.

Considering *T. xiphostachys,* long-billed spiderhunters were observed visiting only twice in a complete day. From four days of observation, no bird was observed on the first day. On the second day, long-billed spiderhunters, as a pair, visited at 12:06. They stayed for 3 min 40 sec, and both foraged on each of seven flowers. On the third day, a long-billed spiderhunter came to the flowers at 09:31, and on the forth day at 09:41 and 13:11. At every observation, every open flower in the inflorescence was visited. On each flower, each bill insertion lasted 7 to 8 sec.

Three species of mistletoes received the visitation of long-billed spiderhunters at different times of the day; *A. dulthieana* was visited from 06:00 to 09:00, *T. sparsa* from 07:00 to12:00, and *T. xiphostachys* from 09:30 to13:00.

Fruit set. In studies of *A. dulthieana* and *T. sparsa* we did not mark individual flowers, thus an accurate ratio of fruit set (fruit/flowers) was unknown. From photographs taken after three months of observation, the fruit set ratio of *A. dulthieana* was estimated at more than 50%, and that of *T. sparsa* more than 60% during non-GF. The inflorescence of *T. xiphostachys* bore 38 fruits out of 46 flowers (6 were taken for measurements), so the fruit set ratio was 82.6%.

During the GF in 1996, *T. sparsa* and *T. xiphostachys* bloomed. At that time, *T. xiphostachys* received only 10% visitation by long-billed spiderhunters while fruit-set ratio decreased to 4.2%. Thus, *T. sparsa* received half the visitation of long-billed spiderhunters recorded during non-GF, but yellow-eared spiderhunters frequently visited the flowers. Therefore, total visitation ratio per flower in the GF was 1.4 times that of the non-GF. The fruit set was 82%, much higher than in the non-GF period.

12.4 Bird Pollination of Zingiberaceae

Before GF in 1996, from July 1994 to June 1995, and from April to June 1996, 44 ginger species (Zingiberaceae and Costaceae) were collected. The flower visitors of 29 species were identified. Flower characteristics were analyzed statistically. Among them, eight species, *Amomum roseisquamosum, Etlingera inundata, E. punicea, Hornstedtia reticulata, H. leonurus, H. minor, E. velutina* and *Plagiostachys strobilifera,* were found to be spiderhunter-pollinated species, while other species were bee-pollinated, either by *Amegilla* or halictids (Sakai et al. 1999b).

Floral characteristics. The flowers of spiderhunter-pollinated species had a long floral tube (>31 mm with the exception of *H. reticulata* and *H. leonurus*).

Eltingera punicea produced showy flowers with a long yellow lip fringed with red on a scandent inflorescence. The flower secreted considerable nectar and sugar (474 mg sugar /day /inflorescence). Flowers of *A. roseisquamosum* were similar internally to the typical bee-pollinated species when dissected, but their outward appearance was quite different. Their lip was tightly rolled up to form a tube holding the anthers beyond the end of the corolla, which made it difficult for insects to insert their proboscides. *Plagiostachys strobilifera* had the smallest flowers with the smallest stigma width and anther length, and the least total nectar sugar per flower among spiderhunter-pollinated gingers. However, its flower number was the largest of all the ginger species. Bees could not open the flowers of *P. strobilifera*, but the nectar that overflowed from the shortest floral tube (8.8 mm) of the spiderhunter-pollinated species was consumed by *Amegilla* and stingless bees. Those bees did not transport pollen of the flowers.

Most flowers visited by spiderhunters were generally pink or red, sometimes with a yellow or white nectar guide. *Amomum roseisquamosum*, an epiphytic species, had white flowers surrounded by pink bracts with yellowish edges. Sugar concentration of floral nectar was slightly lower in spiderhunter-pollinated species ($26 \pm 4\%$) than in bee-pollinated species, but the difference was not significant. However, daily sugar production per inflorescence was elevated in spiderhunter-pollinated species, and in sugar (mean = 148 mg/inflorescence/day) was significantly higher than those of bee-pollinated gingers.

Flower visitors. All birds observed visiting the eight ginger species were little spiderhunters, with the exception of a single visit by a copper-throated sunbird on *P. strobilifera* and by a long-billed spiderhunter on *E. inundata*. Birds generally approached the inflorescences carefully and remained for only a short time (10–25 sec), but sometimes lingered and sang loudly. Occasionally they used a warning call to discourage conspecifics from approaching the inflorescence.

12.5 Pollination of Three *Durio* Species

Pollination ecology of three *Durio* species—*D. grandiflorus*, *D. oblongus*, and *D. kutejensis* (Bombacaceae)—was studied 14 May to 24 June, during GF in 1996. Flower characteristics, flower visitors, nectar secretion and timing of foraging were studied in detail, and bagging experiments were carried out (Yumoto 2000).

Floral characteristics. The flower color of *D. grandiflorus* and *D. oblongus* is white, while *D. kutejensis* is dark red (see Plate 7A). The flowers of *D. oblongus* and *D. kutejensis* are cup-shaped, while that of *D. grandiflorus* is a large whorl of fleshy petals (see Plate 7G). The flower base of *D. grandiflorus* is constricted tightly, so that a long slender bill or proboscis is needed to remove nectar. In *D. oblongus*, the style and filaments of flowers form a tough tube. Petals and sepals of all three species are thick, but the petals of *D. grandiflorus* were thinner compared to the other species.

Flowers of *D. grandiflorus* opened around 14:00 and a faint odor. The flower buds of *D. oblongus* began to open at 11:30, but took about 6 hours to open fully. The floral odor from *D. oblongus* was strong, but much less than that of *D. kutejensis*. Flowers of *D. kutejensis* opened around 16:30, emitting a strong smell similar to durian fruits, which could be recognized from more than 100 meters away. Flowers kept their shape for nearly 30 hours in *D. grandiflorus* and 24 hours in *D. oblongus* and *D. kutejensis*.

Nectar secretion. Floral nectar of *D. oblongus* and *D. kutejensis* was abundant but dilute; the nectar flowed out easily when the flowers were tipped over. The nectar sugar concentration of *D. grandiflorus* was high in the day but low at midnight. Nectar volume in the daytime was larger, with greater variation, compared to night when no animal was observed to visit. Nectar concentration of *D. oblongus* was slightly higher at night than at daytime. From 12:00, just after the flower opened, the nectar volume increased steadily from 16:00 to 04:00. Flowers secreted no nectar by 12:00 in the second day, because no harvesting by animal visitors was observed. The variation of nectar volume was greatest at 18:00 and still noticeable at 22:00, but the volume at 06:00 did not show such variation. Nectar concentration of *D. kutejensis* changed only slightly during the 24-hour observation period and was highest at midnight; variation among flowers was low. Nectar volume steadily increased after the flower opened. A wide range of nectar volume was recognized throughout 24 hours, except just after the flowers opened.

Flower visitors. Flowers of *D. grandiflorus* were visited mainly by two species of spiderhunters, the long-billed spiderhunter and the yellow-eared spiderhunter. The number of visits by yellow-eared spiderhunters was much more than that by long-billed spiderhunters. Especially after 09:00, only yellow-eared spiderhunters were observed to visit flowers. Spiderhunters were dusted with pollen on the forehead and the bill. Besides birds, the Malay bird-wing butterfly *Troides amphrysus* was observed to forage on flowers in the afternoon, and stingless bees (*Trigona* spp.) hovered around flowers but never took nectar or collected pollen. No bat was recorded during four nights of observation.

Spiderhunters were also the main visitors to *D. oblongus*. Three spiderhunter species were recorded: the long-billed spiderhunter, yellow-eared spiderhunter, and the spectacled spiderhunter. The yellow-eared spiderhunter was only observed once during 24 hours. Spiderhunters usually perched on branches and imbibed nectar without touching a stigma or anther. However, in 5 of 24 visits by long-billed spiderhunters and 2 of 21 visits by spectacled spiderhunters, birds perched on the staminal column enclosing the style while taking nectar, and pollen was observed on the ventral part of the bird. A sphingid moth was observed on flowers but did not touch stigmata or anthers. No bat was recorded during four nights of observation.

The flower visitors to *D. kutejensis* were more diverse. At 05:30, giant honeybees (*Apis dorsata*) began to forage on flowers and to collect pollen. Nearly 50 individuals of giant honeybees were counted on a flower in 30 minutes. The number of giant honeybees decreased during 07:00 to 09:00 and increased again at 09:00. After 09:00, giant honeybees began to collect nectar. Stingless bees (Meliponini) began to visit flowers at 06:00, mainly foraging nectar. A small number of giant honeybees visited flowers immediately after anthesis at 16:30. Birds also visited the flowers of *D. kutejensis*. Spiderhunters, the long-billed spiderhunter (four times), the spectacled spiderhunter (six times), and the little spiderhunter (once) were observed to visit flowers, and the plain sunbird also visited flowers. But the most frequent bird visitor was the orange-bellied flower pecker. During 11:00 to 14:00, only the orange-bellied flower pecker visited flowers, while spiderhunters were recorded in the morning and in the evening, and the sunbirds were seen only in the evening. All birds touched the stigma and anthers when they removed nectar from flowers. A male orange-bellied flower pecker was caught by mist net, and pollen was observed on its forehead. From 20:00 to 22:00, the cave nectar bat (*Eonycteris spelaea*), was observed, and an individual was caught by mist net. Pollen was collected on the forehead of the captured bat.

Bagging experiments. All three species of *Durio* appeared to be self-incompatible. The continuously bagged flowers and flowers pollinated artificially by geitonogamous pollen bore no fruit. Flowers pollinated artificially by pollen from the other individual bore fruits in a ratio comparable to flowers pollinated naturally (data on *D. oblongus* were not available). The fruit-set ratio of open-pollinated flowers, however, was very low: 0.037 (*D. grandiflorus*), 0.158 (*D. oblongus*), and 0.039 (*D. kutejensis*).

In *D. grandiflorus* the fruit-set ratios for open pollination, cross-pollination, and 'open only at daytime' were not significantly different (χ^2 test). The fruit-set ratios of open-pollinated flowers of *D. oblongus* and those open during daytime were not significantly different (χ^2 test). These data for *D. grandiflorus* and *D. oblongus* suggest strongly that cross-pollination occurred during the day and not at night (the latter would occur if bats were major pollinators). Data of *D. kutejensis* showed that all the fruit-set ratios—open, cross-pollination, open at daytime, open at night and open in the early morning—were almost the same.

Bird-pollination versus bat-pollination. Observations of animals foraging on flowers showed that *D. grandiflorus* and *D. oblongus* were visited only by birds and that no bats visited these flowers. Moreover, effective pollination was shown by experiments to occur at daytime, not at night. Diurnal nectar volumes showed large variation, which suggests active consumption of nectar by animal visitors at flowers.

Another pattern was observed for *D. kutejensis*. Giant honeybees, birds, and bats came to forage on flowers at different times of day and night, and pollination experiments showed that effective pollination occurred at all times. The results suggest that giant honeybees, birds, and bats all contributed to effective pollination in this species.

The typical characteristics of ornithophily and chiropterophily pollination syndromes are normally easy to distinguish; ornithophilous flowers are characterized by vivid colors, absence of odor, a deep tube or spur, and diurnal anthesis;

those of chiropterophily by whitish or creamy color, strong odor, large-mouthed single flowers or brush inflorescences, and nocturnal anthesis (see Plate 7). Interestingly, characteristics of the three *Durio* species observed are intermediate between ornithophily and chiropterophily. Flowers of *D. grandiflorus* are white, faint smelling, large-mouthed, single flowers with diurnal anthesis: two of these characteristics are typical of ornithophilous flowers and two are typical of chiropterophilous ones. Three characteristics of *D. oblongus*, white color, strong odor, and large-mouthed flowers, are typical for chiropterophily, but its diurnal anthesis is of ornithophily. *Durio kutejensis* also has intermediate flowers between chiropterophily and ornithophily; both diurnal and evening anthesis occur, flowers have a very strong odor, and large size (chiropterophilous), while a red color usually shows ornithophily (see Plate 7G).

Faegri and van der Pijl (1979) pointed out that the Bignoniaceae and Bombacaceae have some species that are intermediate between bird- and bat-pollination; *Bombax malabaricum* (*Gossampinus heptaphylla*) is ornithophilous, but incompletely so with open, red, cup-shaped, diurnal flowers. Its sister species *B. valetonii* is chiropterophilous, with pungent-smelling flowers. Baum (1995) discovered that two species of section *Brevitubae* in *Adansonia* (Bombacaceae) are pollinated by nocturnal mammals (fruit bats and lemurs), and that five species in section *Longitubae* are pollinated by long-tongued hawk moths. The studies proved clearly that typological thinking with regard to pollination syndromes is oversimplification and potentially misleading, particularly applied to distinctions between bird- and bat-pollination.

12.6 A New Pollination Syndrome: Squirrel Pollination of *Ganua* (Sapotaceae)

The tree, *Ganua* (herbarium voucher B16, Momose 5025), Sapotaceae, was found on a ridge and bloomed during 16–26 March 1996. This tree had not flowered since 1992. Flower characteristics and flower visitors were studied. The diameter at the breast height of the tree was 22.5 cm and the height was approximately 20 m.

Flower characteristics. The stamen and petals are fused to form a fleshy and berrylike corolla (0.25g in wet weight, 15 mm diameter), and this pseudo-fruit is as sweet as 15% sugar concentration. There are 16 stamens and a pistil in a flower, at the end of a stem. The color of crown is whitish transparent and the anther is whitish yellow. Flowers have no nectar or perceptible odor. The berrylike corolla is easily detachable from the pistil/ovary, which has a very bitter flavor. Anthesis occurred around 05:00 and the longevity of a flower was about one day.

Flower visitors. Observation of foraging animals was carried out 05:00 to 24:00 on 23 March, and from 00:00 to 05:00 on 25 March 1996. Three species of squirrels, *Callosciurus prevostii, Sundasciurus hippurus inquinantus, S. lowii* and a species of flying squirrel, *Petaurista petaurista* were observed to frequently forage on flowers. *Callosciurus prevostii* visited the tree 27 times from 08:00 to 19:00, *S. hippurus* 5 times from 06:00 to 19:00, *S. lowii* 17 times from 06:00 to 18:00, and *P. petaurista* did 12 times during 19:00 to 21:00 and 04:00 to 08:00. Squirrels and flying squirrels were observed to take away the corolla from the twig, hold by the forepaws, and consume it. Pollen was observed attached to the fingers, and fur around the squirrels' mouths. Squirrels avoided eating the bitter pistil/ovary. A squirrel consumed approximately 400 flowers during 25 minutes. Blue-eared barbets also were observed twice to eat four corollas in total during the observation period. No bat or insect was observed on flowers.

Fruit set. The fruit set ratio was estimated as 0.12, based on the records by four tagged twigs at 10 m height (N flowers=256). There was no conspecific blooming tree within at least 300 m distance. The lower branches (2–5 m in height)—which were not visited by any animal during the observation period—received sunlight through the gap but bore no fruit (N>1500). Although no bagging experiment was done, this species seems to need outcrossing to bear fruit.

Most of the known flowers pollinated by mammals are brush-shaped or like large bowls, with protruding style and stamens and offering nectar and pollen as rewards. Our discovery was that squirrel pollination is quite different from other types of pollination. The fused stamen and petals form a berrylike corolla, which is the reward to pollinators. This pollination syndrome was completely unknown prior to our work.

13. Insect Predators of Dipterocarp Seeds

Michiko Nakagawa, Takao Itioka, Kuniyasu Momose, and
Tohru Nakashizuka

In this chapter, we investigate the insect seed predators of dipterocarps. The
Dipterocarpaceae are a dominant family of the upper canopy and emergent canopy layer. Their large seeds scattered in lowland forests are one of the main
consequences of general flowering. The main vertebrate consumers of dipterocarp seeds in old-growth forests are reported to be the bearded pig *Sus scrofa*,
squirrels *Callosciurus prevostii, C. notatus,* the monkey *Presbytis rubicunda,*
and certain birds (Kobayashi 1974; Natawiria et al. 1986; Curran and Leighton
1991; Curran and Leighton 2000; Curran and Webb 2000). These vertebrates
are thought to be generalist seed predators. Although invertebrates also play an
important role as seed predators (Mattson 1978; Janzen 1980; Crawley 1989;
Tanaka 1995; Igarashi and Kamata 1997), there is little information about dipterocarp seed-eating insects, partly because of the difficulty with species iden-
tifications and adequate taxonomic study. We identified insect seed predators of
24 dipterocarps and two species of the family Moraceae during two masting
years; here we analyze host specificity, variation of resource use patterns between the two events, and the association of host ranges with the dominance of
certain seed predators.

13.1 Introduction

The seed is a critical stage in the life history of plants, and mortality at this
stage is known to be highest in the life cycle for most plant species (Hickman

1979; Cavers 1983; Silvertown 1987; Tanaka 1995). Seed predation is therefore one of the most important mortality factors, reported in many studies of pre-dispersal and post-dispersal seed attack by natural enemies (Steven 1983; Auld 1986; Randall 1986; Andersen 1989; Traveset 1991; Crawley 1992; Greig 1993; Hulme and Hunt 1999; Díaz et al. 1999). To escape seed predation, some plants have physical defenses and others produce chemical-rich seeds (Wainhouse et al. 1990; Baumann and Meier 1993; Bennett and Wallsgrove 1994; Grubb and Metcalfe 1996; Grubb and Burslem 1998; Grubb et al. 1998). Masting, or in-termittent production of large seed crops by a population of plants (Kelly 1994), is also thought to be effective for reducing seed predation by both invertebrates and vertebrates (Nilsson 1985; Auld and Myerscough 1986; Nilsson and Woostl-jung 1987; Kelly et al. 1992; Donaldson 1993; Cunningham 1997; Kelly and Sullivan 1997; Sperens 1997; Shibata et al. 1998; Forget et al. 1999).

Lowland mixed dipterocarp forests in Southeast Asia are characterized by general flowering and seeding (GF), which produces supra-annual and irregular fluctuation of flowering and seeding at the community level (Ashton et al. 1988; Appanah 1993; Sakai et al. 1999c; Sakai 2002). If some plants share common seed predators, they could satiate them by producing large seed crops simultaneously. In species-rich forests, seed predation should be discussed not only at the population level, but also considering guilds or even the whole com-munity. Moreover, a significant question has yet to be understood, regarding what the seed predators utilize during non-GF periods. We currently have little understanding of how they respond to such irregular and intense seed produc-tion.

13.2 Seed Predators and Their Predation Pattern

Before our study, more than 40 species, including 23 weevils, 6 scolytids, and 14 micro-Lepidoptera (moths), have been reported as insect seed predators of various Dipterocarpaceae (see Table 13.1). However, the information was limited to mostly one fruiting event. In view of substantial annual variability in the association between tropical insects and their hosts (Roubik et al. 2003), it is necessary to accumulate additional studies about seed predators of dipterocarps. Moreover, these data are not quantitative information but qualitative description. To evaluate the importance and the effect of each seed predator, we should further quantify the pattern of seed predation.

We conducted our study on insect seed predators of 24 species in Diptero-carpaceae (*Dipterocarpus*, *Dryobalanops*, and *Shorea*) and 2 species in Mora-ceae (*Artocarpus*) at Lambir Hills National Park (LHNP) in 1996 and 1998 (see Table 13.2). We sampled fallen seeds on the ground and seeds in seed traps, and reared seed predators in plastic cases kept for more than three months in the laboratory. The seeding of Dipterocarpaceae in 1996 was more intense (more species joined, and a larger amount of seed was produced) than that in 1998 (Sakai 2002).

Table 13.1. Invertebrate seed predators of dipterocarps cited from various sources which were identified to species. Abbreviations in host plants and in literature are D.- *Dipterocarpus*, Dr.- *Dryobalanops*, H.- *Hopea*, N.- *Neobalanocarpus*, P.- *Parashorea*, S.- *Shorea*, and V.- *Vatica*, and L20- Layal and Curran (2000), C20- Curran and Leighton (2000), M96- Momose et al. (1996), T92- Toy and Toy (1992), T91- Toy (1991), T88- Toy (1988), K87-Kokubo (1987), N86- Natawiria et al. (1986), C77- Chan (1977), D74- Daljeet-Singh (1974) and K56- Kalshoven (1956), respectively. *Cydia pulverula* (Tortricidae) in Daljeet-Singh (1974) was mis-identified, and *Andrioplecta shoreae* is the correct species name (Komai 1992, 2001). Seed predators in boldface are also observed in Lambir Hills National Park.

Family	Species name of seed predators	Host plants	Literature
Curculionidae	*Mecysolobus* (= *Alcidoes*) *crassus*	*D. dyeri, Dr. oblongifolia, Dr. aromatica, Dr. lanceolata, one Hopea sp., one Shorea sp.*	M96, K56, D74
Curculionidae	*Mecysolobus* (= *Alcidoes*) *dipterocarpi*	*D. cornutus, S. assamica, S. bracteolata, S. curtisii, S. leprosula, S. macroptera, S. parvifolia, S. faguetiana, S. laevis, S. platyclados, S. acuminata, S. smithiana, H. ferrugine*	D74, N86
Curculionidae	*Mecysolobus* (= *Alcidoes*) *humeralis*	*S. macroptera, S. curtisii, S. laevis, S. platyclados*	D74
Curculionidae	*Dermatoxenus hians*	*Dr. aromatica*	D74
Apionidae	*Nanophyes shoreae*	*S. acuminata, S. laevis, S. lepidota, S. longisperma, S. maxwelliana, S. parvifolia, S. rubra, S. sumatrana, S. dasyphylla, S. leprosula, S. macroptera, S. ovalis, S. pauciflora, S. smithiana, S. singkawang, H. helferi, H. latifolia, H. odorata, S. macrophylla, S. platyclados, D. hasseltii*	T91, T92, D74, C77, N86
Apionidae	*Nanophyes dipterocarpi*	*D. trinervis, D. basseltii*	K56
Anthribidae	*Phloeobius pallipes*	*D. cornutus*	K87

Table 13.1. *Continued*

Family	Species name of seed predators	Host plants	Literature
Scolytidae	*Coccotorypes (= Poecilips) gedeanus*	*Dr. oblongifolia, S. assamica, S. bracteolata, S. macrophylla, S. leprosula, S. paucifolia, S. platyclados, S. curtisii*	D74, C77
Scolytidae	*Coccotorypes (= Poecilips) medius*	*Dr. aromatica, S. resinosa, S. bracteolata, S. macrophylla, S. leprosula*	D74, C77
Scolytidae	*Coccotorypes (= Poecilips) advena*	*Dr. oblongifolia, S. resinosa, S. macrophylla*	D74, C77
Scolytidae	*Coccotorypes (=Poecilips) cinnamomi*	*Dr. oblongifolia, S. acuminata, S. bracteolata, S. macrophylla, S. platyclados*	D74
Scolytidae	*Coccotorypes (=Poecilips) papuanus*	*S. macrophylla, S. platyclados*	D74
Scolytidae	*Coccotorypes (=Poecilips) variabilis*	*S. paucifolia, D. kunstleri*	D74
Tortricidae	*Andrioplecta shoreae*	*D. gracdiflorus, D. dyeri, Dr. oblongifolia, S. acuminata, S. bracteolata, S. macrophylla, S. leprosula, S. platyclados, S. curtisii*	D74
Tortricidae	*Laspeyresia* sp. aff. *magnetica*	*Dr. aromatica*	D74
Pyralidae	*Assara (=Cateremna) albicostalis*	*S. parvifolia, S. bracteolata, B. heimii*	D74
Pyralidae	*Lamorea adaptella*	*S. macroptera*	D74
Heliodinidae	*Stathmopoda* sp. aff. *philaromia*	*S. leprosula*	D74

Table 13.2. List of seeds studied. Literature cited from (1) Ashton (1982); "—" indicates no data. Modified after Nakagawa et al. (2003)

Family	Species[1]	Section[1]	Code names	Number of seeds studied 1996	Number of seeds studied 1998	Size (diameter, cm)[2] longest	Size (diameter, cm)[2] shortest	Wing (long+short)[2]	Tannin (mg/g)
Dipterocarpaceae	Dipterocarpus pachyphyllus	—	DPC	5	1,354	2.0–3.0	2.0–2.8	2+3	—
Dipterocarpaceae	Dipterocarpus geniculanus	—	DGE	18	536	1.0–1.6	1.0–1.6	2+3	1.3±1.7
Dipterocarpaceae	Dipterocarpus crinitus	—	DCR	198	581	1.5–1.8	0.6–0.8	2+3	—
Dipterocarpaceae	Dipterocarpus globosus	—	DGL	68	88	2.5–3.5	2.5–3.5	2+3	116.8±16.77
Dipterocarpaceae	Dipterocarpus palembanicus	—	DPL	77	277	2.5–4.5	1.0–1.5	2+3	0.1±0.1
Dipterocarpaceae		—	DTE	1,681	737	3.0–4.0	1.0–2.0	0	8.1±5.2
Dipterocarpaceae	Dryobalanops aromatica	—	DRA	4,470	77	2.0–3.0	1.0–1.5	5	—
Dipterocarpaceae		—	DRL	1,880	363	1.8–2.5	1.3–2.0	5	—
Dipterocarpaceae	Shorea smithiana	Brachypterae	SBS	9	317	1.8–2.7	1.6–1.8	3+2	34.4±38.0
Dipterocarpaceae	Shorea kunstleri	Brachypterae	SBK	7	—	1.3–2.0	1.1–1.5	3+2	—
Dipterocarpaceae	Shorea macrophylla	Pachycarpae	SPM	—	63	5.5–6.0	2.9–3.2	3+2	—
Dipterocarpaceae	Shorea pilosa	Pachycarpae	SPP	74	—	1.8–2.0	1.7–1.9	3+2	—
Dipterocarpaceae	Shorea xanthophylla	Richetioides	SRX	837	130	1.5–2.0	1.0–1.4	0	—
Dipterocarpaceae	Shorea laxa	Richetioides	SRL	98	398	3.0–3.5	2.0–2.5	0	42.2±3.1
Dipterocarpaceae	Shorea confusa	Anthoshorea	SAC	405	—	1.2–2.0	1.0–2.0	—	—
Dipterocarpaceae	Shorea ochracea	Anthoshorea	SAO	312	—	1.1–1.5	0.9–1.1	—	—
Dipterocarpaceae	Shorea agami	Anthoshorea	SAA	20	—	1.5–2.0	1.2–1.5	3+2	—
Dipterocarpaceae	Shorea falciferoides	Shorea	SSF	1,004	—	7.0–12.5	1.8–2.2	3+2	—
Dipterocarpaceae	Shorea domatiosa	Shorea	SSD	191	79	1.0–3.5	1.3–2.5	3+2	—
Dipterocarpaceae	Shorea argentifolia	Mutica	SMA	—	106	0.9–1.5	0.7–1.0	3+2	—
Dipterocarpaceae	Shorea curtisii	Mutica	SMC	—	34	1.1–1.9	0.7–1.0	3+2	—
Dipterocarpaceae	Shorea parvifolia	Mutica	SMP	3,113	—	0.9–1.6	0.7–0.9	3+2	63.7±3.7[3]
Dipterocarpaceae	Shorea quadrinervis	Mutica	SMQ	314	—	1.5–1.8	0.5–0.7	3+2	—
Dipterocarpaceae	Shorea scabrida	Mutica	SMS	294	—	0.6–1.5	0.6–0.8	3+2	—
Moraceae	Artocarpus tamaran	—	ATA	112	—	—	—	—	—
Moraceae	Artocarpus odoratissimus	—	AOD	31	—	—	—	—	—

Insect seed predators totaled 1425 individuals, including 51 species and 11 families, reared from 24 species of dipterocarps and 2 *Artocarpus* species in the two GFs (see Table 13.3; Nakagawa et al. 2003). Nearly 35% are undescribed species. These seed predators mainly consisted of three taxonomic groups: micro-Lepidoptera (Tortricidae, Pyralidae, Immidae, Sesiidae, Cosmopterigidae, and Crambidae), scolytids (Coleoptera: Scolytidae), and weevils (Coleoptera: Curculionidae, Apionidae, Anthribidae and Attelabidae), similar to those encountered in the previous literature (see Table 13.1 and 13.3). The two-year total numbers of weevils, micro-Lepidoptera, and scolytids consisted of 40.0%, 38.9%, and 21.1%, respectively, of the entire predator assemblage. Weevils were the richest in species number (more than 50%) while species richness of scolytids and micro-Lepidoptera were considerably lower.

Figure 13.1 depicts resource-use patterns of each seed predator ($N \geq 3$) based on the tree genera in 1996 and 1998. In 1996, most predators used more than two genera of seeds; however, 45.8% of seed predators in 1998 predated dipterocarp seeds of only one genus. Although most seed-eating insects in 1996 fed on *Dryobalanops* (see Fig. 13.1A), the seeds of *Dryobalanops* were predated by just four species of seed predators in 1998 (see Fig.13.1B). The species of 13 seed predators in our study were already known. However, their host plants were quite different from the reported species (see Table 13.1 and Fig 13.1). *Nanophyes shoreae* (Apionidae) emerged from not only seeds of *Shorea* and *Hopea* but also *Dryobalanops*, and four scolytids (*Coccotrypes advena, C. cinnamoni, C. gedeanus,* and *C. papuanus*) predated dipterocarp seeds in genera not listed as hosts in previous reports.

Although the plant species studied were not entirely the same both in 1996 and in 1998, we found 11 species of common seed predators ($N \geq 3$) that emerged in both years (see Fig. 13.2). Only *Niphades* sp.1 (Curculionidae) had similar host plants during those years. On the other hand, the dietary pattern of three micro-Lepidoptera changed greatly between the two GF events. In 1998 they did not predate *Dryobalanops* seeds that produced seeds in both years (see Fig. 13.2). Three *Coccotrypes* species used mostly *Dryobalanops* seeds in 1996, while they mainly predated *Dipterocarpus* seeds in 1998. Roubik et al. (2003) also reported variation of association between flower visitors and pollinators and their host flowers in Panama. These shifts of resource-use patterns, perhaps loose niches, by common seed predators indicate the complicated and dynamic relationship between plants and seed predators, and the existence of supra-annual population dynamics of seed predators. Long-term study of seed predators and their host plants is necessary.

There are few reports about life histories of dipterocarp seed predators and their behavior at non-masting years. What do they eat? Where are they? Since three species of dipterocarp seed predators, *Sternuchopsis dipterocarpi, Coccotrypes gedeanus,* and *Andrioplecta shoreae,* also emerged from *Artocarpus* (Moraceae) in 1996 (Fig. 13.1A), it is possible that alternative host plants are common, as discussed by Toy and Toy (1992), and that a seed predator could

Table 13.3. Individual number and code names of insect predators that emerged from dipterocarp and *Artocarpus* seeds in 1996 and 1998. Undiscribed species are marked with a '*'. Modified after Nakagawa et al. (2003).

Seed Predator	Year and Sample Size	
Lepidoptera		
Tortricidae	1996	1998
Olethreutinae	(15180 seeds)	(5140 seeds)
Andrioplecta shoreae Komai; LANSH	349	12
Andrioplecta subpulverula Obraztsov LANSU	7	4
Andrioplecta sp. A; LANSA	0	6
Andrioplecta sp. B; LANSB	0	1
Pyralidae		
Phycitinae		
Assara albicostalis Walker; LASAL	57	7
Gallerlinae		
Lamoria adaptella Walker; LLAAD	60	1
Immidae		
Imma homocrossa Meyrick; LIMHO	0	1
Sesiidae		
Gn. sp. A; LSESP*	0	8
Cosmopterigidae		
Gn. sp. A; LCOSP*	0	11
Crambidae		
Pyranstinae		
Gn. sp. A; LCRSA*	28	0
Gn. sp. B; LCRSB*	3	0
Coleoptera		
Scolytidae		
Coccotrypes advena Blandford; BCOAD	70	18
Coccotrypes cardamomi Schauf.; BCOCA	0	9
Coccotrypes cinnamoni Eggers; BCOCI	18	0
Coccotrypes gedeanus Eggers; BCOGE	34	95
Coccotrypes myristicae Roepke; BCOMY	2	0
Coccotrypes nitidus Eggers; BCONI	17	0
Coccotrypes papuanus Eggers; BCOPA	26	4
Dryocoetiops malaccensis Schedl.; BDRMA	0	1
Hypothenemus areccae Hornung; BHYAR	1	0
Xyleborinus exiguus Walker; BXYEX	2	1
Xyleborinus ferrugineus Ferrari; BXYEX	1	0
Xyleborinus similis Ferrari; BXYSI	0	2
Xyleborinus subdentatus Browne; BXYSU	1	0

Table 13.3. *Continued*

Seed Predator	Year and Sample Size	
Curculionidae		
Mechysolobinae		
Sternuchopsis curranae Lyal; WSTCU	0	66
Sternuchopsis dipterocarpi Marshall; WSTDI	10	18
Sternuchopsis toyi Lyal; WSTTO	26	0
Sternuchopsis shoreaphils Lyal; WSTSH	9	3
Sternuchopsis sp. 1; WSTSA*	0	3
Molytinae		
Niphades sp. 1; WNISP*	18	185
Cryptorhynchinae		
Idotasia sp. 1; WIDSP*	1	4
Imathia sp. 1; WIMSP*	0	20
Ramphinae		
Orchestes sp. 1; WORSA*	40	1
Orchestes sp. 2; WORSB*	0	2
Orchestes sp. 3; WORSC*	1	0
Orchestes sp. 4; WORSD*	0	3
Anthonominae		
Ochyromera sp.2; WOCSP*	0	1
Endaenidius spinipes Kojima et Morimoto; WENSP	6	0
Apionidae		
Nanophyinae		
Nanophyes shoreae Marshall; WNASH	19	14
Meregallia dipterocarpi Marshall; WMEDI	0	7
Meregallia sp. 1; WMESP*	0	4
Ctenomerus sp. 1; WCTSA*	9	0
Ctenomerus sp. 2; WCTSB*	3	5
Manoja sp. 1; WMASP*	0	1
Nanophyes sp. 2; WNASA*	0	2
Gn. sp. 1; WNASP*	0	1
Anthribidae		
Araccerus corporaali Jordan; WARCO	0	48
Phloeobius sp. 1; WPHSP*	0	1
Attelabidae		
Rhynchitinae		
Involvulus sp. 1; WINSA*	1	0
Involvulus sp. 3; WINSB*	0	18
Total	837	588
	(29 spp.)	(37 spp.)

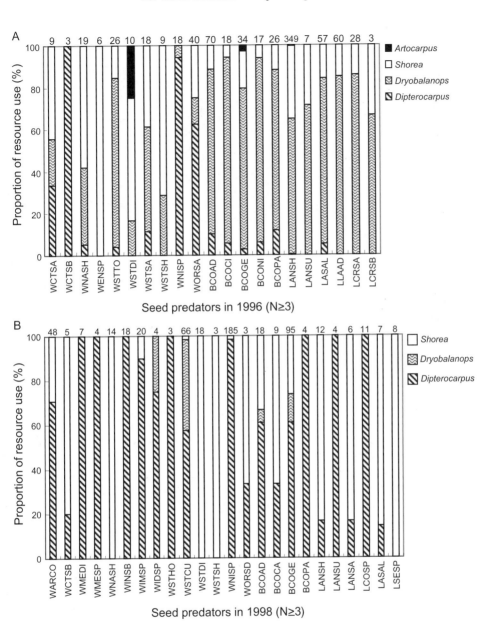

Figure 13.1. Pattern of resource use by dipterocarp seed predators (N≥3), based on plant genus in 1996 (**A**) and 1998 (**B**). The code names of predators are listed in Table 13.3. Numbers of each seed predator are shown above bars (after Nakagawa et al. 2003).

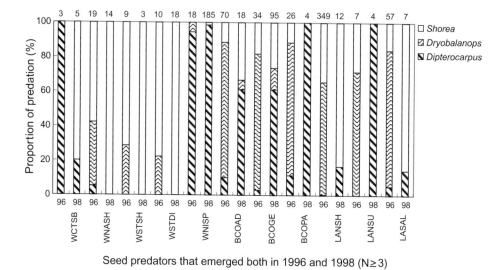

Figure 13.2. Variation of dietary patterns by insect seed predators ($N \geq 3$) that emerged both in 1996 and 1998. The code names of predators are listed in Table 13.3. Numbers of each seed predator are shown above bars (after Nakagawa et al. 2003).

shift its hosts to survive in non-masting years. It is also known that the length of the larval stage varies among individuals of the same species (Maeto 1993). When we dissected dipterocarp seeds after adult insects had emerged from them, to check the remaining seeds, many live larvae of some weevils were found (Nakagawa and Itioka, personal observation). There might be variation in the length of larval or pupal stages, or the larvae of seed predators might take more time to grow than usual, if relatively less specialized on such alternate resources. However, they also could maintain their populations by using more than a single host species.

13.3 Host Specificity of Seed Predators

The predator-satiation hypothesis proposes that masting is an anti-predator adaptation that allows seeds to survive by alternately starving and satiating seed predators (Janzen 1971b; Silvertown 1980). As such, it might be one of the most important hypotheses to explain the ultimate cause of GF (Sakai et al. 1999c; Sakai 2002). In order to posit a synchronous seeding not only at the population level but also among species during GF, it is an essential condition that synchronous-seeding tree species share common seed predators. This would not only produce satiation but also reduce both the risk and cost to an individual tree.

A	
B	**C**

Plate 1. Lambir Hills National Park, Sarawak. **A:** Dawn over Lambir Hills, or, Bukit Lambir in Iban, which means Cock's comb. **B:** In the forest, a *Tristaneopsis* (Myrtaceae)—one of the more distinctive trees in the park. **C:** Fun at the waterfall. After heavy rain, the swollen Sungai Latek forms an impressive cascade.

A	**B**
C	**D**
E	**F**
G	**H**

Plate 2. Scenes from around Lambir Hills National Park. **A:** *Dryobalanops lanceolata* (Dipterocarpaceae) in flower. Underneath the canopy of this tree is the first of the research towers. **B:** A giant bird's nest fern *Asplenium nidus* (Aspleniaceae) viewed from the canopy walkway. **C:** The aromatic flowers of *Goniothalamus* (Annonaceae). Many trees have cauliflorous flowers like these. **D:** The base of a huge *Dryobalanops lanceolata*. This is possibly the biggest tree in the park, with its lowest branch 50 m above the ground. **E:** Rajah Brooke's birdwing *Trogonoptera brookiana brookiana* (Papilionidae). Males (the female is less conspicuous) are common in the park, and the species was first collected and described by Alfred Russel Wallace, here in Sarawak. **F:** A beetle's eye view of an *Amanita* (Amanitaceae) mushroom. Flies and small beetles are attracted by an odor to the mushroom's underside and then become dusted with the spores. **G:** A view of park headquarters from a tall tree. In the foreground is the canteen and car park and beyond are the office and chalets. The laboratories are in the distance, beneath the forest. **H:** Rumah Aji longhouse. The longhouse is approximately 5 kilometers from park headquarters and a source of willing labor and local botanists.

A	**B**	**C**
D	**E**	
F	**G**	

Plate 3. Towers and canopy walkway. **A:** Tower 1 seen from the top of the neighboring tree. Light trap equipment can be seen on the top and mid-levels. This 30 m tower was built in 1992. **B:** The new canopy crane, reaching 80 m high and with a 75 m radius, built in 2000. **C:** Tower 2, built in 1992, stands next to the *Dipterocarpus pachiphyllus* (Dipterocarpaceae), then emerges above the canopy at 42 m. **D:** A younger Dr. Momose and other researchers climb the newly built Tower 2. **E:** The top of Operation Raleigh Tower close to the main waterfall. This tower, built for tourists, is the oldest in the park; as seen here it is also used for research. **F:** The crane standing above the forest with Lambir Hills in the background. **G:** Doctors Momose and Nagamitsu on the canopy walkway, approximately 15 m above the ground.

A	**B**	**C**
D	**E**	**F**
G	**H**	**I**

Plate 4. Researchers in Lambir Hills National Park. **A:** Tamiji Inoue (1947–1997), the founder and leader of the Canopy Biology Project for the first six years. **B:** Peter Ashton on the canopy walkway. Dr. Ashton first established plots in Lambir Hills in the early

1960s and with doctors Hua-Seng Lee, Kazuhiko Ogino, and Takuo Yamakura established the Center for Tropical Science's 52 ha Long-term Ecological Dynamics plot in 1991. **C:** Rapi Abdul Raphman, the first laboratory manager, conducting a phenology census. **D:** Takakazu Yumoto demonstrating single rope climbing techniques near Tower 1. **E:** Takao Itioka sampling insects during the 1996 general flowering, from the crown of a *Shorea falcifaloides* (Dipterocarpaceae). **F:** The honorable foreman who built the towers and walkway, on top of a newly constructed platform. **G:** Mr. Jugok, an experienced local botanist, from Rumah Aji longhouse. **H:** Researchers set up a trap to catch insects. **I:** Sandra Patiño lectures students during a workshop.

A	B
C	D
E	F
G	H

Plate 5. General flowering. In 1996 many trees flowered after several years of little or no reproductive activity. **A:** The canopy in bloom. This tree is *Shorea falcifaloides* (Dipterocarpaceae). **B:** Tamiji Inoue in the canopy of *Dryobalanops aromatica* (Dipterocarpaceae) at dawn. **C:** A flower of *Shorea ochracea* with nectar. **D:** An orchid in the canopy. **E:** The red-winged fruits of *Dipterocarpus geniculatus* (Dipterocarpaceae). **F:** The uncoiled flowers of *D. temphes*. **G:** A profusion of flowers of the thick-stemmed climber *Milletia* (Fabaceae). **H:** A good crop of fruit on *Dryobalanops lancelotata*.

A	B	C
D	E	F	G
H	I	J

Plate 6. The pollinators. **A:** Gall midges (Diptera) on the male infloresence of *Artocarpus integer* (Moraceae). **B:** A pyralid moth (Pyralidae) on the primitive female inflorescence of *Gnetum gnemon* (Gnetaceae). **C:** A giant honeybee (*Apis dorsata*, Apidae) collects pollen from the flowers of *Dillenia excelsa* (Dilleniaceae). **D:** *Apis dorsata* spilling pollen from a flower of *Dryobalanops lancelolata* (Dipterocarpaceae). **E:** A tiny leaf beetle (Chrysomelidae) scrambles out of a flower of *Shorea parvifolia* (Dipterocarpaceae). The populations of these beetles exploded during the general flowering and were the dominant pollinators on many of the strictly general-flowering species. **F:** Beetles *Parastasia* (Scarabaeidae) on the spathe of *Homalomena propinqua* (Araceae). **G:** Leaf beetles on the urceolate calyxes of *Sterculia stipulata* (Sterculiaceae). **H:** The deep purple flower of *Orchindantha inoue* (Loweaceae). This species is pollinated by dung beetles (*Onthophagus*, Scarabaeidae). **I:** A stingless bee, *Trigona erythrogastra* (Apidae) collecting pollen from *Melastoma malabathricum* (Melastomaceae) without touching the stigma. **J:** Female bees (*Amegilla pendleburyi*) visiting a ginger flower *Zingiber longipedunclatum* (Zingiberaceae).

A	B
C	D
E	F
G	H

Plate 7. Vertebrate pollinators. **A:** A Spectacled spiderhunter *Arachnothera flavigaster* (Nectariniidae) thieves nectar from a flower of *Durio oblongus* (Bombacaceae). **B:** The same flower is visited by flies (Diptera) at night. **C** and **D:** A Spectacled spiderhunter pollinates the elongate flowers of *Tritecanthera sparsa* (Loranthaceae). **E:** *Tritecanthera xyphostachys* has a perch, but normally the long-billed spiderhunter (*A. robusta*) perches

on a flower while probing the flowers. **F:** A spectacled spiderhunter sits in the foliage of a third species of Loranthaceae, *Amylotheca duthieana*. Spectacled spiderhunters will defend patches of flowers against other spiderhunters. **G:** Flowers of Durian (*Durio kutejensis*) are normally pollinated by bats but here are visited by large numbers of giant honeybees. **H:** The flowers of *Fragraea racemosa* (Loganiaceae) are visited by bats. The one pictured here is *Macroglossus minimus* (Pteropodidae).

A	B	C
D		E
F		G

Plate 8. Ant-plants in Lambir. **A:** *Crematogaster borneensis* (Myrmicinae) workers use food bodies under the stipules of *Macaranga trachyphylla* (Euphorbiaceae). The exit holes of the *domatia* are clearly visible. **B:** Inside the domatia of another species of *Macaranga* workers of a different *Crematogaster* tend a *Coccus* scale insect (Coccidae), which produces a sweet exudate. **C:** A queen of an undescribed *Crematogaster* tends her eggs inside a sapling of *M. winkleri*. With these eggs she will initiate a colony. **D:** A staghorn fern *Platycerium coronarium* (Polypodiaceae) high in the canopy of *Shorea parvifolia* (Dipterocarpaceae). Inside the bulbous base, formed from modified leaves, an ant *Crematogaster deformis* and two species of cockroach *Pseudoanaplectinia yumotoi* and *Blattella* (Blattellidae) live together. **E:** *C. decamera* workers tend the food bodies of *M. beccariana* on the underside of a young leaf. **F:** A cross-section through the bulb of an ant-plant *Myrmecodia* (Rubiaceae), showing the domatia. **G:** Ant-plants, *Myrmecodia,* are common on the summit of Lambir and also occur as epiphytes high in the canopy.

A	B	C
D		E
F		G

Plate 9. Bees (Apidae). **A:** Stingless bees *Trigona ventralis* and a honeybee (*Apis koschevnikovi*) visiting an artificial honey-water feeder. **B:** The tubular nest entrance of *T. collina*. **C:** A worker *T. collina* collecting pollen from flowers of a grass. **D:** *A. koschevnikovi* workers crowd an artificial feeder during experiments on height fidelity. **E:** An experimental nest box is opened to reveal the combs of *A. koschevnikovi*. **F:** The cannonlike nest entrance of *T. thoracica*. **G:** Dr. Nagamitsu shows the wall of tree resin from an arboreal nest of *T. fimbriata* found in a fallen tree trunk.

A		B
C		D
E		F

Plate 10. Figs in Lambir Hills National Park. **A:** A *hemi-epiphytic* fig seedling, *Ficus sumatrana* (Moraceae), high in the canopy of *Dipterocarpus globosus* (Dipterocarpaceae). As the seedling grows it will drop a root down the trunk until it connects with the ground. **B:** A free-standing hemi-epiphyte, *F. kerkhovenii*, having 'strangled' its host tree. Although often called stranglers, most species do not kill their host. **C:** A cauliflorous fig, *F. schwarzii* with a full crop of developing figs. **D:** A geocarpic fig, *F. uncinata* with figs born on specialized stolons at the base of the stem or under the leaf litter. **E:** A small understory hemi-epiphyte, *F. hemsleyana*. This species always colonizes a position within 3 m of the ground. **F:** *F. aurantiacae*, a bole climber, with its small leaves totally smothering the trunk of its host, bears a single large fig.

A	B	
C	D	
E	F	
G	H	I

Plate 11. Natural history of a fig, *F. schwarzii.* **A:** The receptive male syconium is shown, with short-styled pistilate flowers, separated cup-shaped stigmas, and large immature male flowers around the ostiole. Inset: The pollinating wasp, *Ceratosolen vetustus*, ovipositing on the pistilate flowers. **B:** An almost receptive female syconium; the ostiole is still closed. Pistilate flowers are long-styled but the flattened, bifurcate stigmas join up to form a continuous surface. **C:** Non-pollinating gallers, *Apocryptophagus* sp., oviposit through the wall of a syconium; these are competitors of the pollinator. **D:** Inquilines, *Philotrypesis* sp., oviposit through the wall onto the galls of the pollinator larvae. The larvae out-compete and eventually kill the larvae of the pollinator. **E:** Ants, this one is a *Myrmicaria* sp., patrol the surface of syconia hunting for non-pollinators; they catch ones that are in the process of ovipositing. The tree encourages ants to patrol by offering a sugary reward. **F:** An ovipositing *Apocrypta* sp. This is a parasitoid, probably of *Apocryptophagus*. **G:** The inside of a male syconia at wasp emergence. A group of wingless male wasps cut a tunnel through the ostiole, while female wasps collect pollen from the mature male flowers. **H:** Emerging wasps. The males, having cut the exit tunnel, scatter over the surface to distract the ants. Meanwhile, female wasps emerge very quickly, flying directly from the tunnel. **I:** The female wasps live one day thus must find a receptive tree immediately. Here they hover by the receptive syconia.

A	B
C	D
E	F

Plate 12. Drought and fire in Lambir Hills National Park. **A:** A water-stressed understory shrub showing the leaves browning and curling. **B:** A view over the canopy in Kinabalu National Park, Sabah, during the 1998 drought shows the large number of tree crowns that lost their leaves. **C:** Wilting understory Rubiaceae. Thick-leaved understory plants like this wilt easily in droughts. **D:** During a severe drought there is a buildup of leaf litter from increased leaf fall and reduced decomposition. Fires in the primary forest consume this material, burning slowly but with high intensity. **E:** Occasionally, a large forest tree catches fire. In this case the fire entered via a rotten root (bottom). **F:** The path here has created a firebreak preventing the fire from spreading. To the left one can see how the fire has burnt through the leaf litter and scorched lower leaves, but most stems remain. This fire scorched a ring around the base of stems, thus all but the largest trees later die.

Photograph Credits

Rhett Harrison. Plate 1 A–C; Plate 2 B–D, G; Plate 3 A–C, E, F; Plate 4 F, I; Plate 5 B–G; Plate 10 A–F; Plate 11 A–F; Plate 12 A, C–F.

Tamiji Inoue. Plate 2 A, E–F, H; Plate 3 D, G; Plate 4 A–E, G, H; Plate 5 A, H; Plate 6 B–G, I; Plate 7 G; Plate 8 A–G; Plate 9 F.

Shoko Sakai. Plate 6 A, H, J.

Takakazu Yumoto. Plate 7 A–F, H.

Teruyoshi Nagamitsu. Plate 9 A–C, E, G.

David Roubik. Plate 9 D.

Kazu Kimura. Plate 12B.

Color Plate 3

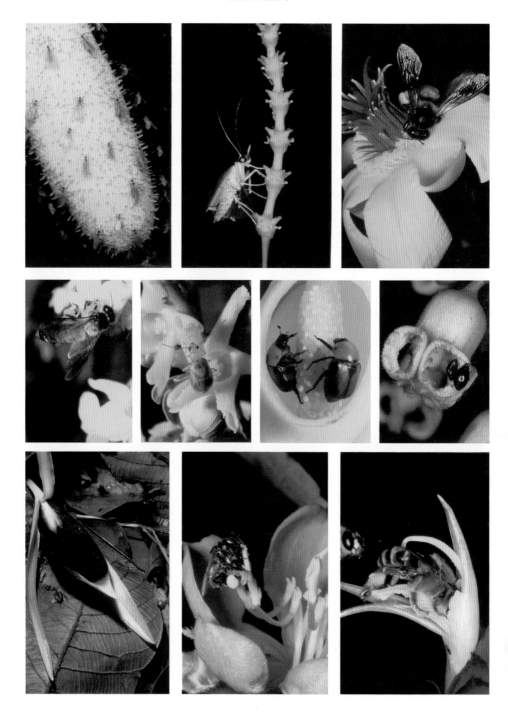

Color Plate 8

Color Plate 9

Color Plate 10

Color Plate 11

Color Plate 12

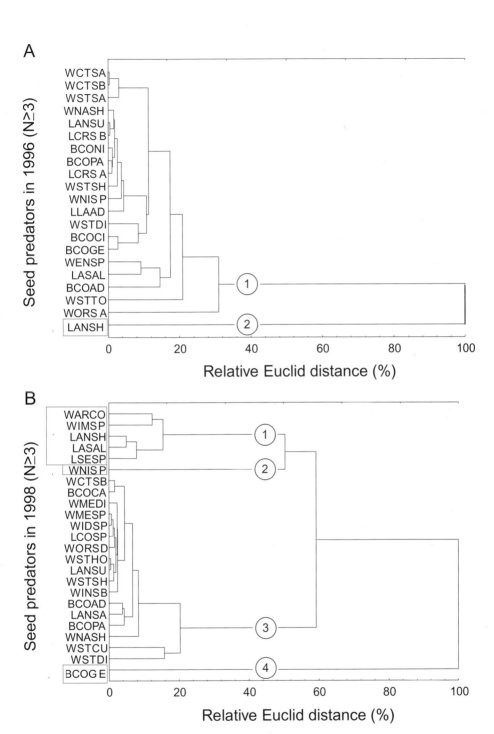

Figure 13.3. Dendrogram of dipterocarp seed predators (N≥3) derived by cluster analysis for 1996 (**A**) and 1998 (**B**). The code names of predators are listed in Table 13.3 (after Nakagawa et al. 2003).

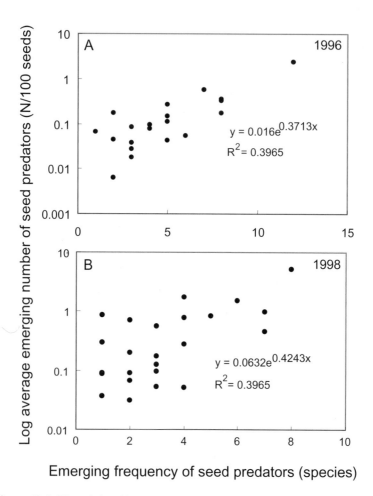

Figure 13.4. The relationship between the average emerging number of each seed predator and the emerging frequency in 1996 (**A**) and 1998 (**B**).

Although host specificity of insect faunas in tropical forests has been thought to be quite high, a growing number of reports show that it is lower than previously thought (Beaver 1979; Basset 1992; Marquis and Braker 1994; Mawdsley and Stork 1997; Barone 1998; Basset 1999; Roubik et al. 2003). Considering sporadic and simultaneous seed production among trees in GF, and the relatively low density of plant species, seed predators have access to a large range of host plants.

As indicated both in our study and in the reviewed literature, insect predators of dipterocarp seeds attack several genera in Dipterocarpaceae (see Table 13.1 and Fig. 13.1): *Andrioplecta shoreae* emerged from more than 10 species; *Coccotrypes advena*, *C. gedeanus*, and *Nanophyes shoreae* also had wide diet ranges. Seeds of *Artocarpus* were predated by three species of dipterocarp seed eaters (see Fig. 13.1A), indicating that some seed predators are polyphagous and have the potential to use various kinds of host plants, although they might have some preference.

Examining the relation between seed predators (N≥3) and their host plants at the guild level with cluster analysis, there were two and four large dietary groups in 1996 and 1998, respectively, and those groups did not correspond to a particular taxonomic group of seed predators (see Fig. 13.3). Furthermore, dominant seed predators, which had a relatively wide dietary range, were not the same between the two years. *Andrioplecta shoreae* (group 2) in 1996 was abundant and used mainly seeds of *Dryobalanops* and *Shorea*. The seed predators in group 1 tended to consume relatively specific seeds and/or to be relatively rare. In 1998, *C. gedeanus* (group 4) had large dietary range and *Niphades* sp.1 (group 2) emerged from mainly one *Dipterocarpus*, but was abundant. Although seed predators in group 3 were similar to those in group 1 in 1996, five species in group 1 mainly predated *S.macrophylla* (see Fig. 13.3).

Figure 13.4 indicates the association between the number of host plant species consumed (diet range) and the mean population of predators per 100 seeds of each tree species (dominance) in 1996 and 1998. Diet range of seed predators was significantly and positively correlated to dominance, which indicates that a *polyphagous* seed predator tends to be a dominant species. Greig (1993) also reported that a polyphagous seed predator of *Piper* (Piperceae), *Sibaria englemani* (Hemiptera), was a dominant predator in Costa Rica, and a polyphagous seed predator, *Stator limbatus* (Coleoptera: Burchidae), was dominant in seeds of *Acacia tenuifolia* (Mimosaceae) in Santa Rosa National Park (Janzen 1980).

Seed predators with a large diet breadth could become dominant species partly because they might have a higher competitive ability than do predators with a narrow diet breadth or because they could digest various resources. Although there was no generalist seed predator found in the whole range of tree species seeding during 1996 or 1998, the fact that host specificity of seed predators was relatively low, that dominant seed predators were likely to have wide range of diet, and that they changed between two GF events, all suggest two conclusions. First, it is possible to satiate insect seed predators by synchronous seeding, and second, to escape seed predation there is possibly a mutualism among tree species.

14. Diversity of Anti-Herbivore Defenses in *Macaranga*

Takao Itioka

14.1 Introduction

Plants evolve various modes of anti-herbivore defenses to reduce damage by phytophagous animals (Feeny 1969; Hartley 1997; Levin 1976; Coley et al. 1985; Rhoades 1985; Ehrlich and Murphy 1988; Mattson et al. 1988; Schultz 1988; Karban and Myers 1989; Tollrian and Harvell 1999). In general, plant species are armed with chemical or physical defenses against herbivores, and sometimes further adjust life histories to accommodate seasonal changes in potential herbivory (Aide 1992). In the tropics, where ants are constantly abundant, many plants have mutualistic relations with them. The ants become, effectively, anti-herbivore agents (Buckley 1982; Beattie 1985; Hölldobler and Wilson 1990; Oliveira and Oliveira-Filho 1991; Davidson and McKey 1993; McKey et al. 1993). Some plants provide extra-floral nectar (EFN) to attract ants; more than 30% of forest tree species offer EFN in Neotropical forests (Oliveira and Oliveira-Filho 1991). The prevalence of EFN for ants suggests widespread advantages of mutualism in the ant-rich tropics. Indeed, only in the tropics have plant species in diverse taxa become evolutionarily specialized to particular ant species. The plants provide nest sites and sometimes nutrients for the ants (Buckley 1982; Beattie 1985; Davidson and McKey 1993; McKey et al. 1993). These *myrmecophytes*, or, ant-plants derive benefit from symbiosis because the ants protect their nest plants from various invaders.

Because ants possess social organization and well-developed communication

(Hölldobler and Wilson 1990), their anti-herbivore defenses can be highly effective. However, symbioses with ants are not necessarily beneficial for plants, because ant defenses take a toll. Plants incur metabolic costs to provide partner ants with nutrients and energy sufficient for defensive activities, and defensive ants thereby diminish resources that may be needed for growth, reproduction (Moles 1994), and non-ant defense tactics (Janzen 1966; Rehr et al. 1973; Folgarait and Davidson 1994, 1995). How do myrmecophytic plants deal with the trade-off between ant defenses versus non-ant defenses, or between ant defenses versus other components in the life history? What environmental or biological factors affect the plant traits involved in the trade-off?

Macaranga (family Euphorbiaceae) is a tree genus of approximately 280 species, distributed from West Africa to South Pacific islands. It has its highest species richness in Borneo and New Guinea. These trees include many obligate myrmecophytic species (Whitmore 1969, 1975; Fiala et al. 1989). More than 16 species, including at least 10 myrmecophytic ones, occur in Lambir Hills National Park (see Plate 8). Since 1994, my colleagues and I have investigated the evolutionary ecology of anti-herbivore defense in *Macaranga* (Itioka et al. 2000; Nomura et al. 2000, 2001; Itino and Itioka 2001; Itino et al. 2001b). This research is briefly described below, in light of factors that affect diversity in anti-herbivore defenses, and provides new perspectives on the *Macaranga-Crematogaster* association.

14.2 Life History at Lambir

Details of basic biological characteristics in myrmecophytic *Macaranga* are well described in Fiala et al. (1989, 1994). The obligate myrmecophytes at Lambir Hills National Park (LHNP) always have mutualistic relationships with the specific partner ant, primarily *Crematogaster*. As shown by Itino (Chapt. 15; Itino et al. 2001a), *Crematogaster* are symbiont ants of all ten *Macaranga*. However, *M. lamellata* also has symbiotic relations with the ant *Camponotus macarangae* (Maschwitz et al. 1996).

At least three species of obligate myrmecophytes have difficulty surviving without the ants (Itioka et al. 2000). In obligate myrmecophytism, plants produce food bodies, which are collected by ants inside the domatia within stipules, and on the leaf surface (see Plate 8A, E). Symbiont ants use food bodies as their main food source, and are often found attacking invading insects and vines of competitive plant species that climb the host plant.

In all obligate myrmecophytes except for *M. hosei*, when seedlings reach 10 cm to 50 cm in height the stems swell and the pith degrades, leaving up to five hollow internodes where ants can build nests. An alate foundress queen of a symbiont ant settles into a stem internode of the seedling. She flies from her parent nest, pulls off her wings just after landing at a *Macaranga* seedling, and then makes an entrance hole on the stem surface. After she enters a stem internode, she confines herself in the hollow stem by filling the hole with plant tissue.

Another alate foundress queen often settles into another stem internode of the plant, thus a seedling may receive two or more foundress queens. Inside a hollow stem internode, a foundress queen feeds her first brood of workers with regurgitated food.

A few months later, workers make holes on the surface of the stem internodes, come out on the plant surface, begin to collect food-bodies, and start to defend host seedlings. Soon after, if the seedling is colonized by other ant queens, which is a common occurrence, battles will take place between residents and invaders. Finally, only one of the ant colonies occupies the seedling. In *M. hosei*, after plants grow to 1.5 m to 2.2 m in height, foundress queens begin to occupy them.

In myrmecophytes, as seedlings or saplings develop, the amount of food bodies they provide allows their symbiont ant colonies to possess sufficient workers for effective anti-herbivore defense. In turn, seedlings with plant-ant colonies of sufficient size are well protected from herbivores. In all three *Macaranga* species, if seedlings reach about one meter in height, their symbiotic relationships with ants usually show the balance of these factors.

In non-myrmecophytes, stems are not hollow but sometimes produce EFN on leaves, stipules or stems. In *M. gigantea*, EFN occurs on stipules and in *M. praestans*, on the young leaves. Non-specific, non-plant-ant species are often found feeding at such EFN.

Our demographic study using cohorts of several *Macaranga* suggests that, usually, at least 8 years, and probably more than 10 years, are required for the initial reproduction of *Macaranga* trees, although the correct ages were unknown (Davies et al. 1998).

Symbiont ant colonies do not reproduce until the colony becomes large. In most myrmecophytic species, reproductives are found only in trees whose sizes are over 3.5 meters in height (Itino et al. 2001b). Male reproductives are sometimes found in young seedlings below about 2 meters in height. In this case the nest plants or the colonies themselves were seriously damaged from plant destruction by humans, mammals or birds, or by treefalls.

As Heckroth et al. (1998) described, coccid Homoptera are always found associated with plant-ant colonies. They occupy the hollow stems of all myrmecophytic species, except in the early growth stage. It is inferred that coccids come into the nests after the ant workers emerge and are not specific to the plant or ant (Heckroth et al. 1998). They imbibe phloem sap of *Macaranga*, producing honeydew, upon which plant-ants feed. It is likely that ants feed on the coccids, but the number of coccids increases as the ant colonies grow.

14.3 Interspecific Variation in Environmental Conditions of Micro-Habitat

As Davies et al. (1998) have demonstrated, there is a remarkable difference in micro-habitat, regarding light, soil, and water conditions among *Macaranga* (see Fig. 14.1). Of all species, *M. gigantea* shows the greatest preference for open

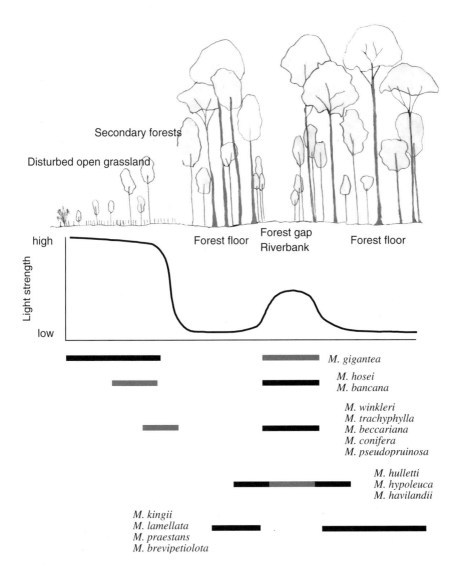

Figure 14.1. Schematic illustration of interspecific differences in micro-habitats of *Macaranga* species in Lambir Hills National Park, in relation to vegetation and light intensity (Davies et al.1998; Itioka et al., unpublished data).

and disturbed areas. This species is commonly found in secondary forests and at forest edges, and its growth rate appears to be higher than that of the others. *Macaranga bancana* and *M. hosei* also appear to require relatively intense light. In contrast, *M. hypoleuca, M. hullettii, M. kingi, M. lamellata, M. praestans* and *M. brevipetiorata* occur in shaded areas beneath the forest canopy. They seem unable to survive in open areas with strong light, probably due to photoinhibition of photosynthesis. *Macaranga winkleri, M. trachyphylla* and *M. beccariana* thrive in intermediate light conditions. They are found primarily in forest gaps and at the forest edge.

The distribution of *Macaranga* species, and their light environments, are not so distinctive and overlap somewhat in LHNP. However, when we also consider water conditions, the variation in micro-habitat becomes clear. Of the light-requiring secondary forest species, *M. bancana* requires the most water. *Macaranga winkleri* and *M. hullettii* also are more common in micro-habitats that have abundant water. The micro-habitats of these three species are close to, or limited to, areas around streams or ponds. In contrast, *M. lamellata* seems to be most tolerant to dry condition and also to drought (T. Itioka, unpublished data).

14.4 Interspecific Variation in Anti-Herbivore Defenses

In LHNP, the distributions of three obligate and forest-gap myrmecophytes—*M. winkleri, M. trachyphylla,* and *M. beccariana*—are convergent. They are usually found in the same gaps. However, Itioka et al. (2000) demonstrated that there is significant variation in the intensity of ant defenses among the three species. The intensity of ant defenses is highest in *M. winkleri*, second highest in *M. trachyphylla*, and lowest in *M. beccariana* (see Fig. 14.2). This conclusion was suggested by ant-exclusion experiments, the aggressiveness of symbiont ants, and the ratio of ant biomass to plant biomass for different tree species.

In all three *Macaranga*, after ant colonies are established, little herbivore damage is found on leaves, except when trees have been seriously damaged. Why, then, are the three *Macaranga* species able to defend themselves almost equally from their herbivores, although the intensity of ant defenses differs significantly? By measuring the effects of *Macaranga* leaves on growth performance of a butterfly larva, Nomura et al. (2000) answered this question. These researchers estimated the intensity of non-ant, that is, physical and chemical, defenses of the three *Macaranga* species. The results showed that *M. winkleri* is strongest and *M. beccariana* weakest in ant defense, but the former is weakest and the latter strongest in non-ant defenses (see Fig. 14.3). Conclusively, the non-ant anti-herbivore defenses were more effective in the *Macaranga* that was less defended by ants. A complementarity in defense seems to result, in these three species, in similar herbivory levels.

Janzen (1966) hypothesized that non-ant defenses in myrmecophytes are reduced because plants cannot afford to maintain both ant defenses and non-ant

☐ Increase of average herbivory damage compared to control (ant-colonized) plants
■ The number of ant workers that aggregated at leaf tips
▨ Ratio of the dry weight of ant colonies to that of above-ground parts of plants

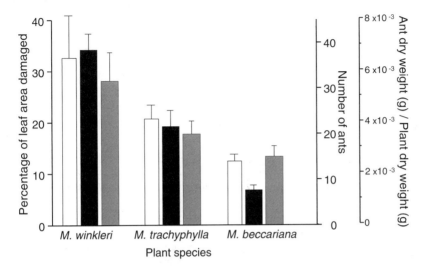

Figure 14.2. Three indices of interspecific differences in ant-defense intensity between *M. winkleri*, *M. trachyphylla*, and *M. beccariana*. *Open bars*: increase of herbivore damage on ant-excluded plants of *Macaranga* compared with the control (ant-inhabiting) plants at 4 weeks after ant exclusion. The increase is expressed by mean difference (*bar*) in percentage of leaf area damaged with standard error (*vertical line*). *Solid bars*: the mean number of ant workers that, in 90 seconds, aggregated around leaf tips (1 cm²). *Shaded bars*: ratio of the dry weight of whole ant nesting nodes to the dry weight of whole above-ground parts (± 1 standard error) of plants in the case of the three *Macaranga* species (g/g), after Itioka et al. (2000).

defenses at the same time. According to the limitation of resources that may be allocated to herbivore defenses, he hypothesized that there is a trade-off between ant defenses and non-ant defenses. That hypothesis may account for causal factors that affect the negative correlation between ant and non-ant defense intensities, among myrmecophytic and non-myrmecophytic *Acacia* species (Rehr et al. 1973).

Nomura et al. (2000) also measured the intensity of non-ant defenses of two non-myrmecophytic species: *M. gigantea* and *M. praestans*. Although light conditions that favor the two species are extremely different, the intensity of not-ant defenses was higher in both, compared to three myrmecophytic species. The more intensive non-ant defenses in non-myrmecophytic species are consistent with Janzen's hypothesis (see Fig. 14.4).

The trade-off hypothesis is also consistent with interspecific variation in intensities of ant and non-ant defenses among the three obligate myrmecophytes

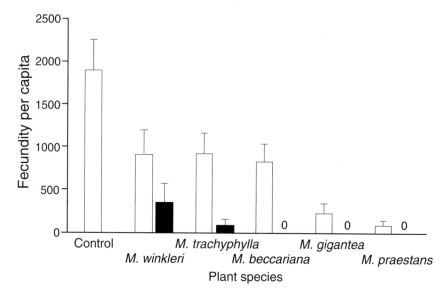

Figure 14.3. Fecundity response of cutworm (*Spodoptera litura*) reared on an artificial diet of dry leaf powder of *Macaranga* species (open bar ± standard deviation) and reared on fresh leaves; control from rearing on artificial diet only (Nomura et al. 2000).

M. winkleri, *M. trachyphylla* and *M. beccariana*, which share similar micro-habitats. The negative correlation between ant versus non-ant defense intensity across the three species can be accounted for by resource allocation. The production of food bodies—expected to affect ant colony growth, ant activity, and hence ant defense intensity—seems to be lower in myrmecophytes with less ant defense (Hatada et al. 2002, Fig. 14.4). This also favors the trade-off hypothesis.

In LHNP several other myrmecophytic species occur. *Macaranga bancana* has a micro-habitat similar to the three abovementioned species. This plant is closely related to *M. trachyphylla* and the plant-ant species is also similar. It is likely that the "place" of this species in the "trade-off space" (see Fig. 14.5) between ant versus non-ant defenses should be close to that of *M. trachyphylla*. The intensity of non-ant defenses of *M. bancana* is similar to that of *M. trachyphylla* (M. Nomura et al. unpublished data). *Macaranga kingii*, *M. hulettii*, *M. lamellatta* and *M. hypoleuca* share relatively shaded micro-habitats, compared to the abovementioned four myrmecophytes. Considering food production, ant aggressiveness, and leaf toughness for these species, all have less intense defense by ants than the four myrmecophytes of high-light habitats (see Fig. 14.5). *Macaranga hosei* is a myrmecophyte but the ant-receiving growth stage is much later than in other myrmecophytes.

Until trees reach 1.5 meters to 2.5 meters in height, they are not colonized by any ants but only attended by non-specific, facultatively-attending ants. During the non-myrmecophyte period in the early growth stage, the stem is not

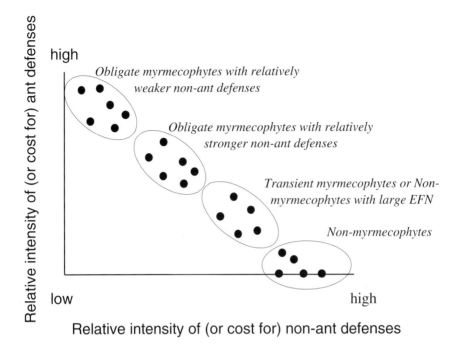

Figure 14.4. Schematic illustration of species distribution on hypothesized trade-off between ant versus non-ant anti-herbivore defenses in *Macaranga*.

hollow and very few food bodies are produced in *M. hosei* (K. Murase et al., unpublished data). In *M. hosei*, physical or chemical anti-herbivore defenses appear to be applied more in the early phase, when ant defense is lacking, than in the later period with plant-ant colonies. The other non-myrmecophytic species have intense non-ant defenses (M. Nomura, unpublished data), irrespective of their requirement for light.

14.5 Factors Affecting Variation in Anti-Herbivore Defenses

Davidson and Fisher (1991) hypothesized that relative investment in ant defenses should be greatest under moderate light conditions, because it is there that the efficiency or optimality of ant defenses is maximized at the average leaf life span there (see Fig. 14.6).

They argued that higher light intensities promote rapid growth, and faster turnover in leaf production. Ant defense by symbiont plant-ant colonies is too costly, compared with other tactics such as ant attraction by EFN to new leaves, because of the faster leaf turnover. Furthermore, the relative cost of ant-defenses decreases as leaf lifespan increases, under lower light intensities. They also ar-

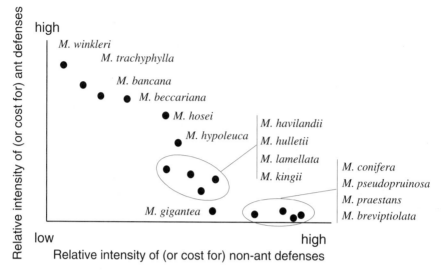

Figure 14.5. Schematic illustration of inferred distribution of *Macaranga* species in Lambir Hills National Park on hypothesized trade-off between ant versus non-ant antiherbivore defenses (after Davies et al. 1998, Itioka et al. 2000, Nomura et al. 2000, Itioka et al., unpublished data).

gued that leaf production rates lessen at lower light intensities, and that relative cost-effectiveness of non-ant defenses decreases while the relative cost of ant defense increases in such situations.

The hypothesis described above seems to be partly supported by the relationship between ant defense intensity and demand for light in *Macaranga* in LHNP. Ant defenses are most intensive in *Macaranga* whose major micro-habitats are in moderate light conditions (see Fig. 14.1, 14.5). Most non-myrmecophytic species exist primarily in high-light micro-habitats (e.g., *M. gigantea*) or shaded areas such as the forest floor (e.g., *M. praestans* and *M. brevipetiolata*). The obligate myrmecophytes with only relatively weak ant defenses, *M. huletti*, *M. hypolecuca*, *M. lamellata* and *M. kingii*, grow in lower light conditions than myrmecophytes with the more intensive ant defenses.

The hypothesis does not, however, completely explain interspecific variation in ant defense intensity according to differences in the demand for light. Within three obligate light-demanding myrmecophytes, *M. winkleri*, *M. trachyphylla* and *M. beccariana*, the micro-habitats overlap (Davies 1998; T. Itioka unpublished data) and the demands for light are not significantly different. Rate of leaf production and the growth rate of individual plant biomass, however, seem to be related to ant defense intensity. Of the three species, the leaf turnover and tree height are both highest in *M. winkleri*, and both lowest in *M. beccariana*. This relation could be accounted for by a difference in cost-effectiveness of ant

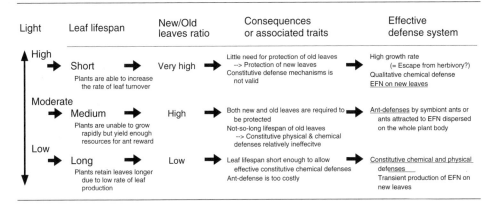

Figure 14.6. Hypothesized association between light intensity in micro-habitats, life history traits, and effective anti-herbivore defense systems.

and non-ant defenses in relation to leaf lifespan (see Fig. 14.6). For trees with long-lived leaves and lower growth rates, metabolic investment in non-ant defenses by chemical and physical mechanisms is not so costly, since those defenses would be useful for a relatively long period (cost per unit time thereby *decreases*). In contrast, ant defenses would be very costly, because a lower output of photosynthesized products, which is correlated with long leaf-life span, is insufficient for maintenance of defensive ant colonies. Thus, different anti-herbivore strategies may diversify and coexist, even under similar light-conditions, if life-history strategies in relation to leaf lifespan and growth rate differ. Life histories may generate various anti-herbivore strategies in which the balance between ant and non-ant defense intensity differs, while in the same micro-habitat. Although Davies (1998) and Davies et al. (1998) examine inter-specific variation in life-history traits—including photosynthetic ability, shade tolerance, growth rate, and age of initial reproduction—further correlations considering life history variation and anti-herbivore defenses suggest that continued study and analysis will be rewarding.

14.6 Species-Specificity in Plant-Ant Partnerships: Maintenance Mechanisms

Species-specificity is remarkably high in the partnership between *Macaranga* and *Crematogaster* (Fiala et al. 1999; Itino et al. 2001a). In LHNP, for some *Macaranga*, almost all trees of a species are colonized by only a single *Crematogaster*. At the same time, one ant species is observed colonizing only a few closely-related *Macaranga*. Focusing on the three forest-gap myrmecophytes—

M. winkleri, *M. trachyphylla* and *M. beccariana*—we find that the respective ant mutualist species are almost completely different (Fiala et al. 1999; Itino et al. 2001a). One proximate factor that helps to maintain species-specificity is chemical recognition by foundress queens (Inui et al. 2001). Using the non-volatile carbohydrates on the surface of leaves and stems of *Macaranga* seedlings as chemical cues, foundress queens correctly recognize their partner species.

The normally species-specific settling by foundress queens into their partner ant-plants has been confirmed in the field, as shown in Fig. 14.7 (Murase et al. 2002).

However, a small proportion of foudress queens settle on a non-partner *Macaranga* myrmecophyte. Because almost all well-developed saplings of obligate myrmecophytes have symbioses only with specific partner ant species, unmatched symbiosis (or 'mistakes') between non-partner ants and plants may lead to a collapse of mutualism when the plant-ant colonies mature. As mentioned

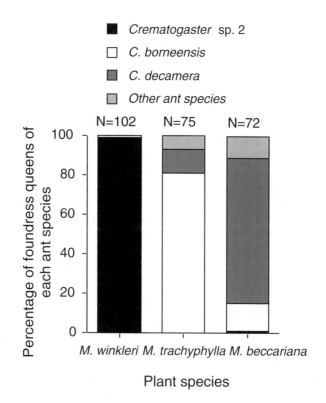

Figure 14.7. Percentage of seedlings settled by foundress queens of different ant species in *M. winkleri*, *M. trachyphylla*, and *M. beccariana* in the field (Murase et al. 2002).

Ant side

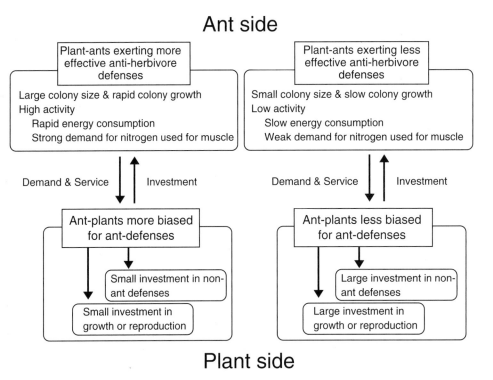

Figure 14.8. Hypothesized demand-and-supply match between *Macaranga* ant-plants and *Crematogaster* plant-ants.

above, intensive ant aggressiveness demands costs that plants can afford only by reducing physical and chemical defenses, or growth (see Fig. 14.8).

Likewise, ant colonies symbioic with plants that invest heavily in ant-defenses may be unable to obtain sufficient nutrition from other species. When the wrong ant settles on a tree normally mutualistic with strong defenders, it may be unable to effectively defend the plant. In this case, the plant cannot compensate for the weak ant defense. In both cases, mismatched myrmecophytism cannot persist. My colleagues and I have just undertaken experiments to examine whether these scenarios are realized or not.

14.7 Anti-Herbivore Defense Variation: Consequences to the Herbivore Community

Macaranga plants affect the community structure of *Macaranga* herbivores (see Fig. 14.9). To feed on *Macaranga* myrmecophytes, herbivores have to overcome

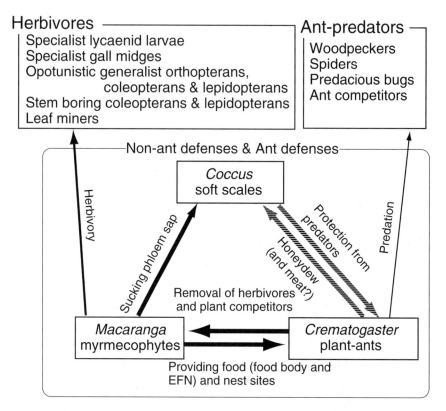

Figure 14.9. Schematic illustration of the herbivore and ant-predator community of the *Macaranga-Crematogaster* system.

ant-defenses, as well as non-ant defenses. Species composition of local herbivores on *Macaranga* plants may differ, depending on the relative importance or intensiveness of ant versus non-ant defense mechanisms. Adaptations by herbivores against ant defenses would be higher in herbivores on plants biased toward ant defenses, compared to those using predominantly non-ant defenses. Thus, contrasting levels of interdependency should select for herbivores having different counter-adaptations.

In LHNP there are notable differences in herbivore composition among *Macaranga* species, and some myrmecophytic plants are fed upon by their respective specialist herbivores, as well as by generalist herbivores (Itino et al. 2001b, and unpublished data, Table 14.2). In addition, the frequency of herbivore damage and herbivore abundance on the plants is noticeably different among species of *Macaranga*. For instance, densities of a leaf gall-making midge, a few species of Phasmidae (stick insects) and a leaf-mining lepidopteran were all relatively higher on *M. beccariana* than on the other myrmecophytic species.

The total density of herbivorous insects was lowest in *M. winkleri* among forest-gap myrmecophytes, but the damage by monkeys and woodpeckers, which feed on ant colonies inside the stems and destroy plant tissues, was most severe in this species. Thus more empirical field data are sure to show even more variety in the *Macaranga* system, and lead to fruitful speculation, hypothesis-building, and experiments.

15. Coevolution of Ants and Plants

Takao Itino

In this chapter, I review the current understanding of the interaction and evolution of *Crematogaster* ants and *Macaranga* plants in Lambir Hills National Park (LHNP). For more detailed biological background, see Chapter 14, Itioka et al. (2000), Itino and Itioka (2001), and Itino et al. (2001a).

15.1 Introduction

Ecological interactions including parasitism, herbivory, and mutualism have generated major diversification in the continual evolution of organic complexity. A reconstruction of the evolutionary history of such interactions requires a *phylogenetic hypothesis*. This explicit statement on the evolutionary history of a group of organisms is useful in testing whether a particular association was inherited by descent (*co-speciation, coevolution,* or *adaptive radiation*) or was produced by opportunistic means (*colonization, ecological fitting*), in which a member of one lineage formed a novel association with another lineage (Farrell and Mitter 1990; Herre et al. 1996).

The symbiotic association between ant plants (myrmecophytes) and ants is a paradigm of mutualism. The ants nest obligatorily in plant stems or other plant cavities (see Plate 12B, C) and benefit from food provided by the host plant (see Plate 12D, E). In return, the plants benefit from protection offered by the ants against herbivores and smothering vines (Janzen 1966). Plant specialization has led to mutual dependence of plants and their partner ants, so in some cases neither party can successfully reproduce without the other (see Chapter 14; Fiala et al. 1989; Itioka et al. 2000). Although these ant-plant symbioses have been regarded as classical cases of coevolved mutualism (Janzen 1966), no evidence of co-speciation has been presented. Rather, recent work (Davidson and McKey 1993; Chenuil and McKey 1996) suggests that frequent evolutionary host shifts by ants have determined the pattern of this association in the American and African tropics.

15.2 Specificity

In LHNP, 11 species of myrmecophytic *Macaranga* have been recorded (Davies 1996), and their microhabitats largely overlap. It is not uncommon to find as many as 5 to 8 species within a small forest gap of 0.1 hectare (Davies et al. 1998).

Historically, the specificity of ants to *Macaranga* has been poorly understood. *Crematogaster borneensis* has been tentatively considered to be a highly variable ant inhabitant of *Macaranga* in Peninsular Malaysia (Fiala and Maschwitz 1990). Recently, however, Itino et al. (2001b) detected four morphological ant species in LHNP, and Fiala et al. (1999) reported nine morphospecies in Southeast Asia. In Lambir, two of the four ants, *Crematogaster* sp. 2 and *C.* sp. 4 were specialists associated with *M. winkleri* and *M. hosei,* respectively (one-to-one associations). On the other hand, both *C. borneensis* and *C. decamera* were associated with four *Macaranga* species (see Fig. 15.1; Itino et al. 2001b).

Key factors maintaining the specific interactions appear to be specialization by *Macaranga* species, according to their natural enemies. Seven *Macaranga* may be organized into three groups (see Table 15.1; Itino and Itioka 2001). The first group (two species) tends to suffer attacks by gall flies (Cecidomyiidae). The second group (four species) is often attacked by leaf herbivores, and the third group (one species) by woodpeckers. The woodpeckers are not herbivores but ant-predators; they break the hollow stem and consume the ants inside. These three groups of *Macaranga* are inhabited by different ant species (see Table 15.1; Itino et al. 2001a): the first group by *C. decamera*, the second by *C. borneensis* or *C.* sp. 4, and the third by *C.* sp. 2. This indicates that each ant species corresponds to a different enemy fauna.

The " 'enemy and ant" specialization in *Macaranga* is concerned with interspecific variation in biotic and non-biotic (chemical/structural) defenses (see Table 15.1; Itioka et al. 2000; Nomura et al. 2000). Interestingly, the degree of biotic and non-biotic defense varies inversely between *Macaranga* species, that is, *Macaranga* species with weaker ant guards tend to have stronger non-ant defenses, as listed in Table 15.1.

Chemical and structural defenses of *Macaranga*, in particular, seem to result in the adaptive specialization of natural enemies. For example, *Macaranga* species susceptible to gall makers (see Table 15.1) may differ in their allelochemical

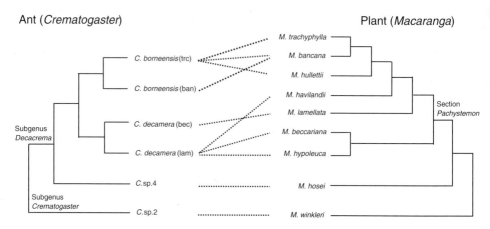

Figure 15.1. Phylogeny of symbiont *Crematogaster* ants and phylogeny of the corresponding *Macaranga* hosts. For *C. borneensis* and *C. decamera,* principal host plants are indicated in the parentheses for each mtDNA lineage. Dashed lines indicate associations. The ant phylogeny is based on mtDNA sequences while the plant phylogeny is based on combined analyses of morphological characters and nrITS DNA sequences (Davies et al. 2001). Only nodes with >50% bootstrapping support are presented as resolved (Itino et al. 2001b, © Blackwell Publishing).

composition from those susceptible to leaf eaters. On the other hand, susceptibility to woodpeckers appears to be related to structural toughness of the stem. With weak stems, *M. winkleri* permits woodpeckers easy access to symbiont ants, while *M. winkleri* is able to grow very fast (Davies et al. 1998), likely at a cost of structural weakness.

Food reward quantity of *Macaranga* to the ants, on the other hand, appears to be crucial for the maintenance of ant specificity. The amount of food reward in *Macaranga* varies among species (see Chapter 14; Itioka et al. 2000; Itino et al. 2001a). Each amount appears suitable for each ant species. For example, a small-sized, nonaggressive ant species (*C. decamera*) is associated with *Macaranga* that provide few food rewards (see Table 15.1; Itioka et al. 2000; Itino et al. 2001a). Even with limited rewards, *C. decamera* can effectively deter the main herbivore, larvae of Cecidomyiidae (Itioka et al. 2000). Although we can imagine a superadapted generalist ant species that can always provide the best services to mutualist plants, irrespective of the amount of food rewards provided, this appears not to be the case. The range of phenotypic plasticity of an ant species is evidently limited.

15.3 Polarization of Food Rewards: Homopterans or Food Bodies?

Primary food rewards of *Macaranga* for their ants differ notably among *Macaranga* species. *M. beccariana* and *M. bancana* provide ants mainly with ho-

Table 15.1.

Macaranga	N	Percentage trees infested by[1]			Principal enemies	Partner ants[2]			Rewards to ants[4]		Chemical/ structural defenses[5]
		gall makers	leaf eaters	woodpeckers		species[3]	aggressive- ness	density	quantity	main item	
M. beccariana	81	24	17	1	gall makers	*C. decamera* (bec)	+	+	+	scale insects	+++
M. lamellata	9	31	5	8	gall makers	*C. decamera* (lam)					
M. bancana	38	0	43	2	leaf eaters	*C. borneensis* (ban)		++	++	scale insects	
M. trachyphylla	112	2	34	4	leaf eaters	*C. borneensis* (trc)	++	++	++	food bodies	++
M. hullettii	9	0	10	0	leaf eaters	*C. borneensis* (trc)					
M. hosei	10	4	23	0	leaf eaters	*C.* sp. 4					
M. winkleri	31	0	9	21	wood-peckers	*C.* sp. 2	+++	+++	+++	food bodies	+

[1] Itino and Itioka (2001)
[2] Itioka et al. (2000), Itioka et al. (2001b)
[3] Because *C. decamera* and *C. borneensis* are divided into different lineages in the molecular phylogeny, and each lineage corresponds to a different specific host, principal host plants are indicated in the parentheses for each lineage (see Fig. 15.1).
[4] Itino et al. (2001a)
[5] Nomura et al. (2000)

mopterans (scale insects) rather than food bodies, as shown in Plates 12D, E and listed in Table 15.1 (Itino et al. 2001a). In contrast, *M. winkleri* and *M. trachyphylla* provide food bodies rather than homopterans. The plant invest- ment made in ants (ant dry weight/plant dry weight) was different among the four *Macaranga* species (Itino et al. 2001a). The Homoptera-dependent *M. bec- cariana* harbored lower biomass of ants than the food-body dependent *M. wink- leri*, suggesting that energy loss is involved in the Homoptera-interposing system, which has one additional trophic level (Itino et al. 2001a).

To understand the evolutionary process of ant-plant symbiosis, it is important to know whether ants use homopterans or food bodies as their main food source. First of all, an evolutionary conflict may exist between ants and plants concern- ing the regulation of ant and homopteran populations. Potentially, ants are able to control the homopteran population by culling, so as to maximize their own fitness, while plants are able to regulate the ant population by adjusting food body production. As optimal ant and homopteran population sizes for plants are usually different from those for ants alone, a conflict must occur. The fact that highly ant-defended *M. winkleri* and *M. trachyphylla* rely primarily on food bodies indicates that ant-plant symbiotic systems tend to exclude homopterans when plants need intensive defenses by ants.

15.4 Co-Speciation

Our survey has confirmed specificity between nine species of *Macaranga* and four morphological species of *Crematogaster* (see Fig. 15.1). Yet, the specificity looks onesided: every *Macaranga* species is principally associated with a single ant morphospecies, whereas two of the four *Crematogaster* morphospecies have several plant associates. We hypothesized that such ant morphospecies might be a mixture of morphologically similar but reproductively isolated cryptic species, or races of a single species that correspond to particular *Macaranga* species.

To test the hypothesis, we examined the mtDNA sequence variation of *Cre- matogaster* ants living in the nine *Macaranga* species (Itino et al. 2001b). The ant samples were collected from seven localities in Bornean and Peninsular Malaysia. A 496-base-pair part of the COI gene of mitochondrial DNA was sequenced for ants from 47 different plants. The ant phylogeny revealed six primary mtDNA lineages, suggesting that the previously detected four morpho- logical species are, in fact, divided into six or more genetically differentiated lineages with different mtDNA types. Four of the six ant lineages have basically one specific partner plant species, while the other two ant lineages have three associates (see Fig. 15.1).

Despite the insufficiency of free-living *Crematogaster* species sampled, the ant phylogeny suggests that *Macaranga*-associated *Crematogaster* have arisen at least twice: once each in the *Crematogaster* subgenera *Decacrema* and *Cre- matogaster*. Based on the assumption that the mean rate of divergence in mtDNA sequences is 2.3% per million years in arthropods (Brower 1994), the age of diversification in plant-ant subgenus *Decacrema* can be estimated as slightly less than seven million years (Itino et al. 2001b).

Given evidence for the high specificity between ants and plants, it is possible to test for co-speciation. The co-speciation hypothesis predicts the topology of ant and plant phylogenies to be congruent and the timing of divergence to be simultaneous. The branching of the ant and plant phylogenies is, in fact, highly congruent (see Fig. 15.1). We rejected the hypothesis that the ant phylogeny is independent of the plant phylogeny (P=0.0011, computed using 10,000 random trees; Page 1993) despite one major disagreement: A member of the *Cremato-gaster decamera* group seems to have once colonized *M. havilandii,* which had presumably been associated with a member of the *Crematogaster borneensis* group.

In addition to the congruence of the ant and plant phylogenies, Tertiary climatic patterns in Borneo and the restriction of myrmecophytic *Macaranga* to aseasonal forests suggest that this clade of *Macaranga* diversified in the late Tertiary (Morley 1998), which corresponds to the diversification period of *Crematogaster* subgenus *Decacrema.* These results suggest that the *Macaranga-Crematogaster* mutualism has been rapidly co-speciating and co-diversifying over the past 6 million to 7 million years (Quek et al. 2004)

The intimate and one-to-one coevolution of *Macaranga* and their associate ants presents a striking contrast to the American and African ant-plant associations, which are less specific (Fonseca and Ganade 1996). In the latter there is no known evidence for co-speciation (Ward 1991; Chenuil and McKey 1996). *Macaranga* is also ecologically unique among myrmecophytes in forming diverse communities of up to eight locally sympatric species, with each species inhabiting a slightly different microsite than the others (Davies et al. 1998). The species-specificity of the ant guards appears to promote the coexistence of multiple *Macaranga* species on a small spatial scale, because each ant species defeats a restricted group of herbivores (Itino and Itioka 2001) and thereby creates a novel enemy-free space (Holt and Lawton 1993) for its host *Macaranga.* We hypothesize that the close and long standing partnership between *Macaranga* and *Crematogaster* is a consequence of the specific and mutually dependent ant-plant symbiotic defense systems.

16. Lowland Tropical Rain Forests of Asia and America: Parallels, Convergence, and Divergence

James V. LaFrankie

This chapter primarily compares the forest at Lambir Hills National Park, Sarawak, with other lowland tropical forests in Asia and America. No two tropical forests are exactly alike. But do differences result from histories, often overshadowed by adaptive convergence toward uniform forest ecology? Or do the actual differences reflect fundamental differences in ecological function?

16.1 Introduction

Lowland equatorial wet forests generally display a uniform structure. In each hectare there are 35 m³ to 45 m³ of basal area, 400 to 600 trees over 10 cm dbh (diameter at breast height), and 1000 to 3000 trees 1 cm to 2 cm dbh (see comparative tallies in Davies and Becker 1996; Richards 1996; Lieberman et al. 1996). The general range of tree diversity among continents is also roughly similar, although depressed in African wet lowland forest (Gentry 1988).

Parallel familial representation is characteristic of lowland wet forest in all continents. Of the 168 families represented by trees of Southeast Asia, all but 24 are shared with tropical America. Generally speaking, genera are not shared among continents. However, the few of the shared genera are often conspicuous in the ecological uniformity displayed by their species, which reinforces the notion of a similarity among all tropical forests. A naturalist from Central America knowing *Anaxagorea panamensis* (Annonaceae) as an abundant small treelet with ballistically dispersed seeds and highly aggregated populations would immediately recognize *Anaxagorea luzonensis* at Mount Makiling in the Philippines or *Anaxagorea javanica* in the Malay peninsula. The genus *Xylopia* (Annonaceae) appears on all three continents, as does *Campnosperma* (Anacardiaceae), which forms characteristically dense populations in wet open sites in both America and Asia. *Rinorea* (Violaceae) is recognized in both America and Asia (and in Africa) by flower, fruit, and leaf features, and by the similar clumped distribution among the ballistically dispersed species. *Ormosia amazonica* (Leguminosae) and *Ormosia sumatrana* are quickly recognized as congeners. Other genera of this sort include *Croton* (Euphorbiaceae), *Dacryodes* (Burseraceae), *Elaeocarpus* (Elaeocarpaceae), *Sterculia* (Sterculiaceae), and many smaller Rubiaceae such as *Psychotria*.

It is more typical that a taxonomic family is represented in each continent by regionally endemic genera that display a recognizable suite of ecological features with regard to habit, physiology, flowers, and fruit. The most obvious examples of such families are the Myristicaceae, Sapotaceae, Clusiaceae, Myrtaceae, Malvaceae, and Lauraceae, and many herbs, pteridophytes, orchids, and gingers. Nutmeg trees of the family Myristicaceae are exemplary of such parallel development. The family is easily recognized from vegetative features alone: alternate, entire leaves, the simple leaf stalk, exstipulate, branches typically arranged around the trunk in the pattern of a wagon wheel, and the trunk exuding a red sap. The flowers and fruit are uniform across the family: small, tri-partite, the plants dioecious, and the fruit a single-seeded dehiscent berry with a brightly arilate seed. In Neotropical forests we find the following genera and numbers of species: *Compsoneura* 9, *Virola* 40, *Osteophloeum* 1, *Iryanthera* 23, and *Otoba* 7. In Asian forest, the family is represented by *Horsfieldia* 100, *Endocomia* 4, *Myristica* 72, *Knema* 60, and *Gymnacranthera* 7.

A second factor that implies homogeneity among the rain forests of the world is the convergent adaptation among unrelated taxa, which has yielded so many singular and well-known examples. Those include the high light- and nutrient-demanding mymecophytes *Macaranga* (Euphorbiaceae) in Asia, and *Cecropia* (Cecropiaceae) in America, the hummingbird/sunbird parallels (Karr and James 1975), the Asian pangolin and American anteaters, and the convergent trophic structure among non-volant mammals (Eisenberg 1981). Finding ecological equivalents is a game inevitably played by any naturalist who visits a new continent.

With an eye on uniform physiognomy, on parallel family development, and on ecological convergence of divergent taxa, Gentry (1988) argued for the global similarity of tropical forests. He saw the same tree families dominating lowland wet forests all over the world, with the principal exception of the dipterocarps of Southeast Asia. He further suggested that the ecological role of diptercarps was taken up in the Neotropics by the legumes, especially the caesalpinoid legumes, and particularly on poor soils and/or in seasonal climates. According to Gentry, "Legume trees play essentially the same role in Neotropical forests as the dipterocarps do in Asia"(Gentry 1993).

Janzen (1977), however, pointed out a conspicuous contrast between the hemispheres. The understory of Asian forests appears dominated by the saplings of canopy trees which are juvenile and thus sterile, whereas his experience in America, especially in wet lowland forests, indicated an understory rich in small treelets that flowered and fruited frequently. He emphasized the impact that this had on herbivores and seed dispersers (and consequently on pollinators and flower visitors, as discussed in other chapters in the present book).

There has been little formal comparison of tropical forest composition among continents, perhaps chiefly because of scant information on the lower strata, of the trees no more than 1 cm to 2 cm in diameter.

16.2 A Comparative Study of Forest Composition

Here the principal comparison of trees focuses on four forests—two in America and two in Asia—where large-scale plots have been established. These are Lambir Hills, Pasoh Forest in Peninsular Malaysia, Yasuní forest in Ecuador, and Barro Colorado Island in Panama. Additional comparisons are made to the well-studied site of La Selva Field Station, Costa Rica. The tabulations and results are from published sources or re-calculated for comparison from the original data. The large-scale plot methods are found in Manokaran et al. (1990) and Condit (1998). Data for the individual sites can be found for Lambir Hills (Lee et al. 2002, 2003); for Pasoh, in Manokaran and LaFrankie (1990); Kochummen et al. (1990); and Manokaran et al. (1992); for Yasuní, in Valencia et al. (2004a-c); for Barro Colorado Island in Condit (1998); and for La Selva, in Lieberman and Lieberman (1987), Clark and Clark (1992), and McDade et al. (1994).

The four forests are structurally similar (see Table 16.1). BCI is the least similar, having lower density and far lower diversity owing chiefly to its geographic position and more strongly seasonal climate.

Asian forests are distinguished by numerous families of large canopy trees that in tropical America either are represented sparsely or not at all (see Table 16.2). Foremost are the Dipterocarpaceae, Fagaceae, Ebenaceae, and Polygalaceae. Conversely, some families are diverse and abundant among Neotropical canopy trees but poorly represented in the wet equatorial forests of Asia. These include Bignoniaceae, Cecropiaceae, and Malpighiaceae. However, it will be quickly seen that these are of a different character than the former families, being either trees characteristically of dry habitats (such as Bignoniaceae, well represented in dry forests of the American tropics) or trees of a somewhat ruderal or early successional habit, such as Cecropiaceae. The Old World tropics do not have a family that is easily paralleled with Bignoniaceae.

Table 16.1. Summary comparison of study sites and permanent plots of trees

	Pasoh Malaysia	Lambir Malaysia	Yasuní Ecuador	BCI Panama
Altitude(m)	100	100	235	100
Annual rainfall (mm)	1850	2300	~3000	2500
Months with <100 mm rainfall	0*	0*	0	3
No. trees/ha				
≥1 cm dbh	6477	7068	6094	4707
1–2 cm dbh	2566	3155	2357	2569
<10 cm dbh	5922	6430	5392	4289
≥10 cm dbh	554	637	702	418
Basal Area (m^2)				
Tree Flora				
Families: Plot	93	93	86	62
Genera: Plot	294	288	299	184
Species: Plot	823	1188	1104	304

* Each year may include periods of up to 20 consecutive rain-free days but they are irregular, yielding monthly means over 100 mm.

16.3 Individual Tree Families

Notes on the characteristics of individual families, given below, will be followed by more general findings and summaries.

Annonaceae. While trees of the family Annonaceae display fundamentally parallel ecology among the continents, a profound divergence between America and Asia is seen among lianas. Asian Annonaceae include the following genera and number of species, strictly lianas: *Tetrapetalum* 2; *Rauwenhoffia* 5; *Cyathostemma* 8; *Dasoclema* 1; *Ellipoeiopsis* 2; *Uvaria* 110; *Ellipia* 5; *Anomianthus* 1; *Artobotrys* 100; *Schefferomitra* 1 (PNG); *Desmos* 25; *Friesodielsia* 50; *Melodorum* 5; *Pyramidanthe* 1; *Mitrella* 5; *Fissistigma* 60. In contrast, the Annonaceae are represented among Neotropical lianas by perhaps 3 species of *Annona*, and perhaps a few scrambling shrubs. Besides being species-rich, the annone climbers are also one of the most abundant liana families, which serves to raise their total abundance and species richness to higher levels in Asia than normally found in the Americas (Appanah et al. 1993).

Arecaceae. While the palm family is species-rich and abundant in most parts of the tropical world, their ecological representation in forests contrasts sharply among continents. In almost all American forests, palms are an important part of the lower canopy and collectively may comprise a large fraction of basal area. Lambir Hills is typical of Asian forests because palm trees are a minor component of the tree flora over 10 cm dbh. While the 52 ha plot includes nearly 1200 species of trees, no more than 25 are palms, and no palm is especially

Table 16.2. Comparison of tree composition in four lowland tropical forests: Pasoh Forest, Peninsular Malaysia; Lambir Hills, Sarawak Malaysia; Yasuní, Ecuador (based on re-calculations from data summarized in Valencia et al. 2004a,b, and LaFrankie et al. (in review); and Barro Colorado Island, Panama). Tallies are based on 50 ha samples (25 ha in Yasuní) but are presented as equivalent 25 ha values. Tallies are made for each species and are summed across families. Saplings are trees 1 cm to 2 cm dbh belonging to a species that exceeds 30 cm dbh. Treelets are trees 1 cm to 2 cm dbh belonging to species that do not exceed 30 cm dbh. Ratio in parentheses indicates saplings over 30 cm dbh recorded per adult tree.

	PASOH				LAMBIR				YASUNÍ				BCI			
	SPP	>30 cm	Sapling (ratio)	Treelet	SPP	>30 cm	Sapling (ratio)	Treelet	SPP	>30 cm	Sapling (ratio)	Treelet	SPP	>30 cm	Sapling (ratio)	Treelet
Total	827	2004	27540 (13.74)	34128	1194	3036	49189 (16.20)	29879	1104	1678	10644 (6.34)	42884	304	2133	21479 (10.07)	42741
Families of canopy trees more abundant in Asia																
Dipterocarpaceae	33	576	5952 (10.33)	5	88	1198	12627 (10.54)	101	.	.	.(.)	.	.	.	.(.)	0
Fagaceae	15	100	684 (6.84)	34	21	28	340 (12.14)	36	.	.	.(.)	.	.	.	.(.)	0
Ebenaceae	23	15	299 (19.93)	2706	35	37	1519 (41.05)	429	5	2	11 (5.50)	20	.	1	.(.)	7
Polygalaceae	10	23	725 (31.52)	2	25	10	157 (15.70)	1374	1	.	.(.)	1	.	.	.(.)	0
Dilleniaceae	3	4	4 (1.00)	37	5	10	494 (49.40)	352	.	.	.(.)	.	.	.	.(.)	.
Memecylaceae	12	3	19 (6.33)	1241	12	5	236 (47.20)	204	.	.	.(.)	.	.	.	.(.)	.
Anisophylleaceae	1	1	68 (68.00)	0	5	5	453 (90.60)	492	.	.	.(.)	.	.	.	.(.)	.
Alangiaceae	4	4	471 (117.75)	1	4	1	87 (87.00)	66	.	.	.(.)	.	.	.	.(.)	.
Ixonanthaceae	2	59	516 (8.75)	0	1	51	1781 (34.92)	0	.	.	.(.)	.	.	.	.(.)	.
Families of canopy trees more abundant in America																
Bignoniaceae	2	.	.(.)	0	.	.	.(.)	.	6	7	15 (2.14)	1043	3	79	90 (1.14)	0
Cecropiaceae	.	.	.(.)	.	.	.	.(.)	.	16	104	297 (2.86)	719	3	52	52 (1.00)	0
Araliaceae	1	.	.(.)	3	1	.	.(.)	4	4	13	35 (2.69)	294	3	39	11 (0.28)	0
Malpighiaceae	.	.	.(.)	.	.	.	.(.)	.	7	4	18 (4.50)	131	2	.	.(.)	25
Boraginaceae	2	0	5 (.)	0	1	0	75 (.)	0	9	4	.(.)	598	3	34	121 (3.56)	224
Nyctaginaceae	.	.	.(.)	.	.	.	.(.)	.	22	17	319 (18.76)	889	2	22	19 (0.86)	30
Hippocrateaceae	.	.	.(.)	.	.	.	.(.)	.	5	16	57 (3.56)	416	.	.	.(.)	.
Shared Families																
Euphorbiaceae	84	98	1999 (20.40)	6101	125	150	3926 (26.17)	7002	34	127	215 (1.69)	2697	11	81	580 (7.16)	592
Legumes																
Mimosoideae	9	23	344 (14.96)	51	9	5	20 (4.00)	163	54	156	984 (6.31)	2371	18	24	481 (20.04)	562
Caesalpinioideae	11	154	597 (3.88)	110	6	43	202 (4.70)	9	15	35	129 (3.69)	1582	6	78	1492 (19.13)	58
Phaseoideae	4	33	151 (4.58)	0	4	11	42 (3.82)	1201	32	48	176 (3.67)	243	14	93	1481 (15.92)	696
Dialium-group	4	31	262 (8.45)	0	5	14	112 (8.00)	48	1	7	41 (5.86)	0	.	.	.(.)	.
Bauhinia-group	.	.	.(.)	.	.	.	.(.)	.	2	.	.(.)	218	.	.	.(.)	.

Table 16.2. Continued

	PASOH				LAMBIR				YASUNÍ				BCI			
	SPP	>30 cm	Sapling (ratio)	Treelet	SPP	>30 cm	Sapling (ratio)	Treelet	SPP	>30 cm	Sapling (ratio)	Treelet	SPP	>30 cm	Sapling (ratio)	Treelet
Meliaceae	43	14	511 (36.50)	1297	55	13	636 (48.92)	932	39	106	490 (4.62)	2299	6	301	3891 (12.93)	1
Rubiaceae	47	15	688 (45.87)	4618	59	18	801 (44.50)	3952	80	58	464 (8.00)	2729	32	144	2654 (18.43)	9173
Annonaceae	42	38	647 (17.03)	4035	54	28	842 (30.07)	2915	40	16	173 (10.81)	1575	9	41	475 (11.59)	3928
Burseraceae	25	123	2203 (17.91)	141	40	234	4463 (19.07)	161	17	61	818 (13.41)	133	6	105	2540 (24.19)	215
Anacardiaceae	32	66	592 (8.97)	548	33	215	3291 (15.31)	453	7	36	187 (5.19)	14	4	36	49 (1.36)	0
Myrtaceae	48	75	897 (11.96)	506	58	186	2002 (10.76)	249	56	7	113 (16.14)	1321	7	14	938 (67.00)	427
Lauraceae	49	25	210 (8.40)	923	78	116	1879 (16.20)	918	81	119	1128 (9.48)	1872	10	114	1379 (12.10)	21
Clusiaceae	34	46	1341 (29.15)	536	50	80	1317 (16.46)	735	19	9	104 (11.56)	197	8	15	204 (13.60)	1517
Sapotaceae	15	30	119 (3.97)	268	33	61	538 (8.82)	398	54	38	469 (12.34)	903	5	41	663 (16.17)	0
Moraceae	25	51	299 (5.86)	33	45	60	595 (9.92)	486	51	90	1272 (14.13)	1173	21	106	593 (5.59)	1210
Myristicaceae	31	39	1884 (48.31)	472	40	58	1541 (26.57)	735	17	162	395 (2.44)	435	3	63	505 (8.02)	0
Verbenaceae	8	1	38 (38.00)	97	9	37	883 (23.86)	16	9	.	.	502	1	.	.	19
Chrysobalanaceae	7	35	113 (3.23)	20	8	18	89 (4.94)	1	19	23	281 (12.22)	177	4	27	1174 (43.48)	0
Apocynaceae	7	20	43 (2.15)	71	7	12	50 (4.17)	5	11	16	48 (3.00)	185	1	72	404 (5.61)	20
Lecythidaceae	4	8	734 (91.75)	1	8	10	161 (16.10)	44	15	154	520 (3.38)	306	1	18	29 (1.61)	0
Elaeocarpaceae	8	1	13 (13.00)	77	11	7	148 (21.14)	22	17	25	367 (14.68)	77	1	8	134 (16.75)	0
Combretaceae	4	1	0 (0.00)	2	3	3	7 (2.33)	0	9	13	33 (2.54)	14	2	18	16 (0.89)	0
Celastraceae	7	11	86 (7.82)	132	14	17	653 (38.41)	463	3	2	6 (3.00)	24	1	1	15 (15.00)	0
Sapindaceae	20	43	2918 (67.86)	1472	19	17	930 (54.71)	180	18	7	68 (9.71)	147	7	1	64 (64.00)	697
Malvaceae	27	88	652 (7.41)	492	47	113	2227 (19.70)	729	29	62	880 (14.19)	3183	16	334	436 (1.31)	147
Flacourtiaceae	12	15	640 (42.67)	89	20	26	1227 (47.19)	670	30	7	36 (5.14)	1428	14	52	212 (4.08)	896
Rutaceae	6	1	45 (45.00)	164	4	12	3 (0.25)	126	12	2	16 (8.00)	155	4	35	126 (3.60)	0
Ulmaceae	4	5	123 (24.60)	392	4	14	306 (21.86)	8	4	15	79 (5.27)	2	2	2	1 (0.50)	22
Olacaceae	8	17	86 (5.06)	384	7	14	548 (39.14)	164	9	19	81 (4.26)	110	2	8	157 (19.63)	21
Families of shrubs and treelets																
Monimiaceae	1	.	.(.)	5	1.		.(.)	11	17	.	.	898	2.	.	.(.)	89
Myrsinaceae	8	.	.(.)	3142	9.		.(.)	960	4	.	.	225	4.	.	.(.)	286
Melastomataceae	3	0	41 (.)	67	9.		.(.)	558	59	8	31 (3.88)	2069	13	2	173 (86.50)	3400
Piperaceae	.	.	.(.)	.					22	.	.	3144	8.	.	.(.)	1787
Solanaceae	.	.	.(.)	.					14	.	.	247	5.	.	.(.)	137
Violaceae	3	.	.(.)	1657	4.		.(.)	153	9	.	.	2644	2.	.	.(.)	14298
Acanthaceae	.	.	.(.)	.			.(.)		2	.	.	55	2.	.	.(.)	3
Capparidaceae	.	.	.(.)	.			.(.)		3	.	.	449	2.	.	.(.)	1149
Pandaceae	3	.	.(.)	456	1.		.(.)	257	.	.	.	.	1.	.	.(.)	.

numerous. However, we must add an important qualifier. Asian lowland forests are very often rich in caespetose palms with large fan leaves—the genus *Licuala* foremost among these. *Licuala* is often abundant to the point of local physical dominance. Individual leaves can be 2 meters across, and the 50-ha plot in Pasoh included more than 17,000 individuals. Furthermore, in most humid forests of the Sunda Shelf we find palms richly and abundantly represented among climbing plants. More than 20 species of rattans occur at Lambir (K. Ickes, unpublished report). While noting the relative absence of large tree palms in Asian forests, we should bear in mind that palms may nevertheless show a globally identical presence in terms of leaf area and/or biomass.

Bignoniaceae. In American tropical forests, Bignoniaceae can become relatively abundant both as trees and as lianas. The family is poorly represented in Asian wet forests. In Pasoh, among the trees were three species in two genera, but these were represented by only a few individual saplings, all of which had died before the 1995 census. In the Lambir Hills plot the family did not appear at all.

Dipterocarpaceae. Dipterocarps are ecologically unique and have no equivalent in America (see Plates 2, 5). They represent the most profound ecological divergence among the continents. (Technically, the family is represented in the Americas by two species of subfamily Pakaraimaeoideae, but it is absent from most of the land area and absent in all lowland forests.) The family chiefly comprises tall trees of large diameter—although most genera include a few species of small stature—and they collectively represent a large fraction of the basal area of forests in the lowland wet areas of the Sunda Shelf. The plant body is highly resinous, the leaves pinnate-veined, evergreen, the roots ectomycorrhizal, associated with a rich basidiomycoflora; the flower is variously sized, bell-shaped and malvalean in form, pollinated by insects, presumably obligately outcrossing (Ashton 1989); the single-seeded fruit, variously sized from very large to small, is either non-dispersed, falling in dense clusters around the mother tree, or dispersed opportunistically via thick calyx wings during rare, high winds, or perhaps dispersed secondarily by animals through scatter hording; the phenology is exemplary of the exaggerated masting habit, with three to five or more years of vegetative growth followed by species-specific synchronous flowering (Ashton et al.1988; LaFrankie and Chan 1991); the saplings are highly aggregated. The geography is also indicative of the ecology. In mixed dipterocarp forest of the Sunda Shelf, individual species often have a restricted distribution. For example, the most abundant species of *Shorea* and *Dipterocarpus* in the 52 ha plot at Lambir are *Shorea acuta* and *Dipterocarpus globosus*, species that are not widespread. Among the five most abundant species of Dipterocarpaceae in Lambir and Pasoh forests, we find no species in common (see Okuda et al., editors, 2003).

Fagaceae. The oaks and chestnuts often dominate forests in the high latitudes. While a few species of *Quercus* are found in the Mesoamerican mountains, where they sometimes dominate (Guariguata and Saenz 2002), they are essentially absent from the Neotropics below 1000 meters, or south of Panama. In contrast to their absence in tropical America and Africa, Fagaceae play a significant role in the lowland forests of Asia. The Lambir plot includes 21 species, while the plot at Pasoh has 15 species, thus representing the ninth-ranked family in basal area (Kochummen et al. 1990). They share with dipterocarps a masting habit and produce large and essentially non-dispersed seeds, presumably subject to scatter-hording by squirrels and small rodents.

Lecythidaceae. This family is easily recognized in all tropical forests by the candelabra of branches bearing dense, spiraling rosettes of oblanceolate leaves, with the twigs often hollow. Lecythids are typically a minor element in the lowland forests of Asia, and also in much of America, but collectively are near dominant in the more seasonal habitats such as near Manaus, Brazil. However, America and Asia are represented by different clades, with different ecologies (Morton et al. 1997). In Southeast Asia the family is represented by the Barringtoniae. The flowers are open powder puffs. Some riparian and littoral species have red flowers opening during the day, but species of the closed forest are all (or, almost all) night blooming, presumably pollinated by bats or large moths. The fruit is invariably indehiscent and fleshy. The following genera and species are typically found: *Barringtonia* 40; *Careya* 4; *Chydenanthus* 1; *Petersianthes* 1; *Planchonia* 5. In America the flowers of Lecythidaceae are diverse in symmetry and the fruit varied, but especially abundant are those with dehiscent capsules, seeds winged or not, or arilate or not: *Gustavia* 40, *Grias* 6, *Asteranthos* 1; *Allantoma* 1; *Cariniana* 15; *Couroupita* 3; *Bertholettia* 1; *Couratari* 19; *Eschweilera* 83; *Lecythis* 26. Lecythidaceae is represented in Lambir by *Barringtonia* and in Yasuní and BCI by genera such as *Gustavia*. While the trees can be recognized as con-familial, the reproductive ecology differs sharply and, in light of the great difference in sapling abundance, the population ecology also likely differs.

Melastomataceae. The family Melastomataceae is nearly cosmopolitan, but recent phylogenetic work (Renner et al. 2001), together with the compositional data from the Center for Tropical Forest Science/Smithsonian Tropical Research Institute large plot network, make clear the strong asymmetry of familial representation. Asian forests are most richly represented by *Memecylon*, which forms a sister clade of the rest of the melastomes. It is the latter that are so well represented in the understory of the Americas. *Melastoma* itself is found in Asia but typically in gaps and wet, open forest. The tribes Kibessieae and Astronieae—the basal-most clade within the melastomes—are sometimes represented, but chiefly at altitudes approaching 1000 meters. The other true melastomes found in Asia include some abundant herbs of the forest floor, such as *Sonnerlila* and *Phyllagathis*. Nowhere in Asia do we find an assemblage of shrubs and treelets comparable to the family's representation in BCI and Yasuní (see Table 16.2).

Piperaceae. Pipers can be found all over Asia, but chiefly as small plants <1 meters tall, or, more often, as weak-stemmed climbing plants. More than one

million trees have been recorded in the CTFS Asian plot network, but not a
single individual of *Piper* has been found. This is in stark contrast to the rich
diversity and abundance of *Piper* on BCI, and in most of wet tropical America.
La Selva is perhaps the premier location for *Piper*, with more than 40 species
of shrubs and treelets.

Sapindaceae. This family is a vexing confusion of small trees. The family is
more often than not represented by the most abundant species at each of the
CTFS-AA forest sites: Pasoh (*Xerospermum noronhianum*); Huai Kha Khaeng,
Thailand (*Dimocarpus longan*); and Palanan, Philippines (*Nephelium ramboutan-
ake*). In Lambir the family is best represented by the many species of rambutan
of the genus *Nephelium*. The American tropical forests sometimes include a few
abundant trees in the Sapindaceae, but they are far more important as lianas,
where they typically rank as one of the most diverse and abundant families.
Sapindaceae are not represented among lianas in Asia. Perhaps associated with
the contrasting habit, the American Sapindaceae that grow as lianas are primarily wind-dispersed, whereas the trees of wet Asian forests almost all bear
seeds with fleshy arils, which attract animals as dispersers.

16.4 Other Ecological Elements of Divergence

No pretense is made to formally review a subject so large and involving such a
heterogenous mass of complex data. Rather, the intention is to note contrasting
ecological features that have recently come to light, through either the CTFS
plot program or other work.

Lianas. Family composition of lianas differs between Asia and America. Other
than legumes, which are among the most species-rich and abundant families in
both hemispheres, the tropical forests of Old and New Worlds share few dominant liana families. In Asia the dominant families are Annonaceae, Arecaceae,
Connaraceae, Celastraceae and Icacinaceae (Appanah et al. 1992; Putz and Chai
1987). The composition of lianas at BCI (Foster and Hubbell 1990), La Selva
(Hammel 1990), and Manu, Perú (Foster 1990) is fairly uniform; dominant families are Bignoniaceae, Sapindaceae, Malpighiaceae, Dilleniaceae, Aristolochi-
aceae, Cucurbitaceae, Menispermaceae, and Passifloraceae. The contrasting
taxonomy is not as important as the contrasting ecology. Whereas in America
the bulk of liana seeds are wind-dispersed, in Asia the majority are fleshy fruited.

Understory treelets. LaFrankie et al. (2002) compared the large permanent forest plots at Lambir and Pasoh with Yasuní in Ecuador and Korup in Cameroon.
They found profound differences in the understory composition, regarding the
proportions of tree species of large and small stature. In the understory of the
American and African forest, roughly 70% of the small trees belong to species
that reach a maximum diameter of less than 10 cm dbh. By contrast, in the two
Asian forests nearly 60% of understory trees are saplings of canopy trees that

reach a maximum diameter of over 20 cm. Florisitc information from other Neotropical sites reinforces this picture. The shrub flora of La Selva, Costa Rica (Hammel 1990), and Manu, Peru (Foster 1990) are similar. Some abundant shrubs and treelets are of families also abundant in Asia. These include Rubiaceae, Myrtaceae, Myrsinaceae, and Arecaceae, but their abundance is often less in the Asian forests. However, the rich and abundant families of Piperaceae, Solanaceae, Acanthaceae, and Melastomataceae are poor or absent in the Asian forest understory.

Canopy saplings. While the Asian understory is generally depauperate in shrubs and treelets, it is very rich in canopy tree saplings. This appears to be a general phenomenon. The contrast with the Neotropics appears to be quantitatively significant and taxonomically broad (see Table 16.2). Lambir is more than 25% richer than Pasoh in saplings among all trees over 30 cm dbh, but it is 60% richer than BCI and 250% richer than Yasuní. The result is not solely due to families of canopy trees exclusive to Asia, and the difference persists even without Dipterocarpaceae. The nutmegs (Myristicaceae) are instructive. The family is well represented among trees over 30 cm dbh in all four forests, but the two Asian forests have 26 saplings per adult at Lambir, and 48 at Pasoh, versus 8 at BCI and only 2 at Yasuní. A similar trend is seen in the Annonaceae. Additional data from other Neotropical sites are few, but studies that are somewhat comparable tend to show similar results. For example, Clark and Clark (1992) surveyed 150 ha for six emergent tree species at La Selva, Costa Rica. All six species had more trees above 10 cm dbh than the number in the class of 1 cm to 4 cm dbh. While the abundance of those trees over 30 cm dbh would place them within the 50 most abundant species at Lambir, the number of saplings 1 cm to 4 cm dbh is nearly ten times less than any such species at Lambir. This finding suggests a profound and general divergence between continents, in the nature of canopy tree regeneration.

Termites. Davies et al. (2003) examined termite communities in tropical forests of the world and found large inter-regional differences in the ecological composition, suggesting large differences in related ecosystem processes at the soil surface.

Mammals. Eisenberg (1981) saw in the small non-volant mammals a clear pattern of convergence among trophic guilds between Asia and America. Nevertheless, there is also evidence of strong divergence among mammals. Bats typically have two to three times more species in Neotropical than in Asian forests, and typically include a large fraction of strictly frugivorous species. Primate ecology also shows strong divergence among continents. Species richness exhibits a strong positive correlation with the area of tropical forest on each continent (Reed and Fleagle 1995). However, while in South America, Africa, and Madagascar, species diversity shows a strong positive correlation with mean annual rainfall, no such relation exists within Asia. Kappeler and Heyman (1996) examined convergence of life history traits of primates. They found basic aspects of primate life history varied significantly among the continents. New World primates are significantly smaller than primates in other regions and lack species larger than about 10 kg; only in Asia do we find strictly frugivorous primates. Asia lacks primarily sap-feeding primate species, whereas the Neotropics lacks primarily folivorous primates, nor do the Americas have solitary species.

Birds. Lists of birds cannot always be sensibly compared, because birds are highly mobile and use various parts of the landscape in different ways. A so-called resident species can mean different things to different observers. Nevertheless, tropical forests sites in America appear to have on the order of 30% more resident bird species than comparable forest in Asia. Robinson and Terborgh (1990) cite one of richest Neotropical locations having 239 resident species, while Pasoh, Malaysia, with 180 species, is 75% of that Neotropical locale (Francis and Wells 2003).

Continental differences in avifauna are perhaps more important than differences in diversity, with regard to abundance and trophic organization. Frugivores make up about 30% of the species in the Neotropics, and they are almost always among the most abundant species caught in mist nets (Robinson and Terborgh 1990). On a per gram basis, frugivores might make up a majority of the avian biomass in the Neotropical forest understory. In contrast, perhaps no bird of the Asian tropical forest understory is an obligate frugivore. Similar differences are likely for nectivorous birds. Although sunbirds and spiderhunters are often considered Asian ecological equivalents of Neotropical hummingbirds, the former are far less diverse, less abundant and evidently vastly less important as pollinators.

Dispersal ecology. Ingle (2003) found that many elements of dispersal ecology in montane forest in Mindanao, Philippines, were contrary to findings that appear uniform across the Neotropics (e.g., Foster et al. 1986; Uhl 1987; Gorchov et al. 1993, 1995; Medellin and Gaona 1999). For example, Ingle found that small wind-dispersed seeds dominated seed traps. Birds dispersed more seeds and species of successional plants than did bats. Ingle emphasized that, contrary to the common view that Neotropical and Asian fruit bats are ecological equivalents, Asian fruit bats belong to a separate suborder and cannot echo-locate. Furthermore, the frugivorous bird communities of Asia and the Neotropics are taxonomically distinct. This point was clearly made by Corlett (1998), who states that "On current evidence, it appears that most seeds in the Oriental Region . . . are dispersed by vertebrate families which are either endemic to the region . . . or to the Old World." Thus, dispersal ecology of the Old and New World Tropics has been evolving along independent lines for millions of years.

Phenology. A great deal is still uncertain about phenological patterns in Asia, especially as one leaves the ever-wet equatorial forests. Nevertheless, it is increasingly clear that Asia is characterized by supra-annual masting (Curran et al. 1999, Sakai et al. 1999c, Chapter 4). American tropical forests are characterized by a diversity of phenological habits, with a large proportion of species that flower annually (Bawa et al. 2003).

The comparative chemistry of Asian and American tropical forest trees is poorly investigated and yet suggests today (as it did to Janzen almost 30 years ago) an area of major divergence among the ecosystems. Contrasting sugars in flowers and fruit, and relative lipid content, could also vary both in absolute quantity and temporal and spatial pattern, in tropical Asia compared to America.

16.5 Lambir Hills Compared to La Selva

These miscellaneous notes can be summarized by a synoptic comparison of Lambir Hills forest with the well-studied wet forest at La Selva, Costa Rica. About half the basal area at Lambir is composed of some 80 species of Dipterocarpaceae, none of which dominates, and the most abundant are geographically restricted and of patchy occurrence. At La Selva, half the basal area is composed of *Pentaclethra macroloba* (Mimosoideae) and three species of palms; all four species are geographically widespread. Palms are well represented at Lambir, but not as trees. Instead, they appear as numerous, stemless understory plants and as abundant climbers. The lianas as a class differ sharply between forests. Other than legumes found in both forests, woody climbers at Lambir are chiefly palms, Annonaceae and Icacinaceae—all of which bear fleshy fruit. At La Selva the principal lianas are Bignoniaceae, Sapindaceae, and Malpighiaceae, most having dry, wind-dispersed fruit and seeds. At Lambir the plants in the understory that have 1 cm to 2 cm dbh include 2000 to 3000 individuals in a hectare, of which nearly 70% are the saplings of trees that will exceed 10 cm dbh at maturity. The similar size class at La Selva is much more sparse, and as far as current evidence shows, more than half the individuals are of species that will flower and bear fruit at tree sizes of less than 10 cm dbh. Related to that compositional difference is a difference in abundance of canopy saplings.

Large emergents at La Selva are about as evenly abundant as emergent species at Lambir, but the number of saplings of such species at Lambir is three to ten times greater than their counterparts in La Selva. Lambir Hills has forest that lies almost entirely sterile for many years before displaying a general flowering, whereas La Selva includes a majority of species that flower and fruit more frequently. Pollinators at La Selva appear to be far more diverse with a higher fraction of bird, bat, and species-specific long-distance pollinators. In contrast, the plants at Lambir (with some exceptions) are predominantly pollinated by less-specific fauna of beetles, Lepidoptera, and bees (see Corlett 2004). The birds and bats at Lambir are about 60% as rich as the community at La Selva and include few if any obligate frugivores.

Thus, while Lambir Hills in Sarawak and La Selva in Costa Rica are both lowland wet tropical rain forests of comparable physiognomy, rich diversity, parallel families, a few shared genera, and many examples of convergence, they nevertheless are profoundly different forests. Almost every aspect of ecological dynamics differs between these American and Asian tropical forests. A better description and understanding of such differences will impact basic ecological conundrums as well as influence management decisions aimed at conservation or timber.

17. Lambir's Forest: The World's Most Diverse Known Tree Assemblage?

Peter S. Ashton

17.1 Reminiscing: How Lambir Hills Became a Site for Researching Forest Ecology

If Lambir Hills National Park (LHNP) were to be dedicated to anyone, it should be to Ian Urquhart. I first met Ian in 1953 at Cambridge, where we were both undergraduates. I was seeking to join the Explorers Club, which he had revived from wartime quiescence after an expedition with two colleagues to one of the driest places in the world: the Danekil country and Lake Thana in Northwest Ethiopia. So it was a surprise on arriving 10 years later in one of the wettest countries, Sarawak, to find Ian with his Sarawakian wife happily settled into the administrative service. Tragically, Ian's career was cut down by cancer, but not before he had had a chance to contribute to the wise stewardship of what was then Sarawak's Fourth Division, with its capital at Miri, where Ian served as Resident. It was from there that he sought my opinion, in 1965, concerning the biological case for conserving at least that part of the then Lambir Forest Reserve, which included the water catchment for the city of Miri. He was a keen outdoors person and naturalist, knew of the waterfall, which became the park's main attraction after the road was built in the 1970s, and had scaled the steep defile to the summit trigonometric survey marker on Bukit Lambir. He had the vision to foresee the hills as a recreation area for local residents. Little did he know that the forest at Lambir Hills, with perhaps 2500 species, is among the richest areas in tree species for its size and almost certainly the richest in the Old World.

My first botanical foray into Lambir Forest Reserve was as Forest Botanist with Joseph Au and forest herbarium staff in 1962. At that time the reserve covered the whole area from the edge of the kerangas, on the Pleistocene raised beach that is now under the Miri airport, west almost to the banks of Sungei Sibuti, and south to a track that had been opened up as far as the Riam and Bulak Setap oil-drilling sites, now under shifting cultivation and oil palm plantations. Access from the north was from the coastal Bakam jeep trail, from which the timber concession of the Yong Khaw company could be entered, or from the south, after a perilous slippery journey assisted at times by winch, along the so-called Riam Road, which later became the main road to Bintulu and ultimately Kuching!

I used to choose priority sites for exploration from the excellent maps of surface geology prepared by the geological service and brought together in the invaluable monograph by Liechti and others (1960). Lambir Hills had high priority for me because only here do the mainly sandstone sediments of the Neogene syncline directly abut, and conformably overlie, the slightly older Setap shales of Northeast Sarawak (see Fig. 17.1).

Similar, though younger, sandy sediments found at Lambir Hills also underlie the Andulau Forest Reserve to the east in Brunei, whose forest was a focus of ecological study when I worked there four years earlier (Ashton 1964; Austin et al. 1972). At that time, I found the Andulau forest to be exceptionally rich in tree species and northwestern Borneo endemics. Further, two-thirds of the species I censused in fifty 0.4 ha plots at Audulau, common as well as rare, were not encountered at my second site 100 kilometers to the east on the Setap shale at Kuala Belalong. I concluded that the contrasting floristics of the two forests was due to their contrasting soils: those at Andulau are deep, sandy, low in nutrients, and have a surface raw humus horizon of varying depth, whereas Kuala Belalong soils are clay-rich, truncated and shallow, and lacking in surface organic matter but higher in mineral soil nutrients. Critics argued that the differences could be due to the distance between sites and the influence of limited seed dispersal, rather than to any habitat differences. Resolution of that problem has important implications for forest mapping, management, and silviculture. The immediate juxtaposition of the two habitats in Lambir Hills provided the means for a test.

Lambir Hills is predominantly sandstone, but the interface to the Sibuti Member of the Setap Shale Formation occurs just within the southern boundary of the former forest reserve, and the current national park. My initial exploration on the younger rocks north of the main ridge of Bukit Lambir, itself of relatively hard Lambir Formation sandstone supporting humic podsols and kerangas forest, revealed gentle topography and deep sandy soils as at Andulau. The tree flora had more kerangas elements, implying lower soil nutrient status and higher sand content. South of the Bukit Lambir ridge, the topography is more rugged than at Andulau and the soils overall somewhat richer in clay, but the flora are strikingly similar. The Sibuti Member is considered transitional between the predominantly sandstone Lambir Formation and the predominantly shale Setap Shale Formation, so that the change from sands to clays is generally gradual. Fortu-

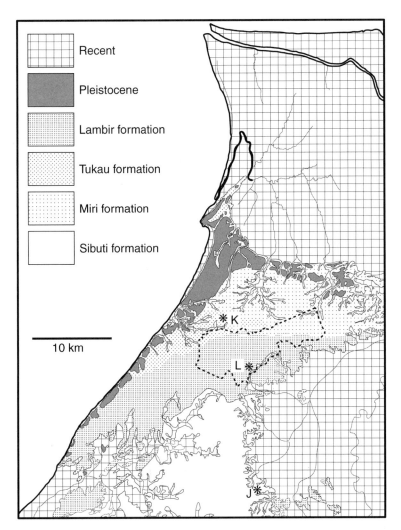

Figure 17.1. Geological map of Lambir Hills and adjacent area (modified after Liechti et al. 1960). *Thick dotted line:* area of LHNP; note isolation of podsol raised beaches (Pleistocene, *shaded*) and sandstone areas (Lambir, Tukau, and Miri formation, *dotted*) by the peat swamps (recent, *hatched*) to north and east, and by Sibuti shale (Sibuti formation, *unshaded*) to south. Approximate positions of the 1963–65 plot clusters are indicated by letters.

nately, though, one kilometer to the west of where the park headquarters now stand along the former Riam road, I found a ridge leading north to Bukit Lambir along which the transition is quite sharp. I returned with the ecological team of the Forest Department Sarawak, led by Kuching herbarium curator Ilias bin Pa'ie. Altogether twenty-five 1.5 acre (0.6 ha) plots were laid out in primary mixed-dipterocarp forest in the Lambir Hills area, and trees greater than 12 inches in girth (9.6 cm diameter) were censused: 4 in the Yong Khaw concession north of Bukit Lambir on leached sandy humult ultisols (see Fig. 17.1, site K) and 6 on the aforementioned ridge south of Bukit Lambir, one of which was on clay udult ultisols over shale and the rest on sandy humult ultisols (site L). Because of the very small area of shale within the forest reserve, a further 5 were placed on clay udult ultisol soils over the Sibuti shales in the headwaters of the Bakong river which drains the southern flank of Bukit Lambir, which is now under cultivation (site J), and 10 farther south again on the same soils in the then Bok-Tisam Protected Forest, now under oil palm cultivation (see Fig. 17.2, Site E). Each plot cluster, north to south, was 10 km to 15 km apart over a total transect of 50 km. Soils were sampled and analyzed (see Chapter 3).

Analysis of the floristic data from these and 80 other such plots, situated between Lambir Hills and the Santubong peninsula in western Sarawak over a distance of 500 kilometers, has set the forests at Lambir Hills in a regional context (see Fig. 17.2; Potts et al. 2002). It showed the preeminence of soil

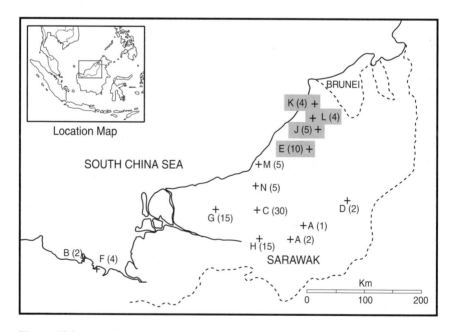

Figure 17.2. Map of northwest Borneo, indicating site locations of 105 mixed diptero-carp forest plots (Potts et al. 2002). Numbers in parentheses indicate number of plots at each site. Lambir Hills plot sites are highlighted.

factors over distance (and, by inference, dispersal constraints) in determining the floristic composition of mixed dipterocarp forest communities in the heterogeneous dissected tertiary landscape of NW Borneo (see Fig. 17.3).

Cluster analysis divided all 105 plots first into groupings of those on humult soils and those on udult soils, or associated humults at higher altitudes but on

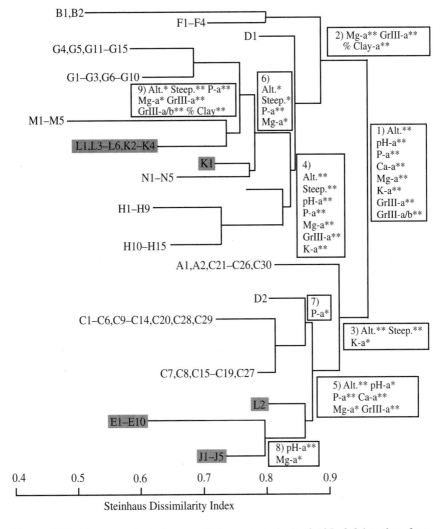

Figure 17.3. Cluster analysis (average-linkage clustering) of 105, 0.6 ha plots from mixed dipterocarp forests of Sarawak and East Malaysia (from Potts et al. 2002). Capital letters of plots in dendrogram indicate sites (see Figure 17.2), mineral soil (concentrated HCl extractable) and topographic factors significantly (*P<0.05; **P< 0.01) related to numbered divisions are listed in boxes. Lambir Hills plots highlighted (after Potts et al. 2002).

the same lithology, irrespective of geographic location. Plots sharing the same locality were mostly grouped together in a subclass within one of these two groupings. Plots within one locality were at least 200 meters apart but the localities of plot clusters were at least 50 kilometers apart—the four separate clusters within the Lambir Hills complex and two sets in western Sarawak excepted. Plots within each cluster (the southern Lambir Hills cluster excepted) also shared the same lithological substrate, which correlated with mineral soil chemical variables, and sometimes with topography. The plots from the Lambir Hills region exemplified the overall pattern. The Yong Khaw and, with one exception, the southern Lambir Forest Reserve plots were in the humult grouping, and all but one Yong Khaw plot shared the same subclass. However, the southern Lambir Hills plot on udult clay soil over Sibuti shales—notwithstanding that it was within 300 meters of the nearest plot on humult soil—was placed in the udult grouping in a subclass comprising it, the Bakong, and all but one of the Bok Tisam plots, which all shared the same substrate. The remaining Bok Tisam plot was on soils of higher sand content and was placed in the humult grouping, associated with plots from the Rejang sedimentary formation of central Sarawak.

17.2 Biogeography of Lambir Hills

Northwest Borneo nevertheless contains the most phyto-geographically fragmented region of the Sunda lowland rain forest. That forest covers the southeastern-most part of Peninsular Thailand, Peninsular Malaysia, Sumatra, western Java, Borneo, and the islands in between (Ashton 1972, 1992, 1995; Wong 1998). The inland hills of northwestern Borneo harbor the widespread Sundaland flora and include distinct elements of three local provinces. Those comprise Sarawak west of the Lupar Valley and western Kalimantan north of the Kapuas river and west of the Kapuas lakes, which also contains a distinct Peninsular Malaysian element (see Fig. 17.4A), the central Borneo hills north to the Rejang drainage, and the Sarawak coastal hills east of the Lupar and west of the Niah rivers (see Fig. 17.4B), and Sabah east to Temburong District, Brunei and south to the Sankulirang peninsula of eastern Kalimantan, including a strong Philippine element (see Fig. 17.4C). Northwest Borneo also includes the richest component of a coastal hill and swamp flora first identified by Corner (1960) as The Riau Pocket (see Fig. 17.5). The area of that floristic province is now recognized as the coastal hills of Perak State, Peninsular Malaysia, some offshore islands of Sumatra, notably Simeuluë to the northwest and the Riau and Lingga Archipelagoes south of Singapore, and northwest Borneo from the mouth of the Kapuas River northeast through western Kalimantan and Sarawak west of the Lupar, and coastal hills from the Oya and Mukah rivers in central Sarawak to Labuan island and Beaufort Hill in southwest Sabah (see Fig. 17.6). The Lambir Hills complex is seen to be pivotal, because it contains the best remaining example of the Riau Pocket flora, and because it is in the transition zone

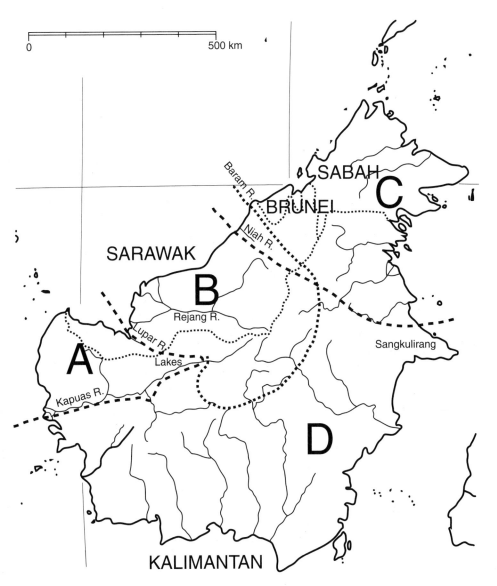

Figure 17.4. The four principal floristic provinces of Borneo and their approximate boundaries. The boundary between provinces A and B follows the Lupar Valley, but others are less sharply defined.

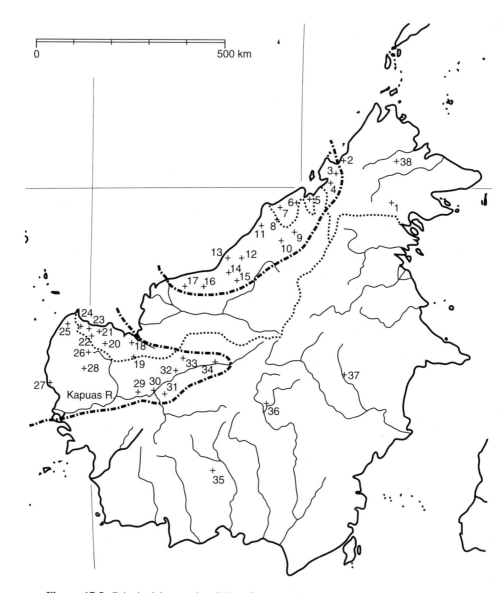

Figure 17.5. Principal known localities of tree species associated with leached sandy clay and sandy soils, based on those in which more than 3 of these soil specialist dip-terocarp species occur. Localities within the Riau pocket, north of the Kapuas River valley (delineated) include the most species (38), of which 24 are endemic to this province of the Riau pocket, and the richest localities (localities 20 Semengoh, 16 Iju Hill, 11 Lambir Hills, 7 Andulau Hills, each with 28 to 32 species); south of the line there are fewer species (12) and no endemics; maximum known richness to the south and east is approximately 8 species (locality 36, Ulu Barito). Lower locality density and richness in Kalimantan only in part reflect lower recording intensity (personal observation); the current status of these sites is unknown, but few likely remain undisturbed.

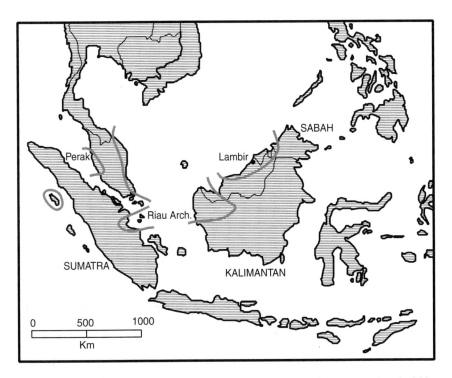

Figure 17.6. The Riau pocket floristic province of West Malesia (areas enclosed within heavy lines on northwestern Borneo, Peninsular Malaysia, and Central Sumatra).

Figure 17.5. *Continued*

Numbers on the map indicate the following localities:

1 Meliau basin (intact); 2 Papar (degraded); 3 Beaufort Hill (degraded); 4 Mengalong forest reserve Sipitang (logged); 5 Biang hill (intact); 6 northern Ladan hills (partially intact); 7 Andulau f. r. (logged); 8 Labi hills (partially intact), Puan (degraded) and Teraja hills (partially intact); 9 Mulu National Park; 10 Batu Belah (logged); 11 Lambir Hills N. P.; 12 Similajau f.r.(degazetted); 13 Nyabau f.r. (degazetted); 14 Segan f.r. (degazetted); 15 Ulu Minah, Kakus (logged); 16 Iju hill (deforested); 17 Ulu Kenyana, Mukah (converted); 18 Ulu Sebuyau (remnants); 19 (Klingkang range (remnants); 20 Semengoh f. r. (10 ha remaining); 21 Kubah N.P. (Mt Matang) (partially intact); 22 Undan hill, Bau (deforested); 23 Meroyong hill (partially intact); 24 Sebandar, Tamin hills, Lundu (remnants); and in Indonesia (status uncertain) 25 Paloh, Sambas; 26 Sanggau; 27 Mempawah; 28 Ngabang, Landak; 29 Sekadau; 30 Sintang; 31 Catit, Melawi; 32 Semitau; (33 Kapuas lakes); 34 Putussibau; 35 Lebak, Sampit; 36 Ulu Barito; 37 Long Bleh, Kahayan; 38 Ulu Telupid ultrabasics.

between the Sabah and the central Bornean floristic provinces—and includes elements of each.

Early in my field studies (Ashton 1964), I observed that the tree floras on different soils manifested distinct regional geographical ranges and distributions. The Lambir Hills flora of deep sandy humult ultisols, deep podsols on Pleistocene raised beaches, and peat swamps is strikingly concentrated in the Riau Pocket. These are the soils with surface acid raw humus that form over substrates low in nutrients and under de-oxygenated, waterlogged conditions, or on freely draining soils prone to periodic drought. The flora of all three habitats reaches its zenith of richness in Asia on the low coastal hills and peat swamps between the Oya River in Sarawak and the Andulau Hills and Belait swamps of Brunei. Host-specific phytophagous arthropods and microorganisms may be expected to share a similar biogeography. Lambir Hills is the jewel in this botanical crown, albeit sadly tarnished by the loss of the rich kerangas forest on the raised beach that now underlies Miri airport, and by the logging of peat swamps immediately to the southeast.

17.3 Establishment of a Large-Scale Ecological Dynamics Plot at Lambir Hills

In 1988, Takuo Yamakura spent a month at Harvard, ostensibly to consult early Japanese forestry and ecology literature in the Arnold Arboretum Library. We became friends, and the plan to jointly establish a large tree demography plot, following the protocols established by Stephen Hubbell and later formalized by the Center for Tropical Forest Science, resulted. With funding from the U.S. National Science Foundation and the U.S. Agency for International Development to Harvard, and the Japanese Ministry of Education (Monbusho) to Osaka City University, the plot was initiated following formal agreement with the Sarawak Forest Department on 5 December 1992. Two re-censuses have now been completed at five-year intervals following the first census. The 52 ha plot was carefully located to include maximum edaphic and topographic diversity. Yamakura and I, with the assistance of Pamela Hall, selected a boundary to include the ecotone between the udult clay and the humult sandy ultisols, with uniform areas of each, as well as the gentle but dissected dip slope and steep landslip-prone scarp slope of the same ridge south of Bukit Lambir along which the earlier 1.5 acre plots had been placed. Some landslips occurred there during the record rainfall of December through March 1963 (see Fig. 17.7). The method of survey and census is detailed by Manokaran et al. (1990) and Condit (1998).

The establishment of the 52 ha plot has enabled detailed empirical and experimental tests of the causes of local floristic patterns in hyperdiverse plant communities. Already, Lee et al. (2002) have shown that the individual floristic communities on specific soils are of a richness similar to the mixed dipterocarp forest of the Peninsular Malaysian lowlands. They also found that two-thirds of the tree flora is specific either to udult or to humult soils, the same proportion

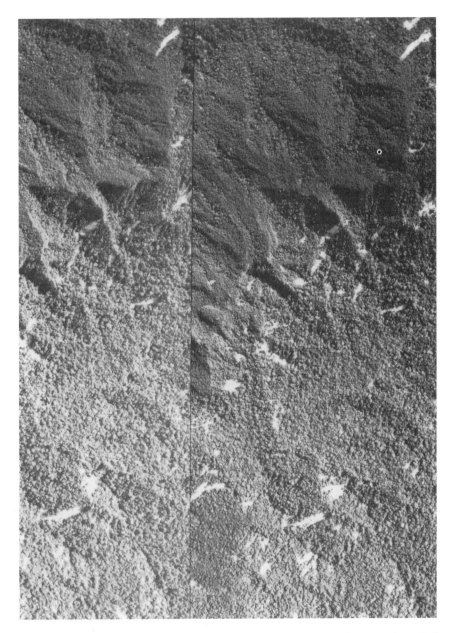

Figure 17.7. Sandstone dips and scarps in the Lambir Hills, site of plots 1 through 6 (see Fig. 17.5). Photographed in 1964 after exceptional rainfall, note landslip scars. Scale 1:25,000.

as estimated by me (Ashton 1964) on a comparison of plots on these soils 100 kilometers apart in Brunei. The opportunity for chance emigration across the ecotone therefore seems seldom to occur. Because tree species of low nutrient soils may be grown with ease as specimens on high nutrient soils in botanic gardens, or on the basic volcanic soils of Kebun Raya Indonesia where even peat swamp species flourish, the failure of humult ultisols specialists to invade the udult soils of the plot implies exclusion by stringent competition.

Tree species richness in the mature phase of mixed dipterocarp forests of northwestern Borneo peaks along the mineral soil nutrient gradient, although skewed toward low nutrient concentrations and within the humult ultisols. This pattern exists at the local scale of the Lambir Hills complex (see Fig. 17.8; Ashton 1998).

However, Lee et al. (2002) showed a different pattern of richness among 1

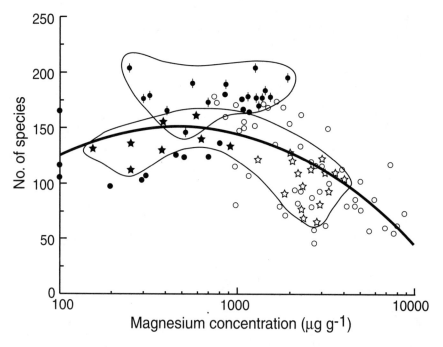

Figure 17.8. Patterns of species richness in mixed dipterocarp forests of northwestern Borneo in relation to total mineral soil Mg concentration (at 20 to 30 cm depth), as represented in 105 plots, each of 0.6 ha. *Solid symbols*: plots on humult ultisols; *open symbols*: plots on udult ultisols and inceptisols; Encircled plots include: *closed circle with vertical bar*, the Bukit Iju rhyolite; *starred plots*, located in the Lambir Hills complex. *Closed star*, Bukit Lambir sandstone; *open star*, plots on the zonal udult soils of the Lower Tinjar and confluent southern Lambir Hills. Iju rhyolite and Lambir sandstone represent ecological islands of ca. 40 km² and ca. 800 km², respectively (modified from Ashton 1998).

ha squares within the 52 ha plot, where richness is highest on steep slopes and ecotones, and lowest on gentle topography both on humult and udult soils. The 105, 1.5 acre plots upon which the general patterns were based were carefully situated to avoid canopy gaps. The pattern of richness within the 52 ha plot indicates that richness due to dynamic or edaphic heterogeneity *within* samples overrides soils-correlated patterns of richness *between* homogeneous samples of the mature phase of the climax forest. Though there appears to be no difference between the richness of mixed dipterocarp forest on udult and humult soils within the 52 ha plot, this is probably because the forest on udult soils, being adjacent to the cleared margin of the trunk road, is exceptionally prone to canopy disturbance, and therefore exceptionally rich in pioneer and early successional species.

Four of the earlier 1.5 acre plots south of Bukit Lambir, including the plot on udult clay soil, have been re-censused at five-year intervals since 1963, with four in Bako National Park, western Sarawak, on very low nutrient sandy humult soils, and five on nutrient-rich andosols over Tertiary basalt at Bukit Mersing, central Sarawak. Ashton and Hall (1992), who analyzed the dynamics within them during the first 20 years, showed a relationship between vertical structure, dynamics, and richness, which is likely to apply within the 52 ha plot and elsewhere at Lambir Hills. They failed to find significant differences in the mean diameter growth rate between mature phase stands on different soils. However, maximum individual growth rates were positively correlated with mineral soil nutrient concentrations; these were manifested by pioneer and successional species, which increasingly entered those mature-phase samples over time. Those species were themselves restricted to particular soils and substrates. Davies et al. (1998) also found that 11 pioneer species within *Macaranga* (Euphorbiaceae) in the Lambir Hills 52 ha plot each have specific soils ranges, and that species of udult soils have faster growth rates than those on humult soils (Davies 2001).

Canopy gaps on udult soils were relatively large, mostly created by group mortality after windthrow, whereas canopy trees on humult soils more often die individually as standing trees (Ashton and Hall 1992; Gale 1997). Differences in the growth rates of successional trees in canopy gaps appear to explain the striking differences in structure between forests on udult and humult soils. Plot census data showed that forests on udult clays have a higher density of very large, presumably emergent trees, mostly dipterocarps, and a lower density of understory trees. Six profile diagrams of forests were prepared on udult soils and 10 on humult soils in Brunei and Sarawak, including one each from Bakong and Bok-Tisam udult soils and one from southern Lambir Hills humult soils (see Fig. 17.9–17.11; Ashton 1964; Ashton and Hall 1992).

The figures illustrate that a continuous emergent canopy, associated with a sparse or almost absent main canopy and sparse understory, often pertains in the mature phase on well-structured udult soils. This contrasts with the heterogeneous and clustered emergent canopy, dense main canopy, and relatively dense understory of other diagrams from both humult and many udult soils (see Fig. 17.12).

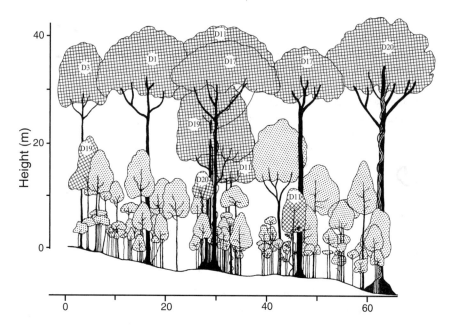

Figure 17.9. Profile diagram of a forest in Bok Tisam Forest Reserve, along hillside, altitude 50 meters (from Ashton and Hall 1992, © Blackwell Publishing). *Circled codes*: dipterocarp species.

It appears that these differences are due to less frequent but larger canopy gaps on well-structured udult soils, such as those over the Setap shales. On such shales, the fast-growing late successional species such as light hardwood dipterocarp *Shorea*—the light red and yellow merantis—successfully compete in gaps, rapidly reaching full height. Thereby, light hardwoods, mostly *Shorea*, dominate the emergent canopy on udult clays. In contrast, on humult soils and udult soils where rooting is unstable, such as poorly structured clays and inceptisols on steep surfaces, the emergent canopy is broken and the main canopy denser. The canopy on low nutrient soils, especially humult soils, is dominated by climax species in a wider range of families, which have regenerated in the predominantly small canopy gaps. Those are mostly slow growing as juveniles, including the heavier hardwood species of *Shorea,* dark red merantis, and selangan batu—sections *Shorea* and *Neohopea* (Ashton 1964, especially his Table 17).

17.4 Conservation Issues

Lambir Hills harbors the only fully conserved example of the richest tree species communities in Eurasia and Africa, and possibly the whole world. Originally, these forests, which are the mixed dipterocarp forests on deep humult ultisols,

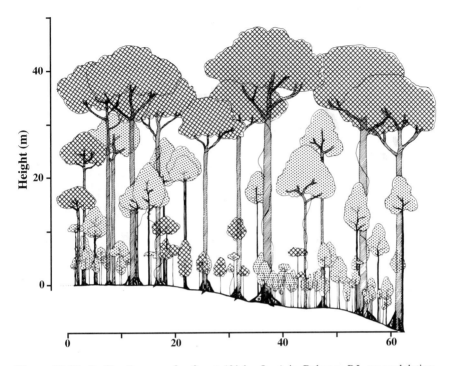

Figure 17.10. Profile diagram of a forest (64 by 8 m) in Bakong, P.J. on undulating land, altitude 110 meters.

occurred on ecological islands on low hills along the northwest coast of Borneo. From southwest to northeast, the main known localities were as follows: Ulu Paloh in extreme northwestern Kalimantan immediately south of the border of westernmost Sarawak, now logged and degraded. Other small areas, now apparently converted to agriculture, formerly existed in west Kalimantan, north of the Kapuas. In Sarawak, Bukit Sebandar, Lundu, now under cultivation; Bukit Undan, Bau District, reduced to a few standing trees; Semengoh Forest Reserve, the primary forest of which is now confined to the natural arboretum of 10 hectares; lower Sungei Sabal Tapang at the base of Gunung Gaharu, Kelingkang Range, now logged and partially converted to shifting agriculture; hills of the Ulu Kenyana, Mukah, now under forestry plantation; Bukit Iju, Arip, Balingian, now under shifting cultivation; Segan Forest Reserve, Bintulu, low hills of clay and sandstone that, with Bukit Iju, were the richest of all, now de-gazetted for oil palm plantation; Nyabau Forest Reserve, now under the oilfield complex; and Lambir Hills, the humult ultisols reduced to 60 km[2] from 350 km[2]. In Brunei, Bukit Puan, Belait District, logged and degraded; the southern and eastern slopes of Bukit Teraja, a small area still intact but not strictly conserved; and Andulau Forest Reserve, Belait, selectively logged but for 300 hectare and a small natural arboretum, all scheduled for strict conservation although there

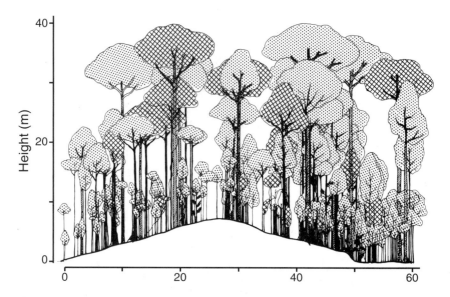

Figure 17.11. Profile diagram of a forest in Lambir Hills, across a ridge, altitude 500 meters.

are rumors that the new Tutong-Kuala Belait trunk road is planned to be driven through it. Finally, in Sabah, Beaufort Hill, a former forest reserve that burned during the 1982–83 El Niño-Southern Oscillation (ENSO) event, is now degraded.

The botanical case for conserving the forest at Lambir Hills is as cast iron as for any comparable area in the world. Indeed, the national park unquestionably qualifies for UNESCO World Heritage status on botanical grounds alone, and it is to be hoped that the authorities will promote that case. Watson (1985) prepared a preliminary yet valuable resource survey of the national park, including management recommendations, which have in part been implemented. Soepadmo and colleagues (1984) provided a list of recorded species, animals as well as plants, on the basis of a collecting expedition undertaken by staff and students of the University of Malaya. Shanahan and Debski (2002) provided an updated list of vertebrates. The park does suffer several disadvantages as a strict conservation reserve, some of which could be overcome by active management.

Although almost the full range of peat swamp communities is conserved in Loagan Bunut National Park, albeit largely selectively logged, it is tragic that none of the kerangas on the raised beaches that follow the coastline north of Lambir Hills and the edge of the Baram peat swamp are conserved. Those at the edge of the peat swamps east of the Lambir Hills, in particular if any remain undegraded, are likely to be particularly rich in tree species, including endemics. The only kerangas of this type currently under any kind of protection is the remains of the fine but now minute stand of *Agathis* (bindang, tolong, kauri) at

Figure 17.12. Dense large-crowned emergent canopy of mixed dipterocarp forest on undulating land overlying Sibuti formation, Ulu Bakong (the area of plots J1–J6), upper part of stereopair; smaller crowned seasonal fresh-water swamp and recent shifting cultivation toward base. 1964. Scale 1:25,000.

Badas Forest Reserve, Brunei, on the far side of the Baram swamps. A botanical census of any kerangas patches on raised beaches remaining by the Sarawak Herbarium staff is recommended, with the aim of evaluating their conservation value.

The park does include within it the only sandstone ridge kerangas conserved in the northeastn Sarawak region, along the summit crests of Bukit Lambir itself. This forest is dominated by *Gymnostoma nobile* (Casuarinaceae) and includes point endemics on the summit rocks.

Whereas the deep humult sandy ultisols that cover most of the park support the tree communities richest in climax species, and hold the highest percentage of local and Borneo endemics of any community in Borneo, the clay udults support a different community. That assemblage is poorly represented in conservation areas throughout the region because of the agricultural potential for plantation crops. It may now be too late to include additional forest on these soils within LHNP, some of which, though selectively logged, appear to persist on the southwest boundary of the park. That community is the richest in central Sarawak and central Borneo endemics. More important is its crucial role as a rich source of food for vertebrates.

The major reason for the low densities of large birds and mammals that currently exist in the park is the persistent illegal hunting (Shanahan and Debski 2002). The very small area of the more productive forests on seasonally flooded alluvium and clay udult ultisols that have been included in the park undoubtedly help somewhat to alleviate the food shortages that are inherent to the humult sandy ultisols and podsols of the area.

Sarawak is exceptional in the tropics by having a National Parks Development Plan first drawn up by botanists and foresters. Ordinarily, vertebrate zoologists play the leading role. Botanical and vertebrate criteria for the selection and design of conservation areas differ in fundamental ways. The first park, Bako, was planned by J.A.R. Anderson, assisted by E.F. Brunig. Anderson then proceeded with the detailed planning of Mulu, Loagan Bunut, and Niah national parks. I prepared preliminary plans for a state network of parks for Sarawak, based on knowledge of tree species' distributions in relation to surface geology and soils. That plan included in addition several further proposals, some of which eventually became legislated as Lambir Hills, Similajau, Kubah, and Gunung Gading national parks, thanks to the tireless efforts of Paul Chai P.K., and others in the Forest Department, Sarawak.

Plants, including tree species, being sedentary are highly site- and especially soil-specific. They often exist and apparently persist in tiny populations of reproductives, occupying minute ecological islands such as mountain peaks and other isolated patches of suitable habitat. In Sarawak, the isolated spots include raised beaches, limestone karst, and on a somewhat larger scale, coastal neogene sandstone hills fragmented by river valleys. It is extraordinary that the richest known individual plant communities in the world occur exclusively on these small hill ranges. In the case of Lambir Hills, the original habitat consisted of only 50 kilometers by 8 kilometers in maximum dimensions (see Fig. 17.1).

The great majority of vertebrates, though specialized in behavior and feeding habits, occur in a wide variety of habitats and plant communities where their nesting and foraging requirements can be met. Because tropical rain forests are generally nutrient-poor and have low productivity of digestible food, most of their vertebrates are wide ranging and require large areas to reasonably ensure survival of their populations. This may be even more pronounced at Lambir Hills, where general flowering and mast fruiting are the rule. Although a botanist would argue that vertebrates represent but a minute part of rain forest biodiversity overall—which is mainly comprised of insects and microorganisms that mostly require no more area for survival than do plant species—one must be mindful that vertebrates play crucial roles in the dispersal of seeds and sometimes pollen. It is, therefore, certain that fully intact rain forest ecosystems and landscapes can only be conserved in large parks, such as Mulu or the Lanjak-Entimau transnational Park. The fact remains, though, that these parks do not contain the full range of plant communities and species represented in their region. The assured conservation of a substantial part of the Sarawak flora—and by implication a vast number of host-specific phytophagous insects, fungi, microorganisms, and also smaller vertebrates—can only be achieved by retaining, in addition to the largest possible forest reserves, a number of smaller parks that include adequately large, representative areas of all principal plant communities and habitats.

Lambir Hills is an example of a small park that is of global significance on account of its hyperdiverse flora, but which may be too small for the conservation of some of the larger vertebrates. In particular, some of the important seed-dispersers including gibbons, flying squirrels, flying foxes, hornbills, and imperial pigeons are threatened. Many tree populations, even families such as Myristicaceae, Bombacaceae, and most Sapindaceae, must depend on the dispersers for their ultimate survival. The problem is further exacerbated by illegal hunting and can only be overcome by introduction of an active management program for animals in danger of local extinction.

Two components for management of endangered vertebrate populations would, therefore, appear to be necessary. First, hunting must be entirely proscribed in the smaller parks such as Lambir Hills, and prohibition enforced. That is only possible in combination with a dynamic program to involve local communities in park management, with clearly defined benefits to participants. Second, sufficient food-rich habitat must be provided. In the case of Lambir Hills this is likely possible now only by plantation management for wild animals. It is often not realized that rain forest plant communities are richest in species and most habitat-diverse where habitat resources, particularly nutrients and water, are relatively limited. Vertebrates, by contrast, are richest in numbers and diversity where nutrients and water are eutrophic and freely available on riverbanks, on frequently flooded clay alluvium, and on nutrient rich water-retaining clay soils. Conservation areas therefore require a combination of habitats. But such essential vertebrate habitats are, not surprisingly, precisely the ones that our own species most wishes to convert for our own use. That is why there is so little Setap shale, and no flood plain, included within Lambir Hills National Park, though extensive areas of both exist adjacent to the south, now mostly degraded and under intermittent cultivation.

One possible solution might be to secure the partnership of local farmers in an agro-forestry scheme in which fruit trees appropriate to the udult clay ultisol habitat—and the food requirements of both threatened vertebrates and people— are planted next to the southern and western park boundaries. Cultivation would need to be on a scale sufficient to sustain the vertebrates yet also permit fruit harvesting for participants, plus yield a reward when vertebrate numbers are being sustained. Planting of hemi-epiphytic figs, which are important resources for vertebrate frugivores and are used by many species in Lambir Hills (Lambert and Marshall 1991; Shanahan and Compton 2001; Terborgh 1986), would provide sustenance for vertebrates and afford opportunities for park visitors to view more wildlife. Illegal harvesting of plant products, notably rattan and timber, continues in the park. Extraction of timber is local and often confined to single trees, though they are often large, for house building. This is inexorably reducing the size of an already perilously small conservation area and continued vigilant policing is critical. Rattan harvesting is pervasive and extensive.

It has to be said that, although the waterfall is an attractive venue for weekend recreation, as is a hike to the summit of Bukit Lambir with its magnificent view of Gunung Mulu and limestone crags, with the vast Baram peat swamp spread like a sea before them, the view from the principal road is unprepossessing. This is in part because more than one track for roads and water pipes have been cut and trees were cleared too far back from the road. The Neogene sandstones are extraordinarily erodable once the surface raw humus has been lost, and catastrophic rainstorms are bound to recur (see Fig. 17.13).

A program of reforestation with trees that occur in nature on these soils is desirable. This will not be simple, because most available nutrients on these infertile soils are retained in the surface raw humus, which has been eroded and must first be reestablished. Fortunately, K. Ogino and his associates have pursued research, in the degraded forest to the north of the park, into re-afforestation and soil restoration, which could provide appropriate guidelines.

17.5 Research Opportunities

Lambir Hills is in several ways ideal for research into tropical rain forests. Now part of the global network of large-scale tree demography and forest dynamic plots of the Center for Tropical Forest Science, it has a replicate in similarly heterogeneous terrain in the similarly perennially wet and uniquely species-rich region of the New World: the Ecuadorian Napo region, in the far western Amazon, at Yasuní. This permits robust testing of generalizations emanating from research at Lambir Hills and comparative data presented in the present volume.

The high site and floristic heterogeneity within the Lambir Hills mixed dipterocarp forest makes it an ideal location for addressing important applied as

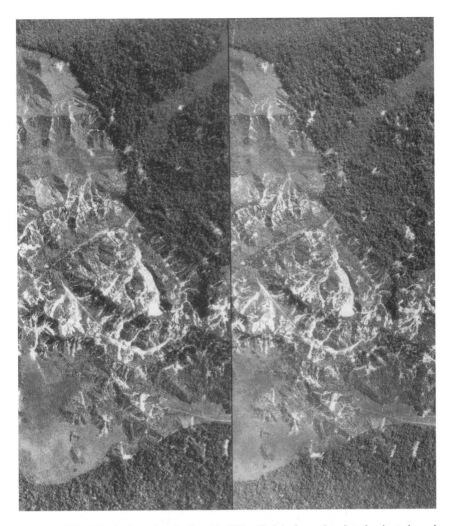

Figure 17.13. Physical erosion in Lambir Hills (Belait formation interlaminated sand and clay), following forest felling and planting (partially contoured) of rubber and after severe rain during the 1963 landas. Note the low incidence of landslips by comparison in the adjacent forested land.

well as theoretical questions, and therefore also ideal for the advancement of solid theoretical foundations—both ecological-silvicultural and socioeconomic—for future forest management and conservation.

Lambir Hills is now well equipped for the necessarily long-term research needed to address these challenges. The 52 ha plot includes a staggering 1200 species, more than half of which are represented by at least 100 individuals. That is roughly the minimum required for the demographic analysis that must underpin both the theory and practice of silviculture. The complex and co-evolutionary interactions, by which competition and natural selection mediate species coexistence in Lambir Hills' stable climate, take place mainly in the leafy canopy and below the soil surface. The canopy walkway, equipped with electricity, and the tower crane together provide opportunities unparalleled at any other rain forest site for a wide range of long-term research in one of the tallest forests remaining in the tropics.

17.6 Site-Specific Forest Management

The potential for sustainable management in the wet tropics can be substantially predicted by the surface characteristics of the forest soil. Where, as is the general case in the zonal yellow-red soils of the tropics, organic matter is sparse in the profile because it rapidly decomposes, soil nutrients are principally stored in the mineral soil and do not depend on sustained forest cover. These include the udult clay ultisols at Lambir Hills. The soil has the capacity to store nutrients applied as fertilizer, albeit often too strong a capacity for the farmer! But over much of Sundaland, and Borneo in particular, there are extensive soils in the well-drained lowlands where the surface bears a blanket of acid organic matter, the 'mor' or 'raw humus.' This denotes a substrate exceptionally poor in soluble nutrients, generally sand or poorly structured clay low in aluminium and iron sesquioxides, and with low capacity to prevent nutrient ions from being leached out in a continuously rainy environment. These are the humult ultisols and tropical humic podsols at Lambir Hills. Here, under natural forest the raw humus accumulates to retain nutrients contributed by the forest above and dropped in litter. Fine roots are concentrated here and, with the aid of symbiotic fungi, are able to release and reabsorb nutrients with extraordinary efficiency; little is leached.

The forest udult soils, we have shown, are rich in relatively fast-growing light hardwoods, including the red merantis, which have been the leading timber on international markets in the past 40 years. The forests are among the easiest to manage sustainably for timber production in the tropics, and the silvicultural methods have been developed and refined by the Forest Research Institute Malaysia as the Malayan Uniform System, or, MUS (Wyatt-Smith 1963, 1995). MUS is monocyclic: that is, trees are removed in a single harvest leaving small juveniles of similar size to produce the next crop. There has been a movement away from monocyclic to polycyclic systems, in which only the larger trees are harvested, leaving medium-sized trees to yield another crop over a shorter felling cycle. Both methods depend for sustainability on the care with which harvesting, the only economically feasible silvicultural intervention, is carried out, because both depend for the new crop on trees established before felling takes place. Even though the udult soils are at a premium for conversion to tree commodity crops, there are vast areas on the folded sedimentary rocks of Borneo where erosion precludes periodic clearing, and the only likely long-term sustainable land-use is the silviculture of native species.

Research in large tree demography plots, such as the CTFS plots at Lambir Hills, Pasoh in Peninsular Malaysia (Okuda et al. 2003), and Palanan in the Philippines can document the phenology and fecundity, and identify the microsite requirements and growth characteristics of species with valuable timber, thereby providing a foundation for refining management protocols.

The humult soils are dangerously erodable, as can be seen along roads in the forestry concession north of Lambir Hills (see Fig. 17.10). This is because the predominating sandy soil and even the young and poorly consolidated rocks beneath rapidly develop canyons once the binding surface roots and raw humus are removed. The surface raw humus is the key: It is, for the most part, bound by the roots of canopy trees, which die and rapidly rot following timber harvesting. Injudicious use of heavy machinery does the rest. A magnificent stand of primary forest gives misleading portent of future productivity, as in the case of the analogous but relatively species-poor seraya (*Shorea curtisii*) forests of the Peninsular Malaysian hills. Maximum growth rates on these soils are low even in intact forests and among pioneer soils. The soils are deficient in nutrients and are drought-prone. Fertilizer leaching makes them marginally economic for agriculture. The solution must be to reestablish conditions favorable for the restoration of surface raw humus, but, even then, the case for economically viable forestry or agriculture is weak in an economy with relatively high labor costs. Nevertheless, rapid rural development around the coasts of Borneo has left vast areas of degraded land bearing these soils, and a major research priority must be to return it to some form of sustainable use. Lambir Hills National Park and the adjacent lands under forest concession and various attempts at agriculture are excellently suited for experimental research to that end.

17.7 Conservation Management

It is now known that the primary mixed dipterocarp forests on humult ultisols along the coasts of northwestern Borneo, and to a lesser extent elsewhere in Sundaland, contain numerous local and endemic tree species. Lambir Hills National Park now comprises the largest surviving example of this forest. The highest research priority must, therefore, be to address maintenance of such fragments. Within that objective, the highest priority must be to find practical means to sustain populations of important and often wide-ranging animals: pollinators, seed dispersers, and top predators in particular. Without them, even these primary forests will suffer extinctions and other long-term degradation. An example of the unpredictable impact that natural forest conversion can have on animal-plant interdependencies is the potentially disastrous effect of greatly increased numbers of wild boar (*Sus* spp.) on absolute and relative survival rates of tree saplings in Pasoh forest, Peninsular Malaysia (Ickes et al. 2001). Wild sows, which benefited from plantations of oil palm in the former Pasoh Forest Reserve, cull 100s of saplings for the nests in which they give birth. These are not research challenges that can be solved in an ad hoc manner, addressing each problem as it arises, because solutions will take longer to achieve than the time before problems will get out of hand. Solutions can only be satisfactorily found by, first, establishing a robust theoretical foundation. Experiments with seedling establishment (Palmiotto 1998; Davies 1998, 2001; Davies and Ashton 1999; Davies et al. 1998) are leading the way.

In their work on pollinators, Momose (1998c), Sakai (1999b), and others have documented the interdependencies between plants and animals. Of particular interest is the competition between plants that share pollinators, particularly when the plant species are related and appear to be ecologically complementary. Interspecific competition among plant species sharing a habitat in biodiverse communities can only be through the mediation of mobile links, except in the case of the commonest species. Strong inferential evidence for such competition has been presented in the case of six co-occurring species of *Shorea* section *Mutica*, which flower in an overlapping sequential succession in Pasoh, and over much of the Peninsular Malaysian lowlands (Ashton et. al. 1988). In the habitat mosaic at Semengoh natural arboretum, Kuching, 11 species co-occur (Ashton 1988), while in the 52 ha plot at Lambir Hills a staggering 14 species are found (see Table 17.1). It seems impossible that so many species can co-exist through control of their numbers by restriction of their fecundity resulting from competition for a common pollinator, leading to sequential flowering peaks. At Lambir Hills, 8 species are humult, and 6 are udult specialists, so that numbers are truly exceptional only at ecotones. Further, at each site, only 3 species exist in densities of reproductives exceeding 2 per hectare in their habitats. The other species exist at less than 1 per hectare: 1 at Pasoh, 5 at Semengoh, and 6 at Lambir Hills.

The number of such 'rare' species is a principal reason for the exceptional richness of such hyperdiverse systems as the forest at Lambir Hills. Some are narrow habitat specialists, or confined to habitats poorly represented in the plot. But there remain many others. Relative infrequency and normally low intensity of catastrophes, such as drought, at Lambir Hills may indicate absence of factors that can force small populations to extinction other than difficulty in securing cross-pollination between scattered individuals: the Allee effect (Allee 1931). Occasional unusually intense droughts do occur at Lambir Hills, though not as severe as in eastern Borneo. They cause increased mortality among juveniles (Delissio et al. 2003) and to varying degrees, among adults (Nakagawa et al. 2000). Interestingly, Potts (2001) showed that rare species showed greater survival during the 1997–98 drought than commoner species, implying that one reason for the number of rare species is that many have a competitive advantage only during rare events. The sequential flowering of *Shorea* species, section *Mutica,* is the subject of a study by Kawasaki and Yamakura (unpublished). They found that species co-occurring at both Pasoh and Lambir Hills flowered in the same sequence. Among them, the facultatively apomictic *S. macroptera*—

Table 17.1. Population density of species of *Shorea*, section *Mutica*, in three Sundaland forests; trees greater than 20 cm diameter per hectare (area sampled indicated in parentheses)

Species	Pasoh Peninsular Malaysia (50 ha)	Semangoh W Sarawak (10 ha)	Lambir, Udult soil (40 ha)	NE Sarawak Humult soil (62 ha)
S. *acuminata*	2.5	—	—	—
acuta	—	—	—	7.7
argentifolia	—	—	0.2	—
curtisii	—	—	—	2
dasyphylla	0.2	3.5	—	—
ferruginea	—	—	9.6	—
hemsleyana	—	0.05	—	—
lepidota	2.7	—	—	—
leprosula	4.1	0.2	0.1	—
macroptera				
ssp. *baillonii*	—	4.6	—	1.7
macroptera	1.8	—	—	—
macropterifolia	—	—	2.3	—
myrionerva	—	—	0.05	—
ovata	—	1	—	1.5
parvifolia	2.9	0.9	3.9	—
quadrinervis	—	4.7	—	2
rubra	—	0.1	—	0.4
rugosa	—	0.1	—	—
scabrida	—	4.7	—	0.2
slootenii	—	1.1	—	0.4

which flowered first when thrip pollinators were at their lowest numbers in Pasoh (Chan and Appanah 1980; Apannah and Chan 1981)—exists at Lambir Hills as two subspecies and with four close relatives, all of which flower more or less synchronously. The subspecies of *S. macroptera* are udult and humult specialists and are found at high population densities. The pattern is similar for two of the related species, and of the final two species one is confined to stream sides; the rarity of the remaining one, *S. slootenii,* is unexplained though it appears to be consistently in low population density throughout its range in northern and eastern Borneo. Such species groups provide interesting material for research into population stability and co-existence.

It has been argued that a major reason for the persistence of sex is that it is the only means of continuously producing rare genotypes, essential in keeping ahead of rapidly evolving pathogens with shorter life cycles (Normark et. al. 2003). If this is so, the overriding prevalence of outcrossing, albeit mostly facultative, among trees in the uniquely equable habitats of biodiverse rain forest implies high levels of species-specific pathogenicity. The implications of the maintenance of outbreeding in populations, in part maintained at low density by density-dependent mortality from pathogens, is highly relevant for conservation management. Importantly, it suggests that conservation of population samples ex situ is less reliable than in natural forests. Large tree demography plots such as the 52 ha plot at Lambir Hills provide the means for the necessarily long-term investigations that will be needed to adequately document the very complex temporal changes in host and pathogen densities that models predict.

No management formula that fails to take account of the social setting of a resource will succeed for long. LHNP is close to a growing city. Communities as diverse as apartment blocks, single-family homes, and longhouses avail of the park. The wide range of land-use that surrounds Lambir Hills National Park, and exists on the same terrain, includes forestry, shifting cultivation, and oil palm plantation. Together, these provide unequaled opportunities to critically examine the socioeconomic opportunities and constraints within which a lasting means of managing a natural resource of truly global uniqueness and importance must be sought.

18. Toward the Conservation of Tropical Forests

Tamiji Inoue
Translated by Kuniyasu Momose

This chapter is translated from *Treasuries of Life: Tropical Rain Forests*, written in Japanese by Tamiji Inoue and published by the NHK Library, Tokyo (1998), as a posthumous work.

18.1 Degradation of Tropical Forests

Half the total area of tropical forests seen at the end of the eighteenth century had disappeared by 1990, and it is estimated that only 10% will remain in 2030 unless social conditions change greatly. That 10%, moreover, will be heavily fragmented forests on steep slopes where people scarcely have access.

Why have tropical forests decreased? In a word, it is because developing countries, to which tropical forests belong, have exploited forests for their economic development. Forests were logged, and wood was exported to developed countries. Some forests were transformed into oil palm plantations or pastures for beef production, which provide oil and meat to export, again, to developed countries through international markets. It is without doubt that destruction of tropical forests is the outcome of the temperate-tropical problem, and that the mass consumption of developed countries has a great influence on degradation of tropical forests. For example, a huge amount of tropical timber was consumed to build modern housing in Japan's urban sprawl after World War II.

It is also true that local people have used the forests to gain daily consumer products on a large scale. Logging roads enable shifting cultivators to easily access forests. This is one cause of forest degradation. Forests are also disappearing by the intensive collection of firewood and wood for charcoal consumed in urban areas. These are linked with population growth and an increasing disparity in wealth within developing countries.

18.2 How Can We Stop the Degradation of Tropical Forests?

Studying tropical forests, we are always overwhelmed by the richness of life and the amazing symbiotic systems found everywhere in these vast forests. Our research has just started, and we are just glimpsing a small fraction of the richness that the tropical rain forests offer.

When being asked what the value of tropical forests is, I always answer: "It is the value of the history of life on land." Only the tropical forests contain the highest biodiversity in the history of the world—formed through the adaptive radiation of angiosperms for at least 100 million years. If tropical forests continue to decrease, this valuable biodiversity, created by the long history of life, will be lost forever.

Biodiversity can be evaluated in various terms, and many writings on biodiversity have been published. Here, I will introduce cases that can contribute to the conservation effort. Being aware that human economic activities are the cause of the degradation of tropical forests, we have to create and implement concrete policies to prevent forest degradation. Otherwise, the end result is an armchair theory. My statement that tropical forests have historical value of 100 million years might reflect nothing more than the self-satisfaction of researchers, unless we construct a system to socialize the historical value of tropical forests and to provide economic systems for its proper usage.

18.3 The Success of Costa Rica

Costa Rica in Central America is famous as a successful example in the conservation of tropical forests and the utilization of biodiversity at the national level. I revisited the country in March 1997 for the first time in eight years to inspect ecotourism. Surprisingly, income through ecotourism jumped to the greatest source of foreign currency in 1996, which exceeded by more than 600 million dollars the income from banana and coffee export. The government of Costa Rica protects areas that are attractive for ecotourists as national parks or nature reserves, which include 11% of the country. Additionally, there are other protected areas managed as field stations by international organizations such as the Organization for Tropical Studies (OTS).

There are various kinds of ecotours. Usually, fewer than ten visitors and one or two guides walk around in a forest for one day. I participated in a tour without revealing that I was a biologist, and I was surprised that field guides were highly educated. Tourists can select various courses from introductory level to those focused on a particular plant or animal group, such as a course for bird watchers. Many people visited the same place repeatedly. A number of people in various professions—including travel agents in the capital, San José, bus drivers, and park guards in national parks—engage in ecotourism to earn their incomes. Tourists are not always simply walking around in a protected area. Some large facilities have been constructed for ecotourists. One example is the aerial tram

(5 kilometers long), located a one-hour drive from San José. They run gondolas carrying up to six passengers through the forest in the same manner as ski lifts. The passengers enjoy watching, within arm's length, epiphytes and humming-birds visiting their flowers. Even elderly persons unable to walk around forests can enjoy nature with this facility. Dr. D.R. Perry, who was among the first to access the canopy of tropical forests over 20 years ago, started this venture business. Annually 80 thousand people visit, and annual sales of the entrance tickets exceed 3 million dollars, he said. His words were impressive: "You make your study by using tax revenue. I obtain research funds from ecotours, con-serving nature at the same time." True to his word, he is not only conserving the forest adjacent to a national park, but his company also provides employment opportunities to people living in villages near the forest.

Ironically, ecotourism depends on the scarcity value of tropical forests caused by destruction of nature. Leading other countries, the government of Costa Rica implemented a policy to manage and use the scarcity value in a sustainable manner. It is certain that ecotourism can be a powerful shot in the arm with an immediate effect for the conservation of biodiversity.

18.4 Exploring Biological Resources

The other project of Costa Rica that is making the country famous is exploration of biological resources by the Instituto Nacional de Biodiversidad (INBio). Vis-iting this institute in a suburb of San José, one must first be surprised by its museum facility, which resembles a car factory, and the strict guards. I visited the entomology section, my specialty, first. People called parataxonomists, which is a newly created professional category, collect specimens from the field to bring into the "factory" of INBio. After learning fundamental knowledge of insect taxonomy and methods of insect collection, they live near the study sites assigned to them. Selected local people with the best knowledge of nature are further educated to become parataxonomists. The preserved specimens brought in are pinned one by one, labeled, and arranged in drawers, sorted by taxa. The process can be compared with car dis-assembly, in which parts are dismantled and returned to the shelves, just the opposite of car assembly in a factory. The whole process is called inventory in terms of biological studies, which originally meant the listing of stock items.

Computers manage the whole process. A bar code is applied to each speci-men, which corresponds to data about the specimen in a computer file. We can immediately find the specimens collected at a certain location and at a certain time in the storage areas, in which specimens are arranged by taxonomic groups. Insect specimens are sorted into species by professional taxonomists with doc-toral degrees. This process takes the longest time. Computer software has been developed to help sort some of the taxa. By answering some questions and watching figures appearing on the computer monitor, the taxonomist can deter-mine the family names. Although only specialists are currently able to use the software, in the near future specimens moving on a belt conveyer can be sorted automatically.

In addition to insects, plants, fungi, and shellfish have been dealt with in INBio. From this year (1997), the all-taxa project, which surveys all organisms, is scheduled to begin. This project was proposed by the world-famous biologist, Dr. D.H. Janzen, who has studied in Costa Rica for several decades. He extended Guanacaste National Park to include an area once used as pasture and started working for forest restoration there. He is seen as the backbone of INBio and has a strong connection with environmentalist politicians, including Al Gore, former vice president of the United States of America.

What makes INBio famous might be biodiversity exploration, rather than the biodiversity inventory projects explained above. Since exploration is impossible without inventory, the existence of both is essential in such activities. Although I was kindly shown around, they did not allow me to take photographs in the biodiversity-prospecting department. At the time of my visit, INBio had contracts with several pharmaceutical companies in Europe and America and allowed them to prospect and screen natural chemical substances and DNA. Besides plants and fungi, they screen insects and their excrement. Frozen samples are vacuum-dried, and chemical compounds are extracted in various solvents. Chemical substances are usually passed on to pharmaceutical companies in this form. INBio also has its own analytical systems. Until now, some useful chemical substances, including one effective on a virus, have been found. The staff complained that people sometimes emphasize only the relationship with pharmaceutical companies in Europe and America, and that INBio is sometimes regarded as merely their subcontractor.

Costa Rica seems to strive to maintain daily incomes from ecotourism and aims for big homeruns from genetic resources. This policy is supported also by the United States, with some large grants provided by the National Science Fundation (NSF).

18.5 Are Non-Timber Products the Answer?

In the past, some environmentalist groups stated that consumption of timber was the primary cause of forest degradation, because logging disturbed forests on an immense scale. However, some examples in Southeast Asia show that logged forests can recover from logging and that a second logging is possible. In the current way of logging, land cover is removed by bulldozers and vegetation of nearly half the area is lost. Even without human assistance, however, seeds are dispersed from remaining trees and vegetation then recovers in the once-cleared area. The next logging is sometimes possible in 50 years. This is 'sustainability' only in terms of wood production, of course. A major part of biodiversity is considered to be lost by such logging activities, but no reliable data have been obtained.

Non-timber products like medicinal drugs are expected to be promising biological resources, which can be exploited without damaging the forests. However, this can cause another type of serious problem when the resources are exploited. An anti-cancer drug was found from American yews and commercialized. Most the American yews have been consumed as ingredients of the drug. Now drug companies have an eye on Asian yews containing similar chemical substances. Since the complete synthesis of the effective chemical does not turn a profit, they exploit and process natural substances from the plants, thereby causing over-exploitation.

If wild plants are used intensively even as crude drugs, they can be collected to extinction. In Indonesia, such excessive collection of plants for crude medicines is already a serious problem. Commercial drugs are so expensive that collection pressures easily exceed the population growth of plants. Therefore, non-timber products should not be gathered from natural populations, and their use cannot be a sustainable usage of the forests, without cultivating the same plants.

18.6 The Necessity of an Integrated Regional Management Plan

From Southeast Asia, we can also find some successful examples in which settlement of conservation areas and the economic activities of surrounding communities were well-matched. In Sulawesi, Indonesia, rainy seasons are so short that rice cultivation becomes impossible if forest degradation causes the loss of water storing capacity. In Domogabone district, a regional development plan was carried out. A national park was settled in the upper streams of the watershed, and an irrigation system was constructed in the lower streams. While one might consider that dams could irrigate the system, dams in the watershed without forests would soon lose their function due to earth and sand eroded from the bare ground. Conservation of the forests provides a most effective natural dam. If not for this regional development plan, the whole watershed would have been used by logging and shifting cultivation, causing destruction of both forests and people's lives, as has happened in many other tropical forest areas.

In an oversimplistic view, tropical forests can be conserved by shutting out human activities from as large a protected area as possible. National parks settled by Europeans in the colonial period are based on this idea. With an unjust plan taking no consideration of the life and economy of local societies, however, national parks are gradually invaded by shifting cultivation and illegal hunting. Such illegal activities are punished everywhere, but local people without alternative ways of life cannot pay penalties. If forests continue to decrease, the price of wood will rise, and remaining forests will suffer more frequent illegal logging. Even in Costa Rica, known as a successful example of forest conservation, this has already started to happen. Construction of road systems for ecotourism makes illegal logging easy.

Tree-planting campaigns are held every year in Japan. Japanese people seem to like tree planting. People from Japanese local governments or nongovernmental organizations also visit and plant trees in Sarawak, Malaysia. Ironically, next door to the people planting trees, secondary forests are being cleared and transformed into oil palm plantations. Demand for palm oil is increasing for cooking in China, and for detergent gentle to the hands and the "environment" in Japan. I do not say that tree planting is useless. For reforestation of unfortunately cleared lands, tree planting is an important technique. Because tree planting in the tropics is not so well-developed a technique as in temperate regions, I agree that it should be studied more. However, as the Forest Research Institute, Malaysia, has clarified, wood production is possible solely depending on natural regeneration, and it is an economically more reasonable way of production. If logging techniques are improved, the intervals of logging could be shortened.

What is most needed now are not specific matters or techniques, but regional development plans involving areas as large as Japanese prefectures (2 to 80 km^2) and active commitments of governments and the citizens of developed countries to such plans. In the area, residents live without conflicts with the conservation of tropical forests. In order to avoid the problems that I pointed out so far, the following zones should be located adequately in the regional development plan:

1. *Nature protection zone.* The area in which human visits are minimized; any activities or research are prohibited.
2. *National parks.* The area for obtaining information and leisure; research or ecotourism are permitted.
3. *Non-timber production zone.* The area in which hunting and gathering, including medicinal herbs and rattans, are permitted, but activities changing the forest structure, such as logging and burning, are prohibited.
4. *Intensive use zone.* The area for production of foods, fruits, medicinal plants, and crops of shifting cultivation.
5. *Industry and residential zone.* The area for processing materials collected from forests. The products include extracted genes or works of art made of wood. In the residential areas there are not only dwellings of workers but also accommodations for all visitors including researchers, artists, and ecotourists.

The time to only discuss some principles of conservation of tropical rain forests has finished. Now we must propose concrete policies for conservation and realize them. We do not have much time.

18.7 Tropical Rain Forests in 2100

One hundred years from now, what will happen to tropical rain forests? Will studies of tropical rain forests be topics in paleontology, like studies of dinosaurs? Will a descendant of Spielberg make a great profit on the film "Tropical Park"? Otherwise, will the smartest lifestyle admired by young people be living near a tropical rain forest to make furniture? While taking an evening walk on an aerial walkway in a forest, will we see an orangutan eating fig fruits and be pleased with the peaceful lifestyle that each of us has?

Appendix A. Reproductive traits, floral characters, and pollinators of 270 plant species (73 families) in a lowland dipterocarp forest at LHNP, Sarawak

Plant Family, genus, species & author	1)	2)	3)	4)	5)	6)	Color 7)	Flower shape 8)	Main pollinators 9)
ACANTHACEAE									
Borneacanthus grandiflorus Brem.	1	s	N	h	d	n	white	bilabiate	*Nomia*
Ptyssiglottis dispar Hallier	1	s	N	h	d	n	white	bilabiate	*Nomia*
ACTINIDIACEA									
Saurauia sp.	1	i	N	h	d	n	pink	spiral	*Amegilla*
Saurauia glabra Merr.	G	i	N	h	d	n, p	white	cup	*Xylocopa*
Saurauia ridleyi Merr.	1	b	N	h	d	n	pink	spiral	*Trigona*
ALANGIACEAE									
Alangium ridleyi King	2	s	GF	h	d	n	white	campanulate deeply lobed	*Apis dorsata*
ANACARDIACEAE									
Buchanania sessilifolia (B1.) B1.	3	s	GF	h	d	n, p	greenish	spiral	d — Canaceidae, Syrphidae, several families of Coleoptera, Diptera
Gluta laxiflora Ridl.	3	s	GF	h	d	n, p	white	spiral	*Trigona*
ANNONACEAE									
Cathostemma aff. *hookerii* King	L	i	N	h	d, n	s	purple	cup	Curculionidae
Enicosanthum coriaceum (Ridl.) Airy Shaw	3	b	N	h	d, n	c, s	white	chamber	Scarabaeidae, Chrysomelidae
Enicosanthum macranthum (King) Sinclair	3	b	N	h	d, n	s	white	chamber	Curculionidae
Fissistigma paniculatum (Dunal) Merr.	L	s	N	h	d, n	s	yellow	chamber	Curculionidae
Friesodielsia glauca (Hk. f. et Th) van Steenis	L	s	N	h	d, n	s	yellow	chamber	Nitidulidae, Curculionidae
Friesodielsia filipes (Hk. f. et Th.) van Steenis	L	s	N	h	d, n	s	yellow	chamber	Nitidulidae, Curculionidae

223

Appendix A. *Continued*

Plant Family, genus, species & author	1)	2)	3)	4)	5)	6)	Color 7)	Flower shape 8)		Main pollinators 9)
Goniothalamus uvarioides King	1	b	N	h	d, n	s	white	chamber	bs	Nitidulidae
Goniothalamus velutinus Airy Shaw	1	b	N	h	d, n	s	yellow	chamber	bs	Nitidulidae, Curculionidae
Meiogyne cylindrostigma (Burck) van Heusden	2	i	N	h	d, n	s	pink	chamber	bs	Nitidulidae, Curculionidae
Monocarpia euneura Miq.	3	i	N	h	d, n	s	yellow	chamber	bs	Nitidulidae, Curculionidae
Polyalthia cauliflora Hk. f. et Th.	2	b	N	h	d, n	s	yellow	chamber	bs	Nitidulidae, Curculionidae
Polyalthia hypogaea King	2	b	N	h	d, n	s	white	chamber	bs	Curculionidae
Polyalthia motleyana (Hk. f.) Airy Shaw	1	i	N	h	d, n	c, s	white	chamber	bs, bf	Chrysomelidae
Polyalthia rumphii (Bl.) Merr.	2	i	N	h	d, n	s	yellow	chamber	bs	Curculionidae
Polyalthia, sp. nov.	1	s	N	h	d, n	s	yellow	chamber	bs	Curculionidae
Polyalthia sp. nov. 2	1	i	N	h	d, n	s	purple	chamber	bs	Curculionidae
Popowia pisocarpa (Bl.) Endl.	2	s	N	h	d, n	p, c	yellow	urceolate	Thripidae	Thripidae
Pyramidanthe prismatica (Hk. f. et Th) Sincliar	L	s	N	h	d, n	s	yellow	chamber	bs	Chrysomelidae
Sphaerothalamus insignis Hk. f.	1	b	GF	h	d, n	s	red	chamber	bs	Curculionidae
Uvaria aff. *elmeri* Merr.	L	b	N	h	d, n	p, s	white	spiral	Blattellidae	Blattellidae
Uvaria sp. nov.	L	s	N	h	d, n	s	yellow	chamber	bs	Nitidulidae, Curculionidae

Appendix A. *Continued*

Plant

Family, genus, species & author	1)	2)	3)	4)	5)	6)	Color 7)	Flower shape 8)	Main pollinators 9)
APOCYNACEAE									
Urceola sp.	L	s	N	h	d	n	white	cup	d	Vespidae, Halictidae, several families of Hymenoptera, Diptera
ARACEAE									
Homalomena propinqua Schott	1	b	N	m	d, n	staminode, p	white*	chamber*	bp, bf	Scarabaeidae, Chrysomelidae
ASCLEPIADACEAE									
Gongonema sp.	L	s	N	h	d	n	yellow	spiral	d	Chrysomelidae, Cleridae, several families of Coleoptera
BOMBACACEAE									
Coelostegia griffithii Benth.	3	i	N	h	d, n	p	brown	cup	bp	Elateridae, Chrysomelidae
Durio grandiflorus (Mast.) Kost et Soegeng	3	i	GF	h	d	n	white	spiral, brush	Spiderhunter (*Arachnothera robusta*)
Durio griffithii (Mast.) Bakh.	2	i	GF	h	d	n, p	white	spiral	Nomia
Durio kutejensis (Hassk.) Becc.	3	i	GF	h	d, n	n	red	cup, brush	Bird
Durio oblongus Mast.	3	i	GF	h	d	n	white	cup, long stamen tube	Spiderhunter (*Arachnothera robusta*)
BURMANNIACEAE									
Burmannia lutescens Becc.	1	s	N	d	d	n	white	cup	Culicidae

Appendix A. *Continued*

Plant

Family, genus, species & author	1)	2)	3)	4)	5)	6)	Color 7)	Flower shape 8)	Main pollinators 9)	
BURSERACEAE										
Canarium denticulatum Bl.	3	s	N	d	d	p	yellow	spiral	d	Chrysomelidae, several families of Hemiptera, Hymenoptera
Dacryodes laxa (Benn.) H. J. Lam	3	s	GF	d	d	p	red	spiral	*Trigona*	
Dacryodes incurvata (Engl.) H. J. Lam	3	s	GF	d	d	n, p	white	spiral	*Apis dorsata, Apis koschevnikovi*	
Santiria griffithii (Hk. f.) Engl.	3	s	N	d	d	n, p	orange	cup	*Trigona, Apis dorsata*	
Santiria laevigata Bl.	4	s	N	d	d	n, p	green	spiral	*Trigona*	
Triomma malaccensis Hk. f.	5	s	GF	d	7–7+	n, p	white	spiral	d	Chrysomelidae, several families Hemiptera,
CELASTRACEAE										
Lophopetalum glabrum Ding Hou	3	s	N / G'	h	d	n, p	orange	spiral	*Trigona* / d	Elateridae, Chrysomelidae, several families of Coleoptera, Diptera
COMPOSITAE										
Mikania micrantha H., B. et K.	G	s	N	h	d	n, p	white	brush*	*Trigona*	
Vernonia arborea Ham.	G	s	N / G'	h	d	n, p	white	brush*	*Ceratina, Braunsapis Nomia, Apis dorsata*	
CONVOLVULACEAE										
Erycibe sp. 1	L	s	GF	h	d	p	yellow	cup	*Trigona*	
Erycibe sp. 2	L	s	GF	h	d	p	yellow	cup	bp	Mordellidae

Appendix A. *Continued*

Plant Family, genus, species & author	1)	2)	3)	4)	5)	6)	Color 7)	Flower shape 8)	Main pollinators 9)
CORNACEAE									
Mastixia rostrata Bl.	2	i	GF	h	d	n, p	white	spiral	bp, Chrysomelidae
COSTACEAE									
Costus globosus Bl.	1	i	N	h	d	n, p	orange*	bilabiate	*Amegilla*
Costus speciosus (Koenig.) J. E. Sm.	G	b	N	h	6-18	n, p	white, yellow*	bilabiate	*Amegilla*
CRYPTERONIACEAE									
Crypteronia cumingii Endl.	3	s	N	h	d	p	white	spiral, brush	*Ceratina, Braunsapis*
DILLENIACEAE									
Dillenia exelsa (Jack) Gilg.	G	s	GF	h	5-17	p	yellow	spiral	*Apis dorsata*
Dillenia suffruticosa (Griff.) Mart.	G	s	N	h	d	p	yellow	spiral	*Xylocopa*
Tetracera akara (Brum. f.) Merr.	L	s	GF	h	d	p	white	spiral	*Apis koschevnikovi*
DIPTEROCARPACEAE									
Dipterocarpus geniculatus Vesque	5	s	GF	h	18:30-18	p, c	orange	campanulate deeply lobed	*Apis dorsata*, bp (Scarabaeidae)
Dipterocarpus globosus Vesque	5	s	GF	h	18:30-18	p, c	white, purple	campanulate deeply lobed	*Apis dorsata*
Dipterocarpus pachyphyllus Meijer	5	s	GF	h	n	n	white	campanulate deeply lobed	Geometridae
Dipterocarpus tempehes Slooten	5	s	GF	h	5-4:30	p	yellow	campanulate deeply lobed	*Apis dorsata*
Dryobalanops aromatica Gaertn. f.	5	s	GF	h	5-12	p	white	campanulate deeply lobed	*Apis dorsata*
Dryobalanops lanceolata Burck	5	s	N	h	5-12	p	white	spiral	*Trigona*
			G'					spiral	*Apis dorsata*

Appendix A. *Continued*

Plant

Family, genus, species & author	1)	2)	3)	4)	5)	6)	Color 7)	Flower shape 8)	Main pollinators 9)	
Hopea pterigota Ashton	3	s	N	h	d, n	c, (p)	yellow	spiral, cup	bf	Chrysomelidae, Curculionidae
Hopea sphaerocarpa (Heim.) Ashton	1	s	GF	h	d, n	p	red	spiral, cup	*Trigona*, Sciaridae	
Shorea agami Ashton	5	s	GF	h	d, n	c, (p)	yellow	spiral, cup	bf	Chrysomelidae, Curculionidae
Shorea beccariana Burck	5	s	GF	h	d, n	c, (p)	yellow	spiral, cup	bf	Chrysomelidae, Curculionidae
Shorea bullata Ashton	5	s	GF	h	d, n	c, (p)	yellow	spiral, cup	bf	Chrysomelidae, Curculionidae
Shorea confusa Ashton	5	s	GF	h	7-18	n, p, c	purple	spiral, cup	bf	Chrysomelidae, Curculionidae
Shorea balanocarpoides Symington	4	s	GF	h	d, n	c, (p)	yellow	spiral, cup	bf	Chrysomelidae, Curculionidae
Shorea falciferoides Foxw.	5	s	GF	h	d, n	c, (p)	white	spiral, cup	bf	Chrysomelidae, Curculionidae
Shorea ferruginea Dyer ex Brandis	5	s	GF	h	d, n	c, (p)	yellow	spiral, cup	bf	Chrysomelidae, Curculionidae
Shorea havilandii Brandis	3	s	GF	h	d, n	c, (p)	white	spiral, cup	bf	Chrysomelidae, Curculionidae
Shorea macrophylla (D. Vr.) Ashton	4	s	GF	h	d, n	c, (p)	pink	spiral, cup	bf	Chrysomelidae, Curculionidae
Shorea macroptera Dyer	5	s	GF	h	d, n	c, (p)	pink	spiral, cup	bf	Chrysomelidae, Curculionidae
Shorea ochracea Symington	5	s	GF	h	6:30-6	n, p, c	yellow*	spiral, cup	bf	Chrysomelidae, Curculionidae
Shorea parvifolia Dyer	5	s	GF	h	17:30-17:30+	c, (p)	yellow	spiral, cup	bf	Chrysomelidae, Curculionidae
Shorea patoiensis Ashton	4	s	GF	h	n	c, (p)	white	spiral, cup	bf	Chrysomelidae, Curculionidae

Appendix A. *Continued*

Plant

Family, genus, species & author	1)	2)	3)	4)	5)	6)	Color 7)	Flower shape 8)	Main pollinators 9)
Shorea pilosa Ashton	5	s	GF	h	d, n	c, (p)	yellow	spiral, cup	bf	Chrysomelidae, Curculionidae
Shorea smithiana Symington	5	s	GF	h	d, n	c, (p)	yellow	spiral, cup	bf	Chrysomelidae, Curculionidae
Shorea superba Symington ex Wood	5	s	GF	h	d, n	c, (p)	white	spiral, cup	bf	Chrysomelidae, Curculionidae
Shorea xanthophylla Symington	4	s	GF	h	d, n	c, (p)	yellow	spiral, cup	bf	Chrysomelidae, Curculionidae, Cleridae
Vatica micrantha Slooten.	4	s	N	h	d, n	c, (p)	yellow	spiral, cup	bf	Chrysomelidae, Curculionidae
Vatica aff. *parvifolia* Ashton	4	s	N	h	d, n	c, (p)	white	spiral, cup	bf	Chrysomelidae, Curculionidae
EBENACEAE										
Diospyros dicotyoneura Hiern	3	i	GF	d	d, n	p	white	urceolate	bp	Staphylinidae, Nitidulidae
ELAEOCARPACEAE										
Elaeocarpus nitidus Jack	G	s	N	h	d	n, p	white	spiral	*Ceratina, Braunsapis*	
Elaeocarpus stipularis Bl.	G	s	N	h	d	n, p	white	cup	*Braunsapis, Trigona, Apis koschevnikovi*	
EUPHORBIACEAE										
Agrostistachys longifolia (Wight) Benth. ex Hk. f.	1	b	N	d	d	n, p	yellow*	brush*	*Trigona*	
Aporusa nitida Merr.	2	s	N	d	d	p	yellow*	brush*	*Trigona*	
Aporusa prainata King ex Gage	3	s	N	d	d	p	yellow*	brush*	d	
Aporusa sarawakensis Schott	2	s	N	d	d	p	yellow*	brush*	*Trigona*	

Appendix A. *Continued*

Plant

Family, genus, species & author	1)	2)	3)	4)	5)	6)	Color 7)	Flower shape 8)	Main pollinators 9)	
Baccaurea racemosa (Reinw.) Muell. Arg.	2	b	N	d	d	p	yellow*	brush*	d	several families of Cantharidae, Muscidae, Diptera
Cephalomappa beccariana Baill.	4	s	GF	m	d	p	yellow*	brush*	'	Curculionidae, Chrysomelidae, several families of Coleoptera, Diptera
Cleistanthus pseudopodocarpus Jabl.	3	s	N	m	d	n, p	yellow	cup	*Trigona*	
Cleistanthus sumatranus (Miq.) Muell. Arg.	3	s	G' GF	m	d	n, p	yellow	cup	*Apis dorsata* *Trigona*	
Cleistanthus venosus C. B. Rob.	3	s	GF	m	d	n, p	yellow	cup	*Apis dorsata*	
Dimorphocalyx denticulatum Merr.	2	i	N	d	d	p	red	spiral	*Trigona*	
Drypetes longifolia (Bl.) Pax ex Hfman	2	b	N	d	d	n, p	white*	spiral	d	Scarabaeidae, Chrysomelidae, several families of Coleoptera, Diptera
Drypetes xanthophyloides Airy Shaw	2	s	GF	d	d	n, p	white*	spiral	d	Mordellidae, several families of Coleoptera, Hemiptera, Diptera
Endospermum peltatum Merr.	G	s	GF	d	d	n, p	yellow	brush*	*Trigona, Apis koschevnikovi*	
Homalanthus populneus (Geisel) Pax	G	s	N	m	d	p	yellow*	brush*	d	*Rhopalomelissa* (Halictidae), *Allodape* (Apidae), *Hylaeus* (Colletidae), Eumenidae, several families of Diptera, Hymenoptera

Appendix A. *Continued*

Plant Family, genus, species & author	1)	2)	3)	4)	5)	6)	Color 7)	Flower shape 8)	Main pollinators 9)
Koilodepas laevigatum Airy Shaw	3	s	N	m	d	p	yellow*	brush*	*Trigona*
Macaranga brevipetiolata Airy Shaw	2	s	GF	d	d	n, p	green*		*Trigona*
Macaranga winkleri Pax et Hfinan	G	s	N	d	d	n, p	green*		*Trigona*
Mallotus griffithianus (Hk. f.) Muell. Arg.	2	s	N	d	d	p	yellow*	brush*	*Trigona*
Mallotus penangensis Muell. Arg.	3	s	N	d	d	p	yellow*	brush*	*Trigona*
Mallotus wrayi King ex Hk. f.	2	s	N	d	d	p	yellow*	brush*	*Trigona*
Moultonianthus leembrugianus (Boerl. et Koord.) van Steenis	3	s	GF	m	d	p	white*	spiral	*Trigona*
Tapoides villamilii (Merr.) Airy Shaw	3	s	N	m	d	n, p	white	cup	d
Trigonopleura malayana Hk. f.	G	s	GF	d	d	n, p	white	cup	*Apis dorsata*
FAGACEAE									
Lithocarpus lucidus (Roxb.) Rehd.	G	s	GF	m	d	p	yellow*	brush	d
Lithocarpus ferrugineus Soepadmo	4	s	N	m	d	p	yellow*	brush	d
FLACOURTIACEAE									
Casearia grewiaefolia Vent	1	s	N	h	d	n	red	cup	Eumenidae

Main pollinators (selected entries):
- *Tapoides villamilii*: Anthicidae, Scarabacidae, Elateridae, several families of Coleoptera,
- *Lithocarpus lucidus*: Braconidae, *Trigona*, several families of Hymenoptera, Diptera
- *Lithocarpus ferrugineus*: Chrysomelidae, Elateridae, several families of Coleoptera,

Appendix A. *Continued*

Plant

Family, genus, species & author	1)	2)	3)	4)	5)	6)	Color 7)	Flower shape 8)	Main pollinators 9)
Hydnocarpus borneensis Sleum.	2	s	GF	h	d, n	n	yellow*	spiral	d; Chrysomelidae, several families of Coleoptera, Diptera
GESNERIACEAE									
Didissandra sp.	1	s	N	h	d	n	white	bilabiate	*Amegilla*
GNETACEAE									
Gnetum cuspidatum Bl.	L	b	N	d	morning	droplet	green*	spiral	fly (Lauxaniidae)
Gnetum gnemon L.	1	s	N	d	n	droplet	green*	spiral	Pyralidae, Geometridae
Gnetum leptostachyum Markgraf	L	b	N	d	morning	droplet	green*	spiral	fly (Muscidae, Conopidae, Drosophilidae)
GRAMINEAE									
Dinochloa sp.	L	s	N	h	d	p	pink*		*Trigona*
GUTTIFERAE									
Calophyllum sp.	2	s	GF	h	d	n, p	white	spiral	*Trigona*
Garcinia penangiana Pierre	2	s	GF	h	d	n, p	white	spiral	*Trigona*
Garcinia nervosa Miq.	2	i	GF	h	d	n, p	white	spiral	d; Curculionidae, Chrysomelidae
Mesua grandis (King) Kosterm.	2	s	GF	h	d	p	white	spiral	*Apis koschevnikovi*
Mesua oblongifolia (Ridl.) Baill.	2	s	GF	h	d	p	white	spiral	*Apis koschevnikovi*
HYPERICACEAE									
Cratoxylum sumatranum (Jack) Bl.	G	s	N	h	d	n, p	pink	spiral	*Trigona*

Appendix A. *Continued*

Plant

Family, genus, species & author	1)	2)	3)	4)	5)	6)	Color 7)	Flower shape 8)	Main pollinators 9)
HYPOXIDACEAE									
Curculigo villosa Wall.	G	b	N	h	d	n	pink	spiral	*Trigona*
ICACINACEAE									
Gomphandra cumingiana (Miers) F. Vill.	2	s	N	d	n	n, p	white	campanulate	*Trigona*
Iodes sp.	L	s	GF	h	n	p	green	spiral	bp, Nitidulidae, Chrysomelidae
IXONANTHACEAE									
Allantospermum borneense Forman	4	s	GF	h	n	n	white	spiral	*Apis dorsata*
LAURACEAE									
Endiandra clavigata Kosterm.	3	s	N	h	d	p	yellow	cup	d, Curculionidae, several families of Coleoptera,
Litsea sp.	2	s	N	d	d	n, p	white	cup	d, Chrysomelidae, Vespidae, several families of Diptera
LECYTHIDACEAE									
Barringtonia sarcostachys Bl. (Miq.)	1	s	N	h	n	n	white*	brush	Sphingidae
LEGUMINIOSAE									
Afzelia borneensis Harms	2	s	GF	h	d	n	red, white*	Caesalpinia-like	d, Chrysomelidae, several families of Diptera,
Bauhinia sp.	L	s	GF	h	d	n	orange	Caesalpinia-like	Several families, Lepidoptera (butterfly)
Caesalpina sp.	G	s	N	h	d	n, p	orange	Caesalpinia-like	*Xylocopa*
Callerya niewenhuisii (J.J. Sm.) Schot.	L	b	N	h	d	n, p	orange	papilionaceous	*Megachile*

Appendix A. *Continued*

Plant

Family, genus, species & author	1)	2)	3)	4)	5)	6)	Color 7)	Flower shape 8)	Main pollinators 9)
Callerya vasta (Kosterm.) Schot.	4	s	N	h	d	n, p	purple	papilionaceous	*Megachile*
Fordia sp.	2	s	N	h	d	n, p	pink	papilionaceous	*Megachile*
Parkia singularis Miq.	3	s	GF	m	n	n	white	tublar, brush	Bat
Parkia speciosa Hask.	3	s	GF	m	n	n	white	tublar, brush	Bat
Sindora beccariana Backer ex de Wit	4	s	N	h	d	n	brown	Caesalpinia-like	*Trigona, Megachile*
Sindora cf. *irpicina* de Wit	4	s	GF	h	d	n	brown	Caesalpinia-like	*Trigona, Nomia, Vespidae*
Spatholobus ferrugineus (Zoll. et M.)	L	s	GF	h	d	n	red	papilionaceous	*Apis dorsata, Apis koschevnikovi*
Spatholobus macropterus Miq.	L	s	GF	h	d	n	white	papilionaceous	*Apis dorsata*
Spatholobus multiflorus Ridder	L	s	GF	h	d	n	red	papilionaceous	d Curculionidae, Chrysomelidae, several families of Hemiptera, Coleoptera
LOGANIACEAE									
Fagraea racemosa Jack ex Wall.	G	s	N	h	n	n	white	bilabiate	Bat
LORANTHACEAE									
Amylotheca duthieana (King) Danser	E	s	N	h	d	n	orange-red	tubular	Spiderhunter (*Arachnothera robusta*)
Macrosolen cochinchinensis (Lour.) van Tiegh.	E	s	GF	h	d	n	yellow	burst open	Birds (*Dicaeum trigonostigma, Chloropsis sonnerati*)

234

Appendix A. *Continued*

Plant									
Family, genus, species & author	1)	2)	3)	4)	5)	6)	Color 7)	Flower shape 8)	Main pollinators 9)
Tritecanthera xyphostachys Tirgh.	E	s	N	h	d	n	pink	tubular	Spiderhunter (*Arachnothera robusta*)
LOWIACEAE									
Orchidantha inouei Nagam. and S. Sakai	1	b	N	h	d, n	deciet	purple*	bilabiate	Scarabaeidae
MARANTACEAE									
Phacelophrynium maximum (Bl.) K. Schum	1	b	N	h	d	n	white	bilabiate	*Amegilla*
Stachyphrynium cylindricum (Ridl.) K. Schum.	1	b	N	h	d	n	white	bilabiate	*Nomia*
Stachyphrynium griffithii (Bak.) K. Schum.	1	b	N	h	d	n	white	bilabiate	*Nomia*
MELASTOMATACEAE									
Melastoma beccariana Cogn.	G	s	N	h	8-12	n	purple	Caesalpinia-like	*Xylocopa*
Melastoma malabathricum L.	G	s	N	h	8-12	n	purple	Caesalpinia-like	*Xylocopa*
Memecylon sp. 1	3	s	N	h	d	p	white	spiral	d Curculionidae, Chrysomelidae, Anthicidae
Memecylon sp. 2	G	s	N	h	d	p	purple	spiral	*Trigona*
Pternandra multiflora Cogn.	G	s	N	h	d	n, p	yellow	spiral	d Acrididae, several families of Coleoptera,
MELIACEAE									
Aglaia palembanica Miq.	2	s	N	d	d	p	yellow	cup	d
Dysoxylum cauliflorum Heim.	3	b	N	h	d	n, p	white	campanulate	*Nomia*
Walsura sp.	3	s	N	h	d	n	white	cup	*Apis koschevnikovi*

235

Appendix A. *Continued*

Plant

Family, genus, species & author	1)	2)	3)	4)	5)	6)	Color 7)	Flower shape 8)	Main pollinators 9)
MENISPERMACEAE									
Diploclisia kunstleri (King) Diels	L	b	N	d	d	n, p	white	cup	d	Syrphidae, several families of Diptera, Coleoptera, Hymenoptera
MORACEAE									
Artocarpus integer L.	4	b	GF	m	n	inflorescence	green*		Cecydomyiidae
Artocarpus odoratissimus Blanco	G	s	GF	m	n	liquid	brown*		d	Drosophilidae, Geometridae, Nitidulidae, several families of Diptera, Lepidoptera, Coleoptera
MUSACEAE									
Musa campestris Becc.	G	s	N	m	d	n	Pink	tubular	Spiderhunter (*Arachnothera robusta*)
MYRISTICACEAE									
Gymnacranthera contracta Warb.	4	s	N	d	d, n	p	white	campanulate	bp	Curculionidae, Chrysomelidae
Horsfieldia grandis (Hk. f.) Warb.	3	s	N	d	d,n	p	yellow	chamber with slits	Thripidae
Kema tridactyla Airy Shaw	1	i	N	d	d, n	p	yellow	chamber with slits	bp	Curculionidae
Knema cinerea (Poir.) Warb. var. *sumatrana* (Miq.) Sinclair	4	s	GF	d	d, n	p	yellow	chamber with slits	bp	Staphylinidae
Knema latifolia Warb.	2	i	N	d	d, n	p	yellow	chamber with slits	bp	Curculionidae

Appendix A. *Continued*

Plant

Family, genus, species & author	1)	2)	3)	4)	5)	6)	Color 7)	Flower shape 8)	Main pollinators 9)
MYRSINACEAE									
Ardisia macrophylla Reinw.	3	i	N	h	d	n	white	spiral	Sunbird (*Nectarinia jugularis*)
Labisia pumila (Bl.) Benth. et Hk. f.	1	b	GF	h	d	n, p	pale purple	spiral	*Nomia*
MYRTACEAE									
Eugenia (=SAN73069)	4	s	GF	h	d	n, p	white*	brush	*Trigona*
Eugenia sp. 4 (KYO)	4	s	N	h	d	n, p	white*	brush	d; Chrysomelidae, Elateridae, several families of Coleoptera, Diptera
Eugenia subrufa King	4	s	GF, G'	h	d	n, p	white*	brush	d *Apis dorsata*; Chrysomelidae, Curculionidae, several families of Coleoptera, Lepidoptera, also *Apis dorsata*
OLACACEAE									
Scorodocarpus borneensis Becc.	4	s	N	h	d	n	white	campanulate	*Braunsapis, Trigona*
Strombosia ceylanica Gardner	3	s	N	h	d	n	white	cup	d; Culicidae, several families of Diptera, Hemiptera
OPILIACEAE									
Lepionurus sylvestris Bl.	1	s	N	d	d	n, p	white	spiral	*Trigona*
ORCHIDACEAE									
Coelogyne foerstermanii Ridl.	E	s	GF	h	d	deciet	white	bilabiate	*Apis dorsata*

Appendix A. *Continued*

Plant Family, genus, species & author	1)	2)	3)	4)	5)	6)	Color 7)	Flower shape 8)	Main pollinators 9)
Dendrobium setifolium Reichb. f.	E	s	N	h	d	deciet	white	bilabiate	*Trigona*
Neuwiedia borneensis	1	s	N	h	d	deceit	white	bilabiate	*Trigona*
PALMAE									
Caryota sp.	2	s	N	h	d	p	yellow	cup	*Trigona*
Licuala sp.	1	b	N	h	d	n, p	white	cup	*Trigona*
PENTAPHRAGMATACEAE									
Pentaphragma lambirensis Kiew	1	s	N	h	d	n	orange	cup	*Amegilla*
Pentaphragma viride Stapf et Green	1	s	N	h	d	n	white	cup	*Nomia*
PIPERACEAE									
Piper vestitum C. DC.	1	s	N	h	morning	n	white*		d	Muscidae, Drosophilidae, Anthicidae, Curculionidae
Piper sp.	1	s	N	h	d	p	yellow*		bp
POLYGALACEAE									
Polygala venosa Juss. ex Poir.	1	b	N	h	d	n	purple	papilionaceous	*Amegilla*
PROTEACEAE									
Heliciopsis artocarpoides (Elmer) Sleum.	2	b	N	h	d	n, p	white	campanulate, deeply lobed	*Trigona*
RHAMNACEAE									
Ventilago malaccensis Ridl.	4	s	N	h	d	n, p	yellow	spiral	*Trigona*
RHIZOPHORACEAE									
Carallia brachiata (Loer.) Merr.	3	s	N	h	d	n, p	green	spiral	d	Chrysomelidae, Canthari-dae, several families of Coleoptera, Hymenoptera, Diptera
ROSACEAE									
Parastemon urophyllum A. DC.	4	s	N	h	d	n, p	white	cup, zygomorphic	*Trigona, Apis koschevnikovi*

Appendix A. *Continued*

Plant

Family, genus, species & author	1)	2)	3)	4)	5)	6)	Color 7)	Flower shape 8)	Main pollinators 9)
RUBIACEAE									
Chasalia sp.	1	s	N	h	d	n, p	white	cup	*Trigona*
Ixora brevicaudata Brem.	1	s	N	h	d	n	orange	tubular	Several families of Lepidoptera (butterfly)
Ixora stenophylla (Korth.) Kuntze	1	s	N	h	d	n	orange	tubular	Several families of Lepidoptera (butterfly)
Ixora urophylla Brem.	1	s	N	h	d	n	orange	tubular	Several families of Lepidoptera (butterfly)
Ixora woodii Brem.	1	s	N	h	d	n	orange	tubular	Several families of Lepidoptera (butterfly)
Ophiorrhiza axillaris Ridl.	1	i	N	h	d	n, p	white	cup	*Trigona*
Pavetta cf. petiolaris Craib. ex Brem.	2	s	GF	h	d	n	white	tubular	Sunbird (*Nectarinia jugularis*)
Pravinia sp.	G	i	N	h	d	n	purple	tubular	Sunbird (*Nectarinia jugularis*)
Steenisia pleurocarpa (Airy Shaw) Bakh. f.	1	s	N	h	d	n, p	white	spiral	*Trigona*
Uncaria longiflorea Merr.	L	s	N	h	d	n	white	tubular	d Otitidae, Chalcididae, several families of Diptera, Hymenoptera
Urophyllum hirsutum Hk. f.	2	i	N	d	d	n, p	white	cup	*Trigona*
Urophyllum pelacalyx Ridl.	2	i	N	d	d	n, p	yellow	cup	*Trigona*

Appendix A. *Continued*

Plant Family, genus, species & author	1)	2)	3)	4)	5)	6)	Color 7)	Flower shape 8)	Main pollinators 9)
Xanthophytum brookei Aexelius	1	s	N	h	d	n, p	white	cup	*Trigona*
RUTACEAE									
Glycosmis sp.	1	i	N	h	d	p	yellow	spiral	*Trigona*
SAPINDACEAE									
Allophyllus cobbe (L.) Laeusch.	1	s	N	h	d	p	white	spiral	d Curculionidae
Nephelium cuspidatum Bl.	3	s	GF	ad	d	n, p	white	brush	*Trigona*
Pometia pinnata Forst.	4	s	N	am	d	n, p	white	brush	*Braunsapis*
SAPOTACEAE									
Ganua (B16)	4	i	GF	h	d,n	c	white	berry-like	Mammal (*Callosciurus prevostii caroli, Sundasciurus hippurus inauinanius, S. lowii; Petaurista petaurista rajah*)
Ganua beccariana Pierre ex Dubard	3	i	GF	h	d, n	c	white	berry-like	Bat
Madhuca (T1)	3	i	GF	h	morning	n, p	white	campanulate, brush	*Xylocopa*
Madhuca (Y141)	3	i	GF	h	d	n	white	campanulate, brush	Sunbird (*Nectarinia jugularis*)

Appendix A. *Continued*

Plant Family, genus, species & author	1)	2)	3)	4)	5)	6)	Color 7)	Flower shape 8)	Main pollinators 9)
Palaquium beccarianum (T. et B.) Pierre	3	s	GF	h	d	n	white	campanulate, brush	Spiderhunter (*Arachnothera robusta*)
Palaquium sp.	3	s	GF	h	d	n	white	burst open, cup	Bird (*Loriculus galgulus*)
Payena acuminata (Bl.) Pierre	4	s	GF	h	d	n	white	cup	*Apis dorsata*
SIMAROUBACEAE									
Quassia borneensis Nooteboom	4	s	N	h	d	n, p	yellow	spiral	d	Staphylinidae, Elateridae, several families of Diptera, Coleoptera
STERCULIACEAE									
Heritiera borneensis (Merr.) Kosterm.	5	s	GF	m	d, n	p	white*	campanulate	bp	Curculionidae, Chrysomelidae
Heritiera sumatrana Kosterm.	3	s	GF	m	d, n	p	red*	campanulate	bp	Curculionidae, Chrysomelidae
Pterocymbium tubulatum (Mast.) Pierre	5	s	GF	m	7–14	n	white*	campanulate	*Apis koschevnikovi, Trigona*
Scaphium borneensis (Merr.) Beumee	4	s	GF	m	7–14	n	white*	campanulate	*Apis dorsata*
Scaphium longipetiolatum (Kosterm.)	5	s	GF	m	7–14	n	red*	bilabiate	*Trigona, Nomia, Megachile*
Sterculia laevis Wall.	2	s	N	m	d, n	p	red*	urceolate	bp	Chrysomelidae
Sterculia stipulata Korth.	1	s	N	m	d, n	p	red*	urceolate	bp	Chrysomelidae
THEACEAE									
Eurya acuminata (DC.) Merr.	G	s	N	m	d	n, p	white	cup	*Trigona*

241

Appendix A. *Continued*

Plant

Family, genus, species & author	1)	2)	3)	4)	5)	6)	Color 7)	Flower shape 8)	Main pollinators 9)
TILIACEAE									
Grewia sp.	2	s	GF	h	d	n, p	white	cup	*Trigona*
Grewia latistipula Ridl.	3	s	N	h	d	n, p	yellow	cup	*Trigona, Ceratina, Nomia*
Grewia stylocarpa Warb. ex Perkins	3	s	N	h	d	n, p	yellow	cup	d	Vespidae, several families of Hymenoptera, Diptera, Coleoptera
Schoutenia glomerata King	3	s	GF	h	d	n, p	white	cup	d	Chrysomelidae, Staphylinidae, several families of Hymenoptera, Diptera, Coleoptera
TRIGONIACEAE									
Trigoniastrum hypoleucum Miq.	G	s	GF	h	d	n	white	papilionaceous	*Nomia*
TRIURIDACEAE									
Sciaphila secundiflora Thw. ex Bth.	1	s	N	h	d	n	white	spiral	Culicidae, Calliphoridae
ULMACEAE									
Trema tomentosa (Roxb.) H. Hara	G	s	N	h	d	n, p	white*	spiral	*Trigona*
URTICACEAE									
Pipturus argentenus (Forst.) Wedd.	G	s	N	d	n	p	white*	spiral	*Trigona*
Dendrocnide stimulans (L. f.) Chew	1	s	N	d	n	p	white*	spiral	*Trigona*
VERBENACEAE									
Callicarpa pentandra Roxb.	G	s	N	h	d	n, p	purple	brush	*Trigona*
Callicarpa havilandii (King et Gamble) H. J. Lam	G	s	N	h	d	n, p	purple	brush	*Trigona*

Appendix A. *Continued*

Plant

Family, genus, species & author	1)	2)	3)	4)	5)	6)	Color 7)	Flower shape 8)	Main pollinators 9)	
Clerodendron phyllomega Steud.	1	s	N	h	d	n	red	brush		Several families of Lepidoptera (butterflies)
Sphenodesma triflora Wight	L	s	GF	h	d	n, p	purple, purple*	brush	*Nomia*	
Stachytarpheta indica (L.) Vahl	G	s	N	h	d	n, p	yellow	bilabiate	*Braunsapis, Ceratina, Trigona*	
Teijsmanniodendron simplicifolia Merr.	2	s	N	h	d	n, p	purple	bilabiate	d	Curculionidae, several families of Coleoptera, Hemiptera
Vitex pubescens Vahl	G	s	N	h	d	n	purple	bilabiate	*Xylocopa*	
Vitex vestita Wall.	G	s	N	h	d	n	yellow	bilabiate	*Nomia*	
XANTHOPHYLLACEAE										
Xanthophyllum sp.	1	s	N	h	d	n, p	white	papilionaceous	*Megachile*	
Xanthophyllum velutinum Chod.	3	s	GF	h	d	n, p	yelowish	papilionaceous	*Xylocopa, Megachile*	
ZINGIBERACEAE										
Alpinia glabra Ridl.	1	s	N	h	d	n, p	red*	bilabiate	*Amegilla*	
Amomum calyptratum Nagam. et S. Sakai	1	b	N	h	d	n, p	orange*	bilabiate	*Amegilla*	
Amomum coriaceum R. M. Sm.	1	b	N	h	d	n, p	white*	bilabiate	*Nomia*	
Amomum durum S. Sakai et Nagam.	1	b	N	h	d	n, p	white*	bilabiate	*Nomia*	
Amomum gyrolophos R. M. Sm.	1	b	N	h	d	n, p	orange*	bilabiate	*Amegilla*	
Amomum oliganthum K. Schum.	G	b	N	h	d	n, p	orange*	bilabiate	*Amegilla*	

Appendix A. *Continued*

Plant									
Family, genus, species & author	1)	2)	3)	4)	5)	6)	Color 7)	Flower shape 8)	Main pollinators 9)
Amomum dimorphum I. Turner	G	b	N	h	d	n, p	white*	bilabiate	*Nomia*
Amomum angustipetalum S. Sakai et Nagam.	1	b	N	h	d	n	white*	bilabiate	*Amegilla*
Amomum roseisquamosum Nagam. et S. Sakai	3	b	N	h	d	n	pink*	bilabiate	Spiderhunter (*Arachnothera longirostra*)
Amomum somniculosum S. Sakai et Nagam.	1	b	N	h	10-16	n, p	white*	bilabiate	*Nomia*
Boesenbergia grandiflora (Val.) Merr.	G	b	N	h	6-18	n, p	white*	bilabiate	*Nomia*
Boesenbergia gracilipes (K. Schum.) R. M. Sm.	1	b	N	h	6-15	n, p	white*	bilabiate	*Nomia*
Boesenbergia sp.	1	b	N	h	6-18	n, p	white*	bilabiate	*Nomia*
Elettaria longituba (Ridley) Holt.	G	b	N	h	6-18	n, p	white*	bilabiate	*Nomia*
Elettariopsis kerbyi R. M. Sm.	1	b	N	h	6-	n, p	white*	bilabiate	*Nomia*
Elettariopsis sp. 1	1	b	N	h	d	n, p	white*	bilabiate	*Amegilla* / *Nomia*
Elettariopsis sp. 2	1	b	N	h	6-	n, p	white*	bilabiate	*Nomia*
Etlingera velutina Ridl.	G	b	N	h	6-18	n	red*	bilabiate	Spiderhunter (*Arachnothera longirostra*)
Etlingera punicea (Roxb.) R. M. Sm.	G	b	N	h	d	n	red*, yellow*	bilabiate	Spiderhunter (*Arachnothera longirostra*)

Appendix A. *Continued*

Plant

Family, genus, species & author	1)	2)	3)	4)	5)	6)	Color 7)	Flower shape 8)	Main pollinators 9)
Globba brachyanthera K. Schum.	1	s	N	h	d	n, p	orange*	bilabiate	*Amegilla*
Hornstedtia reticulata (K. Schum.) K. Schum.	G	b	N	h	d	n	red*	bilabiate	Spiderhunter (*Arachnothera longirostra*)
Hornstedtia minor (Bl.) K. Schum.	G	b	N	h	6–18	n	red*	bilabiate	Spiderhunter (*Arachnothera longirostra*)
Plagiostachys crocydocalyx (K. Schum.) Burtt et Smith	G	b	N	h	d	n, p	white*	bilabiate	*Amegilla*
Plagiostachys glandulosa S. Sakai et Nagam.	G	b	N	h	d	n, p	white*	bilabiate	*Amegilla*
Plagiostachys strobilifera (Bak.) Ridl.	G	s	N	h	6–18	n	pinc*, yellow*	bilabiate	Spiderhunter (*Arachnothera longirostra*)
Zingiber longipedunculatum Ridl.	G	b	N	h	12–18	n, p	white*,red*	bilabiate	*Amegilla*

[1] **Plant habit** 1: forest floor; 2: understory; 3: subcanopy; 4: canopy; 5: emergent, G: gap; L: liana, E: epiphyte

[2] **Flower position** s: on crown surface; i: inside crown; b: below the crown (cauliflorous)

[3] **Reproductive phenology** N: has flowers in non-general-flowering periods only, or both in non-general- and general-flowering periods; G': has flowers in general-flowering periods only; G': pollinators differ in general-flowering periods

[4] **Sexual expression** h: hermaphrodite; d: dioecy; m: monoecy; ad: androdioecy; am: andromonoecy

[5] **Flowering time** d: daytime; n: night time

[6] **Reward** n: nectar; p: pollen; c: corolla; s: stigmatic secretion

[7] *: color of calyx, stamen, stigma, bract, inflorescence; no mark: color of corolla

[8] *: shape of inflorescence; no mark: shape of corolla or calyx

[9] bs: beetles mainly feeding on stigmatic secretions; bf: beetles mainly feeding on floral tissues; bp: beetles mainly feeding on pollen; d: diverse insects

245

Appendix B. List of *Ficus* found in and around Lambir Hills National Park, Sarawak and associated ecological information

1	2	3	4	5	6
Subgenus *Urostigma*					
Section *Urostigma*					
F. caulocarpa Miq.	M	H	3	*Platyscapa fischeri* Wiebes	RH568
F. virens Ait. v. glabella (Bl) Corner	M	H	3	*Platyscapa coronata* (Grandi)	RH68, 76, 202
Section *Conosycea*					
Subsection *Conosycea*					
Series *Validae*					
F. sp. nov. (near annulata)	M	H	0	*Deilagaon megarhopalum* Wiebes*	RH67, 134, 173
Series *Drupaceae*					
F. cucurbinita King	M	H	5		RH83, 88, 89, 92, 94, 176
F. drupaceae Thunb.	M	H	2	*Eupristina (E.) belgaumensis* Joseph	RH571
Series *Indicae*					
F. kerkhovenii Val.	M	HF	11	*Eupristina (E.) leightoni* Wiebes*	NA531, 4777, 4784, RH35, 97, 112, 113
Series *Zygotricheae*					
F. bracheata Wall. ex Miq.	M	H	1		RH570
F. consociata Bl.v. murtoni King	M	H	4	*Waterstoniella malayana* Wiebes	RH53, 109, 200
Series *Crassirameae*					
F. palungensis Weiblen	M	HC	18		RH55, 545
F. stupenda Miq.	M	H	7	*Waterstoniella masii* (Grandi)	RH70, 87, 157
F. subgelderj Corner	M	H	20	*Waterstoniella sp.**	MO712, RH58, 108, 133, 138, 215, 217, 542
F. subtecta Corner	M	H	1		RH572
F. xylophylla Wall. ex Miq.	M	H	19	*Waterstoniella grandii* Wiebes	RH155
Subsection *Dictyoneuron*					
F. binnendykii Miq.	M	H	0		
F. bomeensis Kochummen	M	H	O	*Waterstoniella borneana* Wiebes	
F. delosyce Corner v. obtusa Corner	M	H	24	*Waterstoniella delicata* Wiebes	MO546, RH198, RR150
F. dubia Wall. ex King	M	H	9	*Waterstoniella sp.*	RH11,121, 129, 174
F. glaberrima Bl.	M	H	2	*Waterstoniella williamsi* Wiebes	RH549
F. pellucido-punctata Griff.	M	H	3	*Waterstoniella brevigena* Wiebes	RH573

Appendix B. *Continued*

1	2	3	4	5	6
F. pisocama Bl.	M	H	4	*Waterstoniella* sp.	RH130, 136, 518
F. retusa L.	M	H	8	*Waterstoniella javana* Wiebes	RH574
F. soepadmoi Kochummen	M	H	9		RH196
F. sumatrana Miq.	M	H	0	*Waterstoniella sumatrana* Wiebes	RH69, 82, 122, 140, 171, 180, YA116
F. sundaica Bl.	M	H	0	*Waterstoniella sundaica* Wiebes	RH575
Subsection *Benjamina*					
F. benjamina L.	M	H	0	*Eupristina (P) koningsbergeri* Grandi*	NA332, 4850, RH32, 56, 84, 102, 103, 104, 123, 203
F. callophylla Bl.	M	H	1		RH98, 99, 543
F. subcordata Bl.	M	H	2	*Eupristina (E) philippinensis* Wiebes	NA368, RH72, 74, 81, 199, YA130
F. stricta Miq.	M	HF	0	*Eupristina (P) cyclostigma* Wiebes	RH33, RR89
F. tristanifolia Corner	M	H	1	*Eupristina* sp.	RH577
Subgenus *Pharmacosycea*					
Section *Oreosycea*					
F. vasculosa Wall ex. Miq. v. *acuminata* Miq.	M	LT	13	*Dolichoris vasculosae* Hill	RH547
Subgenus *Ficus*					
Section *Ficus*					
Subsection *Ficus*					
F. deltoidea Jack v. *deltoidea*	D	S	0	*Blastophaga (B) quadrupes* Mayr	RH519, 536
F. deltoidea v. *arenaria* Corner	D	S	0		RH535
F. deltoidea v. *obtusa* Corner	D	S	0		
F. deltoidea v. *bomeensis* Corner	D	E	0		RH2
Subsection *Erisycea*					
F. androchaete Corner	D	ST	3		RH24
F. aurata Miq.	D	ST	39	*Blastophaga (V) auratae* Wiebes	MA465, RH260
F. brunneo-aurata Corner	D	LT	2	*Blastophaga* sp.*	RH537
F. chartacea Wall. ex King	D	ST	139	*Blastophaga (V) medusa* Wiebes	NY183, RH183
F. fulva Reinw. ex Bl.	D	ST	53	*Blastophaga (V) compacta* Wiebes*	RH13, 52
F. glandulifera (Wall. ex Miq.) King	D	ST	70	*Blastophaga (V) sensillata* Wiebes	RH549
F. glossularioides Burm. f.	D	ST	33	*Blastophaga (V) malayana* Wiebes	

Appendix B. *Continued*

1	2	3	4	5	6
F. lamponga Miq.	D	ST	43		RH548
F. setiflora Stapf	D	ST	280	Blastophaga (V) borneana Weibes	RH153, 224, 227, 258
Section *Rhizocladus*					
F. callicarpides Corner	D	C	2		RH530
F. grossivenis Miq.	D	C	5		RH141, 219, 538
F. lanata Bl.	D	C	0		RR179
F. recurva Bl.	D	C	3	Wiebesia boldinghi (Grandi)	JR179, RH540
F. sagittata Vahl.	D	C	0	Wiebesia flava Wiebes	RH539
F. spiralis Corner	D	C	0		RH178
F. trichocarpa Bl.	D	C	0	Wiebes vechti Wiebes	RH96, 110, 116, 117, 194
F. uncinulata Corner	D	C	0		RH196
F. urnigera Miq.	D	C	0	Wiebes sensillata Wiebes	RH532
F. villosa Bl.	D	C	3	Wiebes minuta Wiebes	IK10, RH153, 501, 525, 533
Section *Kalosyce*					
F. aurantiacea Griff. v. angustifolia Corner	D	C	0	Wiebesia planocrea Wiebes	RH531
F. aurantiacea v. parvifolia Corner	D	C	15	Wiebesia contubernalis Grandi*	RH544
F. punctata Thunb.	D	C	2	Wiebesia punctatae Wiebes	RH101, 107, 119
F. ruginervia Corner	D	C	2	Wiebesia sp.	RH127, 529
F. sarawakensis Corner	D	C	5		RH534
Section *Sycidium*					
F. hemsleyana King	D	H,S	3	Liporrhopalum sp.*	RH527, 541
F. heteropleura Bl.	D	H,S	13	Liorrhopalum dubium Grandi	MO7219, NY268
F. leptogramma Corner	D	H,S	1	Kradibia setigera Wiebes	RH551, YA127
F. midotis Corner	D	HC	0	Liporrhopalum midotis Hill	RH182
F. obscura Bl.	D	H,S	3	Liporrhopalum giacominii Grandi	MO114, 244, RH25
F. parietalis Bl.	D	H	0		RH220, 259
F. rubrocuspidata Corner	D	H	3		
F. sinuata Thunb.	D	H,S	4	Liporrhopalum longicomus Grandi	RH526
F. subulata Bl.	D	HC	3	Liporrhopalum erythropareiae Hill	RH188
F. uniglandulosa Walt v. parvifolia Miq.	D	S	1	Liporrhopalum parvifoliae Hill	RH222

Appendix B. *Continued*

1	2	3	4	5	6
Section *Neomorphe,*					
F. *variegata* Bl. v. *variegata*	D	LT	3	*Ceratosolon* (C) *appendiculatus* (Mayr)	RH550
Section *Sycicarpus*					
F. *beccarii* King	D	ST	78	*Ceratosolon* (R) *humatus* Wiebes	RH17
F. *cereicarpa* Corner	D	ST	0	*Ceratosolon* (R) *pilipes* Wiebes*	RH19, 518
F. *condensa* King	D	ST	1	*Ceratosolon* (R) sp.*	RH43, 44, 47
F. *fistulosa* Reinw. ex Bl.	D	ST	119	*Ceratosolon* (C) *c constrictus* Wiebes	RH223, 546
F. *francisci* Winkier	D	ST	54	*Ceratosolon* (R) *josephi* Wiebes*	
F. *geocharis* Corner	D	LT	40	*Ceratosolon* (R) sp.*	RH18, 521, 522
F. *lepicarpa* Bl.	D	LT	0	*Ceratosolon* (R) *vechti* Wiebes*	RH521
F. *megaleia* Corner	D	S	0		
F. *schwarzii* Koord.	D	LT	0	*Ceratosolon* (R) *vetustus* Wiebes*	RH15, 212, 520, YA212
F. sp. nov. (near *megaleia*)	D	LT	106	*Ceratosolon* (R) sp.*	RH52, 179, 197, 523
F. *stolonifera* King	D	LT	780	*Ceratosolon* (R) sp.*	RH45, 65, 524
F. *treubii* King	D	LT	0	*Ceratosolon* (R) sp.*	RH3, 45, 507
F. *uncinata* (King) Becc. v. *uncinata*	D	S	0	*Ceratosolon* (R) *albulus* Wiebes*	RH129
F. *uncinata* v. *gracilis* Corner	D	S	4		MO1108, YA127
F. *uncinata* v. *pilosa* Corner					RH22
F. *uncinata* v. *subbeccarii* Corner	D	S	9		

[1] *Ficus* arrangement follows Corner (1965). Figs were collected between 1993 and 1998. Surveys include complete collections of a 52 ha plot (dbh > 1cm) and 8 ha plot (dbh > 10cm), surveys of hemi-epiphytes and climbers along trails (approx. 60 ha), and *ad hoc* collections both within the park and the surrounding shifting cultivation, particularly near Lupong Aji.

[2] Sexual system: D=dioecious, M=monoecious

[3] Habit: C=climber, E=epipyte, H=hemi-epiphyte, HF=hemi-epiphyte with freestanding phase, HC=hemi-epiphyte with vine-like climbing habit, S=shrub, ST=small tree (<8 m height), LT=large tree (>8 m height)

[4] Number of individuals in a 52 ha plot in Lambir Hills National Park (dbh > 1cm). Species of Sections *Ficus* and *Sycocarpus*, as pioneers, tend to be very clumped hence low abundance in the plot does not indicate rarity over a wider area.

[5] Pollinators (Agaoninae, Agaonidae), arrangement follows Wiebes (1994) *indicates reared from figs in LHNP

[6] *Ficus* specimens: distributed to Lambir, Kuching, Kyoto, Leiden, Kew, and Harvard. Collectors IK:Ikegami, JR:J Rahman, MA:Maida et al., MO:Momose, NA: Nagamitsu, NY:Nyambong, RH:R D Harrison, RR:R Rahman, YA:Yamauchi

Glossary

adaptive radiation: evolutionary diversification of a lineage into different niches

aerial tram: ski lift in rain forest canopy

aerial walkway: walkway in the forest canopy

aff. (affinis): denoting close taxonomic similarity

aggregated population: with clumped spatial distribution

aggressive dominance: dominance through aggressive behavior

allele: gene variant on one of two parental chromosomes (alleles of individuals occur in pairs, but populations usually possess many alleles)

allelochemical composition: composition of chemicals repelling individuals

allopatric: geographically separated spatial distribution

altitudinal gradient: gradual change along with elevation

Amazonia: South American region comprising the drainage basin of the Amazon River

animal seasonality: annual change in activity and/or abundance of animals

andosol: lightly weathered, base-rich reddish brown soil, derived from volcanic substrate

ant guards: mutualisitc protective ants, usually for plants hosting their colony

ant-defense: see *ant guards*

ant-exclusion experiment: removal of mutualistic ants from host plant

anthophilous beetle: beetle that visits flowers

antimicrobial floral resin: resin from certain flowers that kills bacteria, other microbes

ant-plant: plant that attracts or contains ants that protect it from herbivores or may remove competing plants

APGII: angiosperm phylogeny group, second report on classification

architecture (forest): structural properties of forest areas, such as canopy or lower strata

aril: outer seed cover, sometimes fleshy or brightly colored

arilate seed: see *aril*

armchair theory: speculation with no empirical data or experiment

aseasonal: usually unchanging within years

asynchronous flowering: lack of temporal coordination in flower production

autecology: study of major factors affecting a population

axially: positioned along the axis, usually a plant stem

bagging experiment: pollination experiment excluding visitors from flowers

ballistically dispersed: seeds ejected from splitting fruit

bar code: coded bands on a tape, read by machine to record data

basal area: tree trunk area near base

base pairs: purine and pyrimadine bases connected by hydrogen bonds, joining two strands of DNA in a chromosome

basidomycoflora: fungi, class Basidomycetes bearing spores on basidia

BCI: Barro Colorado Island, forested nature preserve, field station at Gatun Lake, Panama Canal

beta-diversity: a measure of taxonomic richness along a transect

bioassay: test for biological activity usually made to study reaction of an organism to a chemical

biodiversity: measure of taxonomic and ecological richness (see *diversity*)

biodiversity prospecting: systematic sampling of plants and other organisms for potential utility to humankind

biotic: with reference to living things

bisexual: flowers bearing male and female reproductive organs

bootstrapping: statistical method based on random sampling of data collected

bract: leaf-like plant part, often colorful or large, near flower

breakup of Gondwana: fragmentation of the great southern continent by seafloor spreading causing movement of tectonic plates since 230 million years ago

buttress: vertical support structures extending trunk base

caespetose: tufted

campanulate: bell-like shape

canopy: high cover of foliage on tree branches

canopy crane: metal construction crane positioned within forest to gain access to the forest canopy

canopy layers: horizontal vegetation at differing heights

canopy observation system: support for observers within forest canopy

canopy raft: inflatable, lighter-than-air platform giving access to canopy

canopy tree (emergent tree): tree superior in height to understory vegetation, having clear access to sunlight

catastrophic disturbance: relatively uncommon event with predominantly negative short-term consequences for living things

cation: positively charged molecule

cauliflorous: inflorescence or fruit on surface of a branch or trunk

CBPS: Canopy Biology Project, Sarawak

character displacement: evolutionary divergence between organisms, usually to diminish competition

chiropterophily: adaptation to pollination by bats

chromosome: DNA-containing strands, in nuclei, with genetic blueprint for organism

clade: phylogenetic unit, inferred from derived characters (features)

cleptoparasite: parasite that seeks stored food of other female to rear its offspring

climatic oscillation: fluctuating climate, such as rainfall or temperature

clumped distribution: see *aggregated population*

coccid: soft-bodied bug, Homoptera, that produces waxy secretion

coevolution: evolutionary change that is multilateral and often causes speciation or changes in more than one population

coevolutionary process: dynamic interaction of populations culminating in adaptive changes

coexistence: sustained geographic cohabitation by two or more species or individuals

cohort: individuals of same age in habitat, usually of one species

COI gene: cytochrome oxidase 1 subunit of a mitochondrial gene

colonized: evolutionary ecology, denoting new association crossing mutualist lineages

color vision: animal vision including colors, more than shades of black and white

community: individuals or groups within a determined habitat type

community structure: species in community and relative abundance

competition: mutually detracting interaction whereby gain of one participant signifies loss by another

complete partitioning: complete separation in resources used by two or more groups

congeners: species belonging to one genus

conspecific: of same species

continental drift: movement of major land masses through slow expansion of seafloor, pushing apart landmasses

convergent adaptation: adaptive changes that produced similar results

cores: geological or other sample drawn as a section, usually across time

corridors: avenues of continuous habitat

cosmopolitan: widespread, generally worldwide distribution

co-speciate: evolutionary diversification of two significantly interacting populations simultaneously and sympatrically

co-speciation: see *co-speciate*

crashes: abrupt decline, usually in abundance

Cretaceous: geologic age, 120 to 67 million years ago

crop size: number of mature fruits or seeds produced

cross-pollination: pollen transfer to stigma of female flowers of different individual plants of the same species

cryptic species: organisms appearing nearly identical but not of the same species

CTFS: Center for Tropical Forest Science (Smithsonian Tropical Research Institute)

cuesta: ridge of distinct scarp and dip slopes along the broken edge of inclined rock strata

dbh (diameter at breast height): tree trunk dimension

dehisce: release of pollen from anther or mature seeds from fruit

dehiscent berry: fleshy fruit that separates from infructescence, having few of many seeds

density-dependent: ecological factor, e.g., mortality, directly related to density of individuals in local population

diapause: quiescent or resting stage, suspension of growth or reproduction, usually induced and terminated by environmental cue

differential mortality: mortality that differs among recognized kinds of individuals

dioecious (dioecy): plant reproductive system of separate individuals with either male, pollen-bearing flowers, or female flowers bearing ovules and stigmata

diploid: bearing a set of maternal and paternal chromosomes

dipterocarp: tree, family Dipterocarpaceae, dominant in many aseasonal and seasonal Asian forests, most important timber source

disperser: animal that carries and distributes seeds

disturbed area: habitat altered to the point of losing species richness and diversity

diurnal anthesis: flowers opening during the day

diversity: measure of taxonomic richness or evenness of dominance by taxa

domatia: plant structures, usually hollow stems, wherein ant colonies live

dry season: annual period of relatively low rainfall

echo-locate: bat perception of objects reflecting their sonic waves

ecological fitting: adaptive interactions among community members, one or more of which did not evolve there and are therefore newcomers

ecological release: increased success in the absence of competitors or natural enemies

ecosystem: the environment and its component species populations, geology, climate, and processes

ecotourism: business from tourists seeking relatively undisturbed, natural habitats

ectomycorrhizal: symbiotic fungi growing mycelia on the surface of other organisms, often roots of tropical plants

edaphic gradient: changing soil conditions

emergent platform: platform positioned at the top of large tree

emergent: tree higher than surrounding trees in mature forest

endemic (endemism): geographically restricted natural occurrence

ENSO (El Niño-Southern Oscillation): heating of western Pacific Ocean, followed by movement of warm air and sea surface conditions eastward, causing abundant rain or drought

environmental cue: perceived, external natural factor prompting a response

epiphyte: plant growing on rocky substrate or another plant

escape hypothesis: theory that specialist natural enemies promote species richness by favoring scarcity (see *Red Queen hypothesis*)

eusocial: insect colonies whose reproductive castes and different generations co-exist

evapo-transpiration: water vapor from leaf surfaces

everwet: perhumid: always under wet or moist conditions

exaptation: trait that serves in a new context as an adaptation, also pre-adaptation

explosive flower opening: forceful, abrupt opening of flower petals

extrafloral nectar: structure for extrafloral nectar (EFN) production, not in flowers

familial: in the same taxonomic family

fan-leaves: large, spreading leaves, greater than 3 m, typified by some palms

feeding organs: mouthparts or structures designed to consume resources

feeding patches: areas where food is located and taken

felling cycle: periodicity of tree harvest

female flowers: flowers with stigmata to receive pollen, a style through which pollen tubes grow, and ovules to produce seeds

female trees: fruit and seed bearing individuals, with no pollen production

fig wasps: small wasps, of large parasitic superfamily Chalcidoidea, family Agaonidae, which reproduce by females "parasitizing" fig inflorescences (syconia) where the larvae develop, and pollinate figs after emerging as adults

fitness: relative ability of individual to pass genes to the next generation's gene pool

floristic composition: numbers and kinds of plant species present

flower visitor: animal that takes food, usually nectar or pollen but also in some species oils or resins, from angiosperm flowers which it may pollinate

flowering phenology: see *flowering tempo*

flowering tempo: timing of flower opening

flush (new leaves): production of new leaves

folivores: animals that feed on foliage—leaves or other green parts

food bodies: protein-rich plant products, usually feeding mutualist symbionts

foraging strategy: rules or plans for taking nourishment from dynamic resources that must be located and competed for

forest edge: boundary of forest, where light, heat, moisture, and physical factors change

forest gap: opening within forest, where light may penetrate (see also *forest edge*)

foundress queen: queen of a social insect that establishes nest and colony after mating

free-living: not parasitic

frugivore: fruit-eating animal

fruit drop: synchronized fruit ripening among trees

fruit set: proportion of flowers becoming fully mature fruit with seeds (not fruit *formation*, which does not include ripening or viability)

fruit: usually sweet or fleshy portion of ripened plant ovary or ovaries, containing seeds

fruiting: fruit-bearing condition

fungus: an organism, neither plant nor animal, with mycelia, fruiting bodies, and occasionally producing spores

geitnogamy (geitnogamous): pollen movement among flowers on an individual

gen. et sp. nov.: taxonomic designation of a new species and new genus

gene: sequence of base pairs in DNA molecule with information for construction of specific protein molecule

general flowering (GF): periodic, local community-wide interval, usually two to four months, when most trees bear flowers

generalist: an organism that uses, feeds upon, or interacts, significantly, with multiple species, habitats or resources; comparable to biological diversity (in contrast to species richness) in the utilization of resources or use of habitats (see also *specialist*)

genetic blueprint: DNA-coded polypeptides, translated to phenotype and visible traits, after interaction with the environment, internal (e.g., cellular) and external

genetic program: genotype, from which phenotype produced; see *genetic blueprint*

genetic recombination: rearrangement of organism's genes on each chromosome during meiosis, so that resulting gametes (eggs or sperm) are not identical to each other or to those of parents

geocarpic: inflorescence and fruit borne on specialized stolons close to ground level

geomorphology: study and aspects concerning configuration and evolution of land forms

germination: initial growth of pollen tube from stigma into style, or of a seed in a soil micro-site

GF period: see *GF*

GF—non-GF cycle: complete period of about 2 to 10 years (mean = 4 to 5), from general flowering, or, GF, to the end of following, longer interval when most trees bear no flowers or fruit

giant honeybee: *Apis* of large size (*dorsata, laboriosa, bingihami, breviligua*) nesting on their exposed giant honeycomb on branch or rocky substrate

global warming: worldwide rise in average temperature

gondola: observation vehicle suspended from canopy observation crane cable

Gondwana: southern continent before its breakup at 230 million years ago

guild: group of ecologically close species performing similar community roles

habitat gradient: gradual or linear variation in habitat type

hemi-epiphyte: epiphytic plant that may grow independently as a vine

herbarium curator: manager of dried plant specimens in a repository, for study

herkogamy (dichogamy): separation in space of male and female organs (stamen and carpel, or anther and stigma) within a flower

heterozygosity: index of genetic diversity based on proportion of individuals with different alleles at a locus

homogamous: having simultaneous pollen dehiscence and stigma receptivity

honeydew: sugary secretion of sap-feeding insects

hornbill: large frugivorous bird, *Buceros,* of Paleotropics, characterized by large, thick bill

host-switching: transfer to a different lineage of hosts (horizontal transfer)

humult soil: acidic soil (pH $<$ 4.2), with surface litter accumulation and root-matted raw humus

humult specialists: plants growing most successfully on humult soil

hyper-diverse: with extremely large number of species and even larger number of potential interactions, mega diverse

illipe nut: abundant forest nut after GF period, from dipterocarp tree, *Shorea,* section *Pachycarpae*, gathered for edible oil

inceptisol: young soil in which no leaching or accumulation of organic matter occurs

indehiscent: flower or fruit that does no split at maturity

inflorescence: flowers grouped together on a stem

inquiline: parasite that spends its entire life within the nest of a host

insinuation: tactic whereby forager approaches aggressive forager without attack

insolation: exposure to sunlight

interference competition: competition including purposeful attack or interference

intermast interval: non-GF, time between GFs or masting events

internode: stem area between two nodes

interspecific: between species

intertropical convergence zone (ITCZ): atmosphere along both sides of equator, where convective thunderstorms are frequent and air pressure low

intraspecific synchrony: synchrony of event for individuals of a species

inventory: tabulation of species or taxonomic groups

ironwood: Belian (family Lauraceae, *Eusideroxylon*), very heavy, resistant wood

irradiation: covering with sunlight

Kerangas forest: forest on acidic, undecomposed organic (peat) soil, freely drained of water

kerangas heath forest: forest on nutrient-poor soil; see *Kerangas forest*

keystone resource: thought to determine the survival of many forest species

keystone species: see *keystone resource*

kriging: statistical method, generalizing point-sampled data into map units

La Niña: preceding or in sequence with El Niño year, unusually having abundant rain

leach: removal of soluble material by percolation

leaf area: area composed by leaves

leafing: production of new leaves

leaf-litter: leaves, plant parts, entering soil decomposition cycle

lepidopteran: butterfly and moth order, Lepidoptera

LHNP: Lambir Hills National Park, Sarawak

life cycle: complete life qualities, from birth to death

life history: see *life cycle*

light environment: amount and quality of light

limestone karst: landscape developing on rocks composed mainly of calcium carbonate of biological origin

lineage: phylogenetic group with shared, recent ancestors

lithological substrate: rocky substrate

locomotory physiology: design and functioning of animal appendages for movement

loose niches: species roles, especially interactions, that shift between generations or relatively short intervals and depend on abundance of participants

Malaise trap: intercept trap for flying insects, using netting and collection jars

male phase: male (stamen, with anther and pollen) period of reproductive cycle or flower lifespan

male strobili: conelike structures or sections of male flowers

malvalean: plants in Malvaceae and relatives

mass flowering: community-wide blooming of many tree species, followed by a general lapse in flowering for years; see *general flowering, masting*

mast seeding: community-wide production of many fruits and seeds (masting), following mass flowering

masting habit: see *mast seeding*

masting: characteristic flowering and fruiting only at intervals of more than one year

mega-event: a community-wide phenomenon including most species and biomass

meranti: dipterocarp timber tree, *Shorea*, giant of the forest

Mesoamerica: region between Mexico and Panama

mesothorax: insect body between head and abdomen

microbe: microorganism, neither plant, animal, fungus

micro-habitat: small area (several mm to m) of particular conditions

micro-lep: Microlepidoptera, moths of very small size

microliter (µL) one-thousandth milliliter

microsite: very small area (centimeter-scale or less) where seed or pollen germination occur with success

migration: movement to distant locality and later return

Miocene: geological period 20–3 million years ago

mist net: horizontal net that ensnares flying animals for live capture and study

mixed dipterocarp forest: SE Asian forest dominated by many species of dipterocarp trees

molecular clock: calibration of time from molecular sequence divergence, i.e., a divergence of 2% occurs in 2 million years

monocyclic system: timber harvest that takes place once

monoecious: a plant producing male and female flowers on different portions of an individual

monophyletic: a taxon descended from most closely related taxon of the same or lower rank; group of direct descent from single ancestral population

monsoon: wind system affecting large regions by reversing direction periodically, bringing wet or dry conditions

morphological matching: correspondence in physical dimensions

morphological: pertaining to measure and form

mtDNA (mitochondrial DNA): genes of cell cytoplasm in the organelles, mitochondria

multivariate analysis: branch of statistics concerned with analysis of multiple measurements on a sample of individuals

mutation: change in single gene involving replacement, duplication, or deletion of base pairs

mycelium (mycelia): vegetative (nonreproductive) branching filaments of fungi

myrmecophyte: plant having structures (domatia) or adaptations to obtain ant symbionts; see *ant-plant*

nectar guide: visual, ultraviolet-reflecting marking indicating location of floral nectar to flower visitor

nectar load: nectar harvested by flower visitor

nectar secretion: production of nectar from specialized plant tissue, nectary

negative density-dependent: selection at high abundance, relaxing when organism is rarer

Neogene: geogological time scale from the Miocene until the end of Pliocene (1.6 million years ago)

Neotropics: tropical regions in North, Central, and South America

nest dispersion: nest distribution in space

niche: see *niche differentiation*

niche differentiation: divergence between species or individuals in ecological requirements and performance

non-arboreal: not associated with trees

non-biotic (abiotic): non-living

non-dispersed seeds: seeds not transported by animals to germination sites

non-GF period: interval between community-wide flowering events

non-pollinating flower visitor: animal not pollinating a plant but taking floral rewards, nectar and pollen

non-volant mammals: mammals incapable of flight

normal distribution: bell-shaped distribution of trait frequencies, extremes least frequent and mean most frequent

oblanceolate: shaped like elongate spear tip

obligate specific-pollinator: pollinator that must pollinate one particular plant species in order to effect its own reproduction

obligately outcrossing: plant unable to produce seeds using own pollen

oligarchy: tree comparatively common across vast forest areas

open pollination: natural pollination, unimpeded by bagging or other experimental techniques used to study plant breeding system

ornithophily: adaptive association with birds, usually in reference to pollination

ostiole: tiny opening leading to concealed fig inflorescence

outcrossing: dispersal of pollen to, or pollen reception from, other flowers of same species

ovipositor: tube through which certain female insects deposit eggs

paleographic history: study of ancient documents

paleontology: study of fossils or ancient life forms

paleotropics: tropical regions in Asia and Africa

Paleozoic: geological era when life diversified, 600 million to 230 million years ago

palynology: pollen and spore science

pan-tropical: distributed throughout tropical latitudes (23', 27" of equator)

paradigm: structured idea or concept (model) applied to body of knowledge

parasitoid: parasite feeding or growing upon or within another organism, which it eventually kills

parataxonomist: trained diagnostician of taxa, without conferred academic degree

paternity analysis: inquiry into father of individual, e.g., determining tree donating pollen to form seeds

pathogen: disease-causing, injurious organisms, often bacteria, viruses, fungi

peat swamp forest: inundated forest in which peat (partly carbonized plants) has accumulated

pest pressure hypothesis: see *escape hypothesis*

petal: segment of corolla or floral sheath, often with odor and bright color

phenology: study of periodic biological phenomena, such as timing of plant growth and production of buds, leaves, flowers, and fruit

photoinhibition: shutting down of photosynthetic machinery following intense insolation

photosynthates: organic compounds synthesized in plant cells from chloroplasts converting sunlight into chemical energy, making carbohydrates, releasing oxygen, and using carbon dioxide and water

phenotypic plasticity: varied outcomes of total interactions between organism's genetic makeup and environment

phylogenetic constraint: limitation in adaptive potential due to phylogenetic characteristics

phylogeny: pathway of descent from ancestors

physiological traits: essential life functions and processes of organism

phyto-geographically fragmented: separated geographic portions of regional flora

phytophagous: consumes plant tissue

pier tree: tree used to stabilize cables of canopy walkway

pinnate-veined: resembling featherlike segments arranged on axis

PNG: Papua, New Guinea

podosol (podsol): leached (drained of nutrients by water) soil

point endemics: endemic taxa with very small ranges, habitats

pollen diet: pollen consumed by herbivore, usually a bee

pollination: see *pollinator*

pollination droplet: gymnosperm sexual recombination, the pollen nucleus entering ovule micropyle

pollination syndromes: floral and occasionally other traits (plant stature, leaf color, shape) significant to pollinators; e.g., red, odorless flowers attract birds; pungent-smelling, pale colored flowers open at night attract hawkmoths or bats, scented and showy diurnal flowers attract insects

pollinator: animal that transports pollen to a stigma of flower (often, no further consideration of ovule fertilization and seed production); biological function of pollen transfer from anther to stigma is only successful when pollen tubes grow through style, reach ovule, then fertilize and form mature fruit and seeds

polycarpic: production of more than one cohort during lifespan

polycyclic system: planned timber harvest after intervals

polyphagous: consuming several foods

population cycle: fluctuation or periodic cycle in abundance

population density: number of individuals per unit area

population ecology: study of groups individuals pertaining to a single species and their dynamic characteristics

population: individuals comprising one species and potentially interbreeding

porose anthers: anthers releasing pollen through a pore or slit at apex

predator satiation: food satiation or filling predators, often in reference to seed consumers

proboscis (plural, proboscides): apical feeding or chemosensory structure

protogynous: female function before male function in a flower having stigmata (borne on carpels) and anthers

proximate cause: causal agent witnessed directly, usually chemical and physical, and genetic program

pteridophytes: ferns, club mosses, and horsetails

quadrat: rectangular sampling area

quasi-biannual oscillation: change generally occurring twice in a year

race: taxonomic, geographic subspecies, conceptually tied to animals

random tree: modeling step in phylogenetic study using a computer

rattan: woody climbing palm of economic value, *Calamus, Daemonorops, Plectomia*, with sharp hooks and reaching a length of more than 100 m in the canopy

raw humus horizon: soil layer with blackened, decayed vegetable matter, which increases water retention and provides plant nutrients

recruit: ability of social insect colonies to provide information, beginning within the nest, allowing many nest mates to find a food location

red meranti: see *meranti*

Red Queen hypothesis: participants in antagonistic relationships forever advance and counter-advance, thus keeping the same place (as found in Carroll, L., *Alice in Wonderland*)

reflexed: bending backward

refuge (refugium, refugia): spaces or habitats where organisms retreat when threatened

relatively dilute nectar: nectar of sugar concentration 15% to 20% (remainder water)

relict (relictual): indicating a population considerably reduced in range or size

re-population events: colonization or long-distance immigration to re-establish population

reproductive ecology: timing, location, interactions, and coordination of reproduction

reproductive maturity: when plant produces flowers, fruit, and seeds, or "buds off" from root extensions

reproductive success: production of reproductively viable offspring

reproductively isolated: separated by distance or timing from procreating or exchanging genetic material

resource cycling: dynamic transfer of energy, nutrients, and other matter among individuals, populations, and the abiotic environment

resource heterogeneity: irregular composition or occurrence of local resources

resource partitioning: shifting away from competitors, in kind or qualities—physical, temporal, or spatial—of resources

robbing: removal by non-pollinating animals of pollen or nectar by perforation or damage to flower

root climbers: plants growing on trees, sending down roots

rope ascent: single or double rope, methods, for tree climbing

rosette: resembling a rose, with dense arrangement, radial symmetry

ruderal: growing on poor soils

sandstone-derived soils: soils derived from sedimentary rock composed of silica, lime, and quartz grains

sapling: young tree, not at reproductive maturity

scandent: growing along ground

scatter-hoarding: animal concealing buried seeds for later use

secondary dispersal: final site of germination reached by second animal dispersal from original site

secondary forest: regenerating forest consisting of mostly fast-growing pioneers and few shade-tolerant trees

section: taxonomic classification, between subgenus and species

seed predator (seed herbivores): animal that consume seeds, not aiding germination

seed rain: seeds dropping from canopy

seed trap: traps catching falling seeds, fruit, and vegetation litter

selection (natural selection): nonrandom survival and reproduction among individuals, owing to current advantages

selective agent: factor (biotic or abiotic) producing changes in survival and/or reproductive success

self-incompatible: plant breeding system requiring outcrossing; see *obligate outcrossing*

self-seeding: producing seeds without outcross pollen reception

sepal: usually green segments forming calyx of flower, a sheath around petals

sexual recombination: reproduction whereby parental gametes may incorporate different alleles at the same loci and produce offspring different from a parent (see also *genetic recombination*)

sexual selection: selection originating in behavior of one sex which affects fitness of opposite sex, and trait evolution, e.g., male fighting apparatus, pollen tube growth rate

shifting cultivation: slash-and-burn cultivation, clearing and burning forest plots

sister clades: formed exclusively by two groups resulting from splitting of ancestral population or lineage

social forager dominance hypothesis: colonial groups, e.g., bees from a colony, often dominate floral resources

soil nutrient status: nutrient abundance and quality

source population: population producing individuals that colonize an area

sp. nov. (new species): taxonomic classification status, previously unrecognized by science

specialist: organism using or interacting with, significantly, few or only one species, or habitat, in feeding, mutualism, parasitism, or antagonism

speciose: having many species

species richness: number of taxonomic units known in an area

staminode: nonfunctional stamen, usually a filament lacking an anther

stand structure: number, species, and sizes of local trees

stigma (plural: stigmata): floral structures receiving pollen

stingless bee: honey-making, tropical colonial bee, with queen, males, and worker bees with no functional sting, tribe Meliponini

stipule: paired, leaflike appendages at leaf base

stolon: stem growing underground or near ground which makes a new plant

strangler: fig tree that grows upon another tree species, gradually covering and killing host tree

sub-annual: occurring more than once annually

subcanopy: height or forest stratum below the canopy (10 to 20 m)

subgenus (subgeneric): taxonomic category between species and genus

suborder: taxonomic category between order and tribe

subsocial: social but having no cooperative rearing of offspring or predetermined reproductive and non-reproductive individuals

subspecies: species variant recognizably different and living outside ranges of other such variants, but able to interbreed and produce viable offspring

successional plants: species abundant earlier in forest regeneration

Sunda islands: western Malay Archipelago

Sunda Shelf: continental shelf uniting Borneo, Malay Archipelago, Sumatra, and Java

Sundaland: see *Sunda Shelf*

supra-annual: occurring at intervals of more than a year

survivorship: survival or its probability until particular size, age, or life stage

sustainable usage: avoiding overharvesting, planned for continual exploitation

symbiont (symbiosis): close association of two or more organisms, often beneficial

sympatric: co-occurring in one habitat

syncline: sloping downward from opposite sites to meet at common area

synconium: reproductive fig structure resembling, hollow inflorescence, with male and/or female flowers on interior surface

taxon: individuals of same named descent group, e.g., family, genus, species, or groups thereof

taxonomic: study of naming and classification, evolutionary analysis and phylogeny

temporal partitioning: utilization at different times by different consumers

temporal segregation: differences in timing or activity

temporal variation: variation over time

Tertiary: Tertiary Period, 67 million to 2 million years ago

thievery: removal of nectar without contacting floral reproductive organs, without damaging or forcing flower open

thirty-day rolling average: conversion of data series to mean averages for successive 30-day periods

three-month shifting average: conversion of data series to means for three-month periods

topographic heterogeneity: wide variation in the topographic features of a region

topographic homogeneity: slight variation in topographic features

topography: detailed and accurate description of a locality

topology: surface features or geometry

trade-off: reciprocal compensation between beneficial adaptations, impossible to augment together

trapliner: animal foraging on selected route at periodic intervals

treefall: fallen tree and gap opened of forest canopy

treelet: small tree, reproductively mature without reaching canopy

trophic guild: species with similar diets and feeding habits

trophic level: compartment in transfer of nutrients and energy, e.g., producers (plants), herbivores, carnivores, and scavengers of dead matter

udult soil: sometimes acidic (pH > 4.2), with rapid litter decomposition, lacking root-matted surface, raw humus

ultimate cause: currently obscure origin or evolutionary cause of adaptation or trait

ultisol: weathered yellowish red soil in humid climate, where clay minerals are leached to deeper horizon (substrate layer)

ultraviolet light trap: urn-shaped, with fluorescent electric light attractive to insects

understory: forest under closed canopy, less than 20 meters aboveground

unisexual: (flower or individual) having a function of either male or female

urceolate: urn-shaped, like a pitcher

vacuum-dried: artificially dried in vacuum

variety: taxon below level of species or subspecies, applied to plant but not to animal population (see *subspecies*)

vegetative growth: plant growth by elongation of stems and roots

volatile cues: airborne odors resulting in behavior in animals

waif dispersal: occasional dispersal of few individuals to a distant habitat

walkway: a horizontal path for easy access to canopy

waterlogged: continually in water

watershed: drainage basin in which rain water flows into streams, rivers, and so forth

windthrow: tree toppling by strong wind

winged seeds: wind-dispersed, with oar or bladelike appendages

Bibliography

Aide, T.M. 1992. Dry season leaf production: an escape from herbivory. *Biotropica* 24: 532–37.

———. 1993. Patterns of leaf development and herbivory in a tropical understory community. *Ecology* 74:455–66.

Aiello, A. 1992. Dry season strategies of two Panamanian butterfly species, *Anartia fatima* (Nymphalinae) and *Pierella luna luna* (Satyrinae) (Lepidoptera: Nymphalidae). In *Insects of Panama and Mesoamerica: Selected Studies*, edited by D. Quintero Arias and A. Aiello, 573–75. Oxford: Oxford University Press.

Allee, W.C. 1931. *Animal Aggregations: A Study in General Sociology*. Chicago: University of Chicago Press.

Andersen, N.A. 1989. Pre-dispersal seed losses to insects in species of *Leptospermum* (Myrtaceae). *Aust. J. Ecol.* 14:13–18.

Anstett, M.C., M. Gibernau, and M. Hossaert Mckey. 1998. Partial avoidance of female inflorescences of a dioecious fig by their mutualistic pollinating wasps. *Proc. Roy. Soc. London B* 265:45–50.

Anstett, M.C., M. Hossaert McKey, and D. McKey. 1997. Modeling the persistence of small populations of strongly interdependent species: figs and fig wasps. *Conserv. Biol.* 11:204–13.

APGII. 2003. An update of the Angiosperm Phylogeny Group classification for the orders and families of flowering plants: APG II. *Bot. J. Linn. Soc.* 141:399–436.

Appanah, S. 1985. General flowering in the climax rain forests of Southeast Asia. *J. Trop. Ecol.* 1:225–40.

———. 1990. Plant-pollinator interactions in the Malaysian rain forests. In *Reproductive Ecology of Tropical Forest Plants*, edited by K.S. Bawa and M. Hadley, 85–102. Paris: UNESCO.

————. 1993. Mass flowering of dipterocarp forests in the aseasonal tropics. *J. Biosciences* 18:457–74.

Appanah, S., and H.T. Chan. 1981. Thrips: the pollinators of some dipterocarps. *Malaysian Forester* 44:37–42.

Appanah, S., A.H. Gentry, and J.V. LaFrankie. 1993. Liana diversity and species richness of Malayan rain forests. *J. Trop. Forest Sci.* 6:116–23.

Appanah, S., and G. Weinland. 1993. Planting quality timber trees in Peninsular Malaysia. Malayan Forest Records. No. 38. Kuala Lumpur: Forest Research Institute Malaysia.

Armbruster, W.S. 1997. Exaptations link evolution of plant-herbivore and plant-pollinator interactions: a phylogenetic inquiry. *Ecology* 78:1661–72.

Armstrong, J.E., and B.A Durmmond. 1986. Floral biology of *Myristica fragrans* Houtt. (Myristicaceae), the nutmeg of commerce. *Biotropica* 18:32–38.

Armstrong, J.E., and A.K. Irvine. 1989. Floral biology of *Myristica insipida* R. Br. (Myristicaceae), a distinctive beetle pollination syndrome. *Am. J. Bot.* 76:86–94.

Arnold, B., L.C. Mejía, D. Kyllo, E.I. Rojas et al. 2003. Fungal endophytes limit pathogen damage in a tropical tree. *Proc. Nat. Acad. Sci. USA* 100:15649–54.

Ashton, P.S. 1964. Ecological studies in the mixed Dipterocarp forests of Brunei state. In *Oxford Forest Memoirs* 25. Oxford: Clarendon Press.

————. 1972. The tertiary geomorphological history of western Malesia and lowland forest phytogeography. *Transactions of the second Aberdeen-Hull Symposium on Malesian ecology. Hull Geography Department Misc. Series* 13:35–49.

————. 1982. Dipterocarpaceae. *Flora Malesiana Ser. I,* 9: 237–552.

————. 1989. Dipterocarp reproductive biology. In *Ecosystems of the World 14B: Tropical Rain Forest,* edited by H. Leigh and M.J.A. Werger, 219–40. Amsterdam: Elsevier Scientific.

————. 1992. Plant conservation in the Malaysian region. In *Harmony with Nature: Proceedings of the International Conference on Conservation of Tropical Biodiversity,* edited by S.K. Yap and S.W. Lee, 86–93. Kuala Lumpur: Malayan Nature Society.

————. 1993. The community ecology of Asian rain forests, in relation to catastrophic events. *J. Bioscience* 18:501–14.

————. 1995. Biogeography and ecology. In *Tree Flora of Sabah and Sarawak,* edited by E. Soepadmo and K.M. Wong, xliii-li. Kuala Lumpur: Ampang Press.

————. 1998. Dipterocarp biology as a window to the understanding of forest structure. *Ann. Rev. Ecol. Syst.* 19:347–70.

————. 1998. Niche specificity among tropical trees: A question of scales. In *Dynamics of Tropical Communities, BES Symposium,* Vol. 37, edited by D.M. Newbery, N. Brown, and H.H. Prins, 491–514. Oxford: Blackwell Scientific Publishers.

Ashton, P.S., M. Boscolo, J. Liu, and J.V. LaFrankie. 1999. A global programme in interdisciplinary forest research: The CTFS perspective. *J. Trop. Forest Sci.* 11:180–204.

Ashton, P.S., T.J. Givnish, and S. Appanah. 1988. Staggered flowering in the Dipterocarpaceae: New insights into floral induction and the evolution of mast fruiting in the aseasonal tropics. *Am. Nat.* 132:44–66.

Ashton, P.S., and P. Hall. 1992. Comparisons of structure among mixed dipterocarp forests of North-western Borneo. *J. Ecol.* 80:459–81.

Augspurger, C.K. 1980. Mass-flowering of a tropical shrub (*Hybanthus prunifolius*): influence on pollinator attraction and movement. *Evolution* 34:475–88.

————. 1981. Reproductive synchrony of a tropical shrub: experimental studies on effects of pollinator and seed predators on *Hybanthus prunifolius* (Violaceae). *Ecology* 62:775–88.

————. 1983. Phenology, flowering synchrony, and fruit set of six Neotropical shrubs. *Biotropica* 15:257–67.

Auld, T.D. 1986. Variation in predispersal seed predation in several Australian *Acacia* spp. *Oikos* 47:319–26.

Auld, T.D., and P.J Myerscough. 1986. Population dynamics of the shrub *Acacia suaveolens* (Sm.) Willd.: Seed production and predispersal seed predation. *Aust. J. Ecol.* 11:219–34.

Austin, M.P., P.S. Ashton, and P. Greig-Smith. 1972. The application of quantitative methods to vegetation survey. III. A re-examination of rain forest data from Brunei. *J. Ecol.* 60:305–24.

Avise, J.C. 2000. *Phylogeography: the History and Formation of Species.* Cambridge: Harvard University Press.

Baillie, I.C. 1976. Further studies on drought in Sarawak, East Malaysia. *J. Trop. Geogr.* 43:20–29.

Baillie, I.C., P.S. Ashton, M.N. Court, J.A.R. Anderson, et al. 1987. Site characteristics and the distribution of tree species in mixed dipterocarp forest on Tertiary sediments in central Sarawak, Malaysia. *J. Trop. Ecol.* 3:201–220.

Barber, R.T., and F.P. Chavez. 1983. Biological consequences of El Niño. *Science* 222: 1203–10.

Barone, A.J. 1998. Host-specificity of folivorous insects in a moist tropical forest. *J. Anim. Ecol.* 67:400–409.

Barth, F.G. 1985. *Insects and Flowers: the Biology of a Partnership.* Princeton: Princeton University Press.

Basset, Y. 1992. Host specificity of arboreal and free-living insect herbivores in rain forests. *Biol. J. Linn. Soc.* 47:115–33.

———. 1999. Diversity and abundance of insect herbivores foraging on seedlings in a rain forest in Guyana. *Ecol. Entomol.* 24:245–59.

Baum, D.A. 1995. The comparative pollination and floral biology of baobabs (*Adansonia*-Bombacaceae). *Ann. Missouri Bot. Gard.* 82:332–48.

Baumann, W.T., and C.M. Meier. 1993. Chemical defense by anolides during fruit development in *Physalis peruviana. Phytochemistry* 33:317–21.

Bawa, K.S. 1980. Evolution of dioecy in flowering plants. *Ann. Rev. Ecol. Syst.* 11:15–39.

———. 1983. Patterns of flowering in tropical plants. In *Handbook of Experimental Pollination Biology,* edited by C.E. Jones and R.J. Little, 394–410. New York: Van Norstrand Reinhold.

———. 1990. Plant-pollinator interactions in tropical rain forests. *Ann. Rev. Ecol. Syst.* 21:399–422.

Bawa, K.S., and S. Dayanandan. 1998. Global climate change and tropical forest genetic resources. *Climatic Change* 39:473–85.

Bawa, K.S., S.H. Bullock, D.R. Perry, R.E. Coville, and M.H. Grayum. 1985. Reproductive biology of tropical lowland rain forest trees II. Pollination systems. *Am. J. Bot.* 72:346–56.

Bawa, K.S., H.S. Kang, and M.H. Grayum. 2003. Relationships among time, frequency, and duration of flowering in tropical rain forest trees. *Am. J. Bot.* 90:877–87.

Beach, J.H. 1982. Beetle pollination of *Cyclanthus bipartitus* (Cyclanthaceae). *Am. J. Bot.* 69:1074–81.

Beattie, A.J. 1985. *The evolutionary ecology of ant-plant mutualisms.* Cambridge: Cambridge University Press.

Beaver, R.A. 1979. Host specificity of temperate and tropical animals. *Nature* 281:139–41.

Beck, N.G., and E.M. Lord. 1988. Breeding system in *Ficus carica*, the common fig. II. Pollination events. *Am. J. Bot.* 75:1913–22.

Becker, P. 1992. Seasonality of rainfall and drought in Brunei Darussalam. *Brunei Museum J.* 7: 99–109.

Becker, P., and M. Wong. 1993. Drought-induced mortality in tropical heath forest. *J. Trop. For. Sci.* 5:416–19.

Becker, P., C. Lye Ong, and F. Goh. 1998. Selective drought mortality of dipterocarp trees: no correlation with timber group distributions in Borneo. *Biotropica* 30:666–71.

Bennett, E.L., A.J. Nyaoi, and J. Sompud. 1997. Hornbills *Buceros* spp. and culture in northern Borneo: Can they continue to co-exist? *Biol. Conser.* 82:41–46.

Bennett, N.R., and R.M. Wallsgrove. 1994. Secondary metabolites in plant defence mechanisms. *New Phytol.* 127:617–33.

Berg, C.C. 1989. Classification and distribution of *Ficus*. *Experientia* 45:605–11.

Bernier, G. 1988. The control of floral evocation and morphogenesis. *Ann. Rev. Pl. Phys. Pl. Mol. Biol.* 39:175–219.

Bertault, J.G. 1991. Tropical forest fires: nearly three million hectares destroyed in Kalimantan. *Bois Forests Trop.* 230:5–14.

Blicher, M.U. 1994. Borneo illipe, a fat product from different *Shorea* spp. (Dipterocarpaceae). *Econ. Bot.* 48:231–42.

Borchert, R. 1983. Phenology and control of flowering in tropical trees. *Biotropica* 15:81–89.

Borchsenius, F., and J.M. Olesen. 1990. The Amazonian root holoparasite *Lophophytum mirabile* (Balanophoraceae) and its pollinators and herbivores. *J. Trop. Ecol.* 6:501–505.

Borges, R.M. 1993. Figs, Malabar giant squirrels, and fruit shortages within two tropical Indian forests. *Biotropica* 25:183–90.

Both, C., and M.E. Visser. 2001. Adjustment to climate change is constrained by arrival date in a long-distance migrant birds. *Nature* 411:296–98.

Boucek, Z. 1988, *Australsian Chalcidoidea.* Wallingford, Oxon, UK: C.A.B. International.

Bronstein, J.L, P. Gouyon, C. Gliddon, F. Kjellberg, and G. Micharoud. 1990. The ecological consequences of flowering asynchrony in monoecious figs: a simulation study. *Ecology* 71:2145–56.

Bronstein, J.L., and D. McKey. 1989. The fig/pollinator mutualism: A model system for comparative biology. *Experientia* 45:601–604.

Brower, A.V.Z. 1994. Rapid morphological radiation and convergence among races of the butterfly *Heliconius erato* inferred from patterns of mitochondrial DNA evolution. *Proc. Natl. Acad. Sci. USA* 91:6491–95.

Brunig, E.F. 1969. On the seasonality of droughts in the lowlands of Sarawak (Borneo). *Erdkunde* 23:127–33.

Buckley, R.C., ed. 1982. *Ant-plant Interactions in Australia.* The Hague: Dr. W. Junk Publishers.

Bullock, S.H., and J.A. Solís-Magallanes. 1990. Phenology of canopy trees of a tropical deciduous forest in Mexico. *Biotropica* 22:22–35.

Burgess, P.F. 1972. Studies on the regeneration of the hill forests of the Malay Peninsula: The phenology of dipterocarps. *Malaysian Forester* 35:103–23.

Burkill, I.H. 1919. Some notes on the pollination of flowers in the Botanic Gardens, Singapore, and in other parts of the Malay Peninsula. *Gards' Bull. Str. Settl.* 2:165–76.

Burslem, D.F.R.P., P.J. Grubb, and I.M. Turner. 1996. Responses to simulated drought and elevated nutrient supply among shade-tolerant tree seedlings of lowland tropical forest in Singapore. *Biotropica* 28:636–48.

Carthew, S.M. 1994. Foraging behavior of marsupial pollinators in a population of *Banksia spinulosa*. *Oikos* 69:133–39.

Carthew, S.M., and R.L. Goldingay. 1997. Non-flying mammals as pollinators. *Trends Ecol. Evol.* 12:104–108.

Casgrain, P., and P. Legendre. 2001. The R Package for Multivariate and Spatial Analysis, version 4.0 d5– User's Manual. Département de sciences biologiques, Université de Montréal. http://www.fas.umontreal.ca/BIOL/legendre.

Cavers, P.B. 1983. Seed demography. *Can. J. Botany* 61:3578–90.

Chan, H.T. 1977. Reproductive biology of some Malaysian dipterocarps. Doctoral Thesis. Aberdeen: University of Aberdeen.

Chan, H.T., and S. Appanah. 1980. Reproductive biology of some Malaysian dipterocarps. *Malaysian Forester* 44:28–36.

Chapin, F.S., E-D. Schulze, and H.A. Mooney. 1990. The ecology and economics of storage in plants. *Ann. Rev. Ecol. Syst.* 21:432–47.

Chenuil, A., and D.F.F. McKey. 1996. Molecular phylogenetic study of a myrmecophyte symbiosis: Did *Leonardoxa*/ant associations diversify via cospeciation? *Mol. Phylogenet. Evol.* 6:270–86.

Chin, S.P. 1993. *A Laboratory Manual of Methods of Soil Analysis*. Kuching, Malaysia: Research Branch, Agriculture Department, Sarawak.

Chittka, L., A. Shmida, N. Troje, and R. Menzel. 1994. Ultraviolet as a component of flower reflections, and the colour perception of Hymenoptera. *Vision Res.* 34:1489–508.

Chuine, I., and E.G. Beaubien. 2001. Phenology is a major determinant of tree species range. *Ecol. Letters* 4:500–510.

Clark, D.A., and D.B. Clark. 1992. Life history diversity of canopy and emergent trees in a Neotropical rain forest. *Ecol. Monogr.* 62:315–44.

Clark, D.B., D.A. Clark, and J.M. Read. 1998. Edaphic variation and the mesoscale distribution of tree species in a Neotropical rain forest. *J. Ecol.* 86:101–112.

Cockburn, P.F. 1974. The origin of the Sook Plain, Sabah. *Malaysian Forester* 37:61–63.

Cockburn, P.S. 1975. Phenology of dipterocarps in Sabah. *Malaysian Forester* 44:28–36.

Coley, P.D. 1998. Possible effects of climate change on plant/herbivore interactions in moist tropical forests. *Climatic Change* 39:455–72.

Coley, P.D., J.P. Bryant, F.S. Chapin III. 1985. Resource availability and plant antiherbivore defense. *Science* 230:895–99.

Colinvaux, P.A. 1996. Quaternary environmental history and forest diversity in the Neotropics. In *Evolution and Environment in Tropical America*, edited by J.B.C. Jackson, A.F. Budd, and A.G. Coates, 359–405. Chicago: University of Chicago Press.

Compton, S.G. 1993. One way to be a fig. *Afr. Entomol.* 1:151–58.

Condit, R. 1998. *Tropical Forest Census Plots: Methods and Results from Barro Colorado Island, Panama and a Comparison with Other Plots*. Berlin: Springer-Verlag.

Condit, R., P.S. Ashton, P. Baker, S. Bunyawejehewin et al. 2000. Spatial patterns in the distribution of common and rare tropical tree species: a test from large plots in six different forests. *Science* 288:1414–18.

Condit, R., P.S. Ashton, H. Balslev, N. Brokaw, et al. 2003. Tropical tree α-diversity: results from a worldwide network of large plots. *Biol. Skr.*

Condit, R., S.P. Hubbell, and R.B. Foster. 1992. Recruitment near conspecific adults and the maintenance of tree and shrub diversity in a neotropical forest. *Amer. Nat.* 140: 261–86.

Corlett, R.T. 1987. The phenology of *Ficus fistulosa* in Singapore. *Biotropica* 19:122–24.

———. 1990. Flora and reproductive phenology of the rain forest at Bukit Timah, Singapore. *J. Trop. Ecol.* 6:55–63.

———. 1993. Sexual dimorphism in the reproductive phenology of *Ficus grossularioides* Burm. f. in Singapore. *Malayan Nat. J.* 46:149–55.

———. 1998. Frugivory and seed dispersal by vertebrates in the Oriental (Indomalaysian) Region. *Biol. Rev.* 73:413–48.

————. 2004. Flower visitors and pollination in the Oriental (Indomalayan) Region. *Biol. Rev.* 79:497–532.

Corlett, R.T., and J.V. LaFrankie. 1998. Potential impacts of climate change on tropical Asian forests through an influence on phenology. *Climatic Change* 39:439–53.

Corner, E.J.H. 1960. The Malayan Flora. In *Proceedings of the Centenary and Bicentenary Congress of Biology*, edited by R.D. Purchon, 21–24. Singapore: University of Malaya Press.

————. 1964. *Ficus* on Mt. Kinabalu. *Proc. Linn. Soc. Lond.* 175:37–39.

————. 1965. Check-list of *Ficus* in Asia and Australasia with keys to identification. *Gard. Bull. Sing.* 21:1–185.

————. 1967. *Ficus* in the Solomon Islands and its bearing on the post-Jurassic history of Melanesia. *Phil. Trans. Roy. Soc. Lond. B.* 253:23–159.

————. 1970. *Ficus* Subgen. *Pharmacosycea* with reference to the species of New Caledonia. *Phil. Trans. Roy. Soc. Lond. B.* 259:383–433.

————. 1985. Essays on *Ficus. Allertonia* 4:125–68.

————. 1988. *Wayside Trees of Malaya*. Kuala Lumpur: The Malaysian Nature Society, United Selangor Press.

Cranbrook, Earl of, and D.S. Edwards. 1994. *A Tropical Rain forest: The Nature of Biodiversity in Borneo at Belalong, Brunei*. Singapore: The Royal Geographical Society and Sun Tree Publishing.

Crawley, L.M. 1992. Seed predators and plant population dynamics. In *Seeds: The Ecology of Regeneration in Plant Communities*, edited by M. Fenner, 157–91. Wallingford: CABI Publishing.

Crawley, M.J. 1989. Insect herbivores and plant population dynamics. *Ann. Rev. Entomol.* 34:531–64.

Crawley, M.J., and R. Long. 1995. Alternate bearing, predator satiation, and seedling recruitment in *Quercus robur* L. *J. Ecology* 83:683–96.

Cressie, N. 1991. *Statistics for Spatial Data*. New York: Wiley.

Croat, T.B. 1975. Phenological behavior of habit and habitat classes on Barro Colorado Island (Panama Canal Zone). *Biotropica* 7:270–77.

Cunningham, S.A. 1995. Ecological constraints in fruit initiation by *Calyptrogyne ghiesbreghtiana* (Arecaceae): floral herbivory, pollen availability, and visitation by pollinating bats. *Am. J. Bot.* 82:1527–36.

————. 1997. Predator control of seed production by a rain forest understorey palm. *Oikos* 79:282–90.

Curran, L.M., and M. Leighton. 1991. Why mast? The role of generalized insect and vertebrate seed predators on the reproductive biology of Dipterocarpaceae in the Gunung Palung Nature Reserve, West Kalimantan. In Proceedings of the 4th Round Table Conference on Dipterocarps, ed. G. Maury-Lechon. Bogor: UNESCO.

————. 2000. Vertebrate responses to spatiotemporal variation in seed production of mast-fruiting Dipterocarpaceae. *Ecol. Monog.* 70:101–28.

Curran, L.M., and C.O. Webb. 2000. Experimental tests of the spatiotemporal scale of seed predation in mast-fruiting Dipterocarpaceae. *Ecol. Monogr.* 70:129–48.

Curran, L.M., I. Caniago, G.D. Paoli, D. Astianti et al. 1999. Impact of El Niño and logging on canopy tree recruitment in Borneo. *Science* 286:2184–88.

Daljeet-Singh, K. 1974. Seed pests of some dipterocarps. *Malaysian Forester* 37:24–36.

Dalling, J.W., H.C. Muller-Landau, S.J. Wright, and S.P. Hubbell. 2002. Role of dispersal in the recruitment limitations of Neotropical pioneer species. *J. Ecol.* 90:714–27.

Damstra, K.S.J., S. Richardson, and B. Reeler. 1996. Synchronized fruiting between trees of *Ficus thonningii* in seasonally dry habitats. *J. Biogeography* 23:495–500.

Danforth, B.N. 1999. Emergence dynamics and bet hedging in a desert bee *Perdita portalis. Proc. R. Soc. London* 266:1985–94.

Davidson, D.W. 1998. Resource discovery versus resource domination in ants: a functional mechanism for breaking the trade-off. *Ecol. Entomol.* 23:484–90.

Davidson, D.W., and B.L. Fisher. 1991. Symbiosis of ants with *Cecropia* as a function of light regime. In *Ant-Plant Interaction*, edited by C.R. Huxley and D.F. Cutler, 289–309. Oxford: Oxford University Press.

Davidson, D.W., and D. McKey. 1993. The evolutionary ecology of symbiotic ant-plant relationships. *J. Hym. Res.* 2:13–83.

Davies, R., P. Eggleton, D. Jones, F. Gathorne-Hardy, and L. Hernández. 2003a. Evolution of termite functional diversity: analysis and synthesis of local ecological and regional influences on local species richness. *J. Biogeography* 30:847–77.

Davies, S.J. 1996. The comparative ecology of *Macaranga* (Euphorbiaceae). Ph.D. thesis. Harvard University.

———. 1998. Photosynthesis of nine pioneer *Macaranga* species from Borneo in relation to life history. *Ecology* 79:2292–308.

———. 2001. Tree mortality and growth in 11 sympatric *Macaranga* species in Borneo. *Ecology* 82: 920–32.

Davies, S.J., and P.S. Ashton. 1999. Phenology and fecundity in 11 sympatric pioneer species of *Macaranga* (Euphorbiaceae) in Borneo. *Am. J. Bot.* 86:1786–95.

Davies, S.J., and P. Becker. 1996. Floristic composition and stand structure of mixed dipterocarp and heath forests in Brunei Darussalam. *J. Trop. For. Sci.* 8:542–69.

Davies, S.J., S.K.Y. Lum, R.K.G. Chan, and L.K. Wang. 2001. Evolution of myrmecophytism in *Macaranga* (Euphorbiaceae). *Evolution* 55:1542–59.

Davies, S.J., M.N. Nur Supardi, J.V. LaFrankie, and P.S. Ashton. 2003b. The trees of Pasoh Forest: Stand structure and floristic composition of the 50-hectare forest research plot. In *Pasoh: Ecology and Natural History of a Southeast Asian Lowland Tropical Rain Forest,* edited by T. Okuda, N. Manokaran, S.C. Thomas, and P.S. Ashton, 35–50. Tokyo: Springer.

Davies, S.J., P.A. Palmiotto, P.S. Ashton, H.S. Lee, and J.V. LaFrankie. 1998. Comparative ecology of 11 sympatric species of *Macaranga* in Borneo: tree distribution in relation to horizontal and vertical resource heterogeneity. *J. Ecol.* 86:662–73.

Debski, I., D.F.R.P. Burslem, P.A. Palmiotto, J.V. LaFrankie, et al. 2002. Habitat preferences of *Aporosa* in two Malaysian rain forests: implications for abundance and coexistence. *Ecology* 83:2005–18.

Degen, B., and D.W. Roubik. 2004. Effects of animal pollination on pollen dispersal, self-pollination and effective population size of tropical trees: a simulation study. *Biotropica* 36:165–179.

Delissio, L.J., and R.B. Primack. 2003. The impact of drought on the population dynamics of canopy-tree seedlings in an aseasonal Malaysian rain forest. *J. Trop. Ecol.* 19: 489–500.

Delissio, L.J., R.B. Primack, P. Hall, and H.S. Lee. 2003. A decade of canopy-tree seedling survival and growth in two Bornean rain forests: persistence and recovery from suppression. *J. Trop.Ecol.* 18:645–58.

Dial, R., and S.C. Tobin. 1994. Description of arborist methods for forest canopy access and movement. *Selbyana* 15:24–37.

Díaz, I., C. Papic, and J.J. Armesto. 1999. An assessment of post-dispersal seed predation in temperate rain forest fragments in Chiloé Island, Chile. *Oikos* 87:228–38.

Dick, C.W., K. Abdul-Salim, and E. Bermingham. 2003. Molecular systematics reveals cryptic Tertiary diversification of a widespread tropical rainforest tree. *Am. Nat.* 160: 691–703.

Diggle, P. 1983. *Statistical Analysis of Spatial Point Patterns.* London: Academic Press.

Dobson, H.E.M. 1987. Role of flower and pollen aromas in host-plant recognition by solitary bees. *Oecologia* 72:618–23.

Donaldson, S.J. 1993. Mast-seeding in the cycad genus *Encephalartors*: a test of the predator satiation hypothesis. *Oecologia* 94:262–71.

Dressler, R.L. 1993. *Phylogeny and Classification of the Orchid Family*. Portland, Oregon: Dioscorides Press.

Duivenvoorden, J.F. 1996. Patterns of tree species richness in rain forests of middle Caquetá area, Colombia, NW Amazon. *Biotropica* 28:142–58.

Dyer, F.C. 1985. Nocturnal orientation by the Asian honey bee, *Apis dorsata*. *Anim. Behav.* 33:769–77.

———. 2002. The biology of the dance language. *Ann. Rev. Entomol.* 47:917–49.

Dyer, F.C., and T.D. Seeley. 1991a. Nesting behavior and the evolution of worker tempo in four honeybee species. *Ecology* 72:156–70.

———. 1991b. Dance dialects and foraging range in three Asian honey bee species. *Behavioral Ecology and Sociobiology* 28:227–33.

———. 1994. Colony migration in the tropical honeybee *Apis dorsata* F. *Insectes soc.* 41:129–40.

Ehrlich, P.R., and D.D. Murphy. 1988. Plant chemistry and host range in insect herbivores. *Ecology* 69:908–909.

Eisenberg, J. 1981. *The Mammalian Radiations*. Chicago: University of Chicago Press.

Eltz, T., C.A. Bruhl, S. van der Kaars, and K.E. Linsenmair. 2002. Determinants of stingless bee nest density in lowland dipterocarp forests of Sabah, Malaysia. *Oecologia* 131:27–34.

Eltz, T., C.A. Bruhl, S. van der Kaars, V.K. Chey, and K.E. Linsenmair. 2001. Pollen foraging and resource partitioning of stingless bees in relation to flowering dynamics in a Southeast Asian tropical rain forest. *Insectes soc.* 48:273–79.

Endress, P.K. 1994. *Diversity and Evolutionary Biology of Tropical Flowers*. New York: Cambridge University Press.

Engel, M.S. 1998. Fossil honeybees and evolution in the genus *Apis*. *Apidologie* 29:265–81.

———.1999. The taxonomy of recent and fossil honeybees. *Journal of Hymenoptera Research* 8:165–96.

Erwin, T.L. 1983. Beetles and other insects of tropical forest canopies at Manaus, Brazil, sampled by insecticidal fogging. In *Tropical Rain Forest: Ecology and Management*, edited by S.L. Sutton, T.C. Whitmore, and A.C. Chadwick, 59–75. Oxford: Blackwell Scientific Publications.

———. 1988. The tropical forest canopy: The heart of biotic diversity. In *Biodiversity*, edited by E.O. Wilson and F.M. Peter, 123–29. Washington D.C.: National Academy Press.

Faegri, K. and L. van der Pijl. 1979. *The Principles of Pollination Ecology*. 3d revised ed. Oxford: Pergamon Press.

Farrell, B.D., and C. Mitter. 1990. Phylogenesis of insect/plant interactions: Have *Phyllobrotica* leaf beetles (Chrysomelidae) and the Lamiales diversified in parallel? *Evolution* 44:1389–403.

Feeny, P.P. 1969. Inhibitory effects of oak leaf tannins on the hydrolysis of proteins by trypsin. *Phytochemistry* 8:2116–26.

Feinsinger, P. 1983. Coevolution and pollination. In *Coevolution*, edited by J. Futuyma and M. Slatkin, 282–310. Sunderland: Sinauer Associates Inc.

Fellers, J.H. 1987. Interference and exploitation in a guild of woodland ants. *Ecology* 68:1466–78.

Ferrari, S.F., and K.B. Steiner. 1992. Exploitation of *Mabea fistulifera* nectar by marmosets (*Callithrix flaviceps*) and muriques (*Brachyteles arachnoides*) in southeast Brazil. *J. Trop. Ecol.* 8:225–39.

Fiala, B., and U. Maschwitz. 1990. Studies on the South East Asian ant-plant association *Crematogaster borneensis* / *Macaranga*: adaptations of the ant partner. *Insectes soc.* 37:212–31.

Fiala, B., H. Grunsky, U. Maschwitz, and K.E. Linsenmair. 1994. Diversity of ant-plant interactions: protective efficacy in *Macaranga* species with different degrees of ant association. *Oecologia* 97:186–92.

Fiala, B., A. Jakob, U. Maschwitz, and K.E. Linsenmair. 1999. Diversity, evolutionary specialization and geographic distribution of a mutualistic ant-plant complex *Macaranga* and *Crematogaster* in Southeast Asia. *Biol. J. Linn. Soc.* 66:305–31.

Fiala, B., U. Maschwitz, Y.P. Tho, and A.J. Helbig. 1989. Studies of a southeast Asian ant-plant association protection of *Macaranga* trees by *Crematogaster borneensis*. *Oecologia* 79:463–70.

Figueiredo, R.A. de, and M. Sazima. 1997. Phenology and pollination ecology of three Brazilian fig species (Moraceae). *Botanica Acta* 110:73–78.

Flenley, J.R. 1998. Tropical forests under the climates of the last 300,000 years. *Climatic Change* 39:177–97.

Folgarait, P., and D.W. Davidson. 1994. Antiherbivore defenses of myrmecophytic *Cecropia* under different light regimes. *Oikos* 71:305–20.

———. 1995. Myrmecophytic *Cecropia*: Antiherbivore defenses under different nutrient treatments. *Oecologia* 104:189–206.

Fonseca, C.R., and G.P. Ganade. 1996. Asymmetries, compartments and null interactions in an Amazonian ant-plant community *J. Anim. Ecol.* 65:339–47.

Forget, P., K. Kitajima, and R.B. Foster. 1999. Pre- and post-dispersal seed predation in *Tachigali versicolor* (Caesalpiniaceae): effects of timing of fruiting and variation among trees. *J. Trop. Ecol.* 15:61–81.

Foster, R.B. 1982. The seasonal rhythm of fruitfall on Barro Colorado Island. In *The Ecology of a Neotropical Forest: Seasonal Rhythms and Long-term Changes*, edited by E.G. Leigh, Jr., A.S. Rand, and D.M. Windsor, 151–72. Washington: Smithsonian Institution Press.

———. 1990. The floristic composition of the Rio Manu Floodplain Forest. In *Four Neotropical Forests*, edited by A.H. Gentry, 99–112. New Haven: Yale University Press.

Foster, R.B., and S.P. Hubbell. 1990. The floristic composition of the Barro Colorado Island forest. In *Four Neotropical Forests*, edited by A.H. Gentry, 85–98. New Haven: Yale University Press.

Frame, D. 2003. The pollen tube pathway in *Tasmannia insipida* (Winteraceae). Homology of the male gametophyte conduction tube in angiosperms. *Plant Biol.* 5: 290–96.

Francis, C.M., and D.R. Wells. 2003. The bird community at Pasoh: Composition and population dynamics, In *Pasoh: Ecology and Natural History of a Lowland Rain Forest in Southeast Asia*, edited by T. Okuda, N. Manokaran, S.C. Thomas, and P.S. Ashton, 375–94. Tokyo: Springer-Verlag.

Frankie, G.W. 1975. Tropical forest phenology and pollinator plant coevolution. In *Coevolution of Plants and Animals*, edited by L.E. Gilbert and P.H. Raven, 192–209. Austin: University of Texas Press.

Frankie, G.W., H.G. Baker, and P.A. Opler. 1974. Comparative phenological studies of trees in tropical wet and dry forests in the lowlands of Costa Rica. *J. Ecol.* 62:881–919.

Gale, N. 1997. Modes of tree death in four tropical forests. Doctoral Thesis. Denmark: University of Aarhus.

Galil, J. 1973. Pollination in dioecious figs: Pollination of *Ficus fistulosa* by *Ceratosolen hewitti*. *Gard. Bull. Sing.* 26:303–11.

Ganeshaiah, K.N., P. Kathuria, R.U. Shaanker, and R. Vasudeva. 1995. Evolution of style-length variability in figs and optimization of ovipositor length in their pollinator wasps: A coevolutionary model. *J. Genetics* 74:25–39.

Garber, P.A. 1988. Foraging decisions during nectar feeding by Tamarin monkeys (*Sa-*

guinus mystax and *Saguinus fuscicollis*, Callitrichidae, Primates) in Amazonian Peru. *Biotropica* 20:100–106.

Gentry, A.H., ed. 1990. *Four Neotropical Rain Forests*. New Haven: Yale University Press.

———. 1974. Flowering phenology and diversity in tropical Bignoniaceae. *Biotropica* 6:64–68.

———. 1988. Changes in plant community diversity and floristic compostion on geographic and environmental gradients. *Ann. Missouri Bot. Gard.* 75:1–34.

———. 1993. *A field guide to the families and genera of woody plants of northwest South America*. Washington, D.C.: Conservation International, Inc.

Ghazoul, J., K.A. Liston, T.J.B. Boyle. 1998. Disturbance-induced density-dependent seed set in *Shorea siamensis* (Dipterocarpaceae), a tropical forest tree. *J. Trop. Ecol.* 86:462–73.

Gilbert, G.S., S.P. Hubbell, and R.B. Foster. 1994. Density and distance-to-adult effects of a canker disease of trees in a moist tropical forest. *Oecologia* 98:100–108.

Gill, A.E., and E.M. Ramusson. 1983. The 1982–83 climatic anomaly in the equatorial Pacific. *Nature* 306:229–34.

Gillet, J. B. 1962. Pest pressure, an underestimated factor in evolution. In *Taxonomy and Geography: a Symposium*, edited by D. Nichols, 37–46. London: London Systematics Association.

Givnish, T.J. 1999. On the causes of gradients in tropical tree diversity. *J. Ecol.* 87:193–210.

Gottsberger, G. 1970. Beitrage zür Biologie von Annonaceen-Bultzen. *Oestrrichische Botanische Zietschrift* 118:237–79.

———. 1989a. Beetle pollination and flowering rhythm of *Annona* spp. (Annonaceae) in Brazil. *Pl. Syst. Evol.* 167:165–87.

———. 1989b. Comments on flower evolution and beetle pollination in the genera *Annona* and *Rollinia* (Annonaceae). *Pl. Syst. Evol.* 167:189–94.

———. 1990. Flowers and beetles in the South American tropics. *Bot. Acta.* 103:360–65.

Grant, V. 1950. The pollination of *Calycanthus occidentalis*. *Am. J. Bot.* 37:294–97.

Greig, N. 1993. Predispersal seed predation on five *Piper* species in tropical rain forest. *Oecologia* 93:412–20.

Gribel, R., and P.E. Griggs. 2002. High outbreeding as a consequence of selfed ovule mortality and single vector bat pollination in the Amazonian tree *Pseudobombax munguba* (Bombacaceae). *Int. J. Plant Sci.* 163:1035–43.

Griswold, T., F.D. Parker, and P.E. Hanson. 2000. In *Proceedings of the Sixth International Conference on Apiculture in Tropical Climates*, 152–57. Cardiff, UK: International Bee Res. Assoc.

Grubb, J.P., and D.F.R.P. Burslem. 1998. Mineral nutrient concentrations as a function of seed size within seed crops: implications for competition among seedlings and defense against herbivory. *J. Trop. Ecol.* 14:177–85.

Grubb, J.P., and D.J. Metcalfe. 1996. Adaptation and inertia in the Australian tropical lowland rain-forest flora: contradictory trends in intergeneric and intrageneric comparisons of seed size in relation to light demand. *Func. Ecol.* 10:512–20.

Grubb, J.P., D.J. Metcalfe, E.A.A. Grubb, and G.D. Jones. 1998. Nitrogen-richness and protection of seeds in Australian tropical rain forest: a test of plant defence theory. *Oikos* 82:467–82.

Guariguata, M.R., G.H. Kattan, eds. *Ecología y Conservación de Bosques Neotropicales*. *Cartago*. Costa Rica: Libro Universitario Reigonal.

Guariguata, M., and G. Saenz. 2002. Post-logging acorn production and oak regeneration in a tropical montane forest, Costa Rica. *Forest Ecol. Management* 167:285–93.

Guilderson, T.P., and D.P. Schrag. 1998. Abrupt shift in subsurface temperature in the tropical Pacific associated with changes in El Niño. *Science* 281:240–43.

Hadisoesilo, S., M. Meixner, and F. Ruttner. 1999. Geographic variation within *Apis koschevnikovi* Buttel-Reepen, 1906, in Borneo. *Treubia* 31:305–11.

Hall, R., and J.D. Holloway, eds. 1998. *Biogeography and Geological Evolution of Southeast Asia*. Leiden: Backhuys Publishers.

Hallé, F., and O. Pascal, eds. 1992. *Biologie d'une canopée de forêt equatoriale—II. Rapport de mission: radeau des cimes Octobre Novembre 1991, reserve de Canpo, Cameroon*. Paris: Foundation Elf.

Hamilton, W.D. 2001. *Narrow Roads of Gene Land, Vol. 2: The Evolution of Sex*. Oxford: Oxford University Press.

Hammel, B. 1990. The distribution and diversity among families, genera and habit types in the La Selva Flora. In *Four Neotropical Forests*, edited by A.H. Gentry, 75–84. New Haven: Yale University Press.

Harms, K.E., R. Condit, S.P. Hubbell, and R.B. Foster. 2001. Habitat associations of trees and shrubs in a 50-ha Neotropical forest plot. *J. Ecol.* 89:947–59.

Harris, D.J. 2002. The vascular plants of the Dzanga-Sangha Reserve, Central African Republic. *Scripta Botanica Belgium* 23:1–274.

Harrison, R.D. 1996. The ecology of the fig-fig wasp mutualism in a lowland tropical forest in Sarawak, Malaysia. Master's Thesis, Kyoto: Kyoto University.

———. 2000a. Phenology and wasp population dynamics of several species of dioecious fig in a lowland tropical forest in Sarawak. Doctoral Thesis, Kyoto: Kyoto University.

———. 2000b. Repercussions of El Niño: Drought causes extinction and the breakdown of mutualism in Borneo. *Proc. Roy. Soc. Ser. B.* 267:911–15.

———. 2003. Fig wasp dispersal and the stability of a keystone plant resource in Borneo. *Proc. Roy. Soc. Ser. B.* 270 (Suppl. 1):76–79.

Harrison, R.D., R. Banka, R. Yumuna, I.W.B. Thornton, and M. Shanahan. 2001. Colonization of an island volcano, Long Island, Papua New Guinea, and an emergent island, Motmot, in its caldera lake. II. The vascular flora. *J. Biogeography* 28:1311–37.

Harrison, R.D., A.A. Hamid, T. Kenta, J.V. LaFrankie, et al. 2003. The diversity of hemiepiphytic figs in a Bornean lowland rain forest. *Biol. J. Linn. Soc.* 78:439–45.

Harrison, R.D., and N. Yamamura. 2003. A few more hypotheses for the evolution of dioecy in figs. *Oikos* 100:628–35.

Harrison, R.D., N. Yamumura, and T. Inoue. 2000. Phenology of a common roadside fig in Sarawak. *Ecol. Res.* 15:47–61.

Hartshorn, G.S., and B.E. Hammel. 1994. Vegetation types and floristic patterns. In *La Selva, Ecology and Natural History of a Neotropical Rain Forest*, edited by L.A. McDade, K.S. Bawa, H.A. Hespenheide, and G.S. Hartshorn, 73–89. Chicago: University of Chicago Press.

Hatada, A., T. Itioka, R. Yamaoka, and T. Itino. 2002. Carbon and nitrogen content of food bodies in three myrmecophytic species of *Macaranga*. *J. Plant Res.* 115:179–84.

Heckroth, H., B. Fiala, P.J. Gullan, A.H. Idris, and U. Maschwitz. 1998. The soft scale (Coccidae) associates of Malaysian ant-plants. *J. Trop. Ecol.* 14:427–43.

Heinrich, B., and P.H. Raven. 1972. Energetics and pollination ecology. *Science* 176: 597–602.

Heithaus, E.R. 1979. Flower-feeding specialization in wild bee and wasp communities in seasonal Neotropical habitats. *Oecologia* 42:179–94.

Hepburn, H.R., D.R. Smith, S.E. Radloff, and G.W. Otis. 2001. Infraspecific categories of *Apis cerana*: morphometric, allozymal and mtDNA diversity. *Apidologie* 32: 3–23.

Herre, E.A. 1987. Optimality, plasticity and selective regime in fig wasp sex ratios. *Nature* 329:627–29.

———. 1989. Coevolution of reproductive characteristics in 12 species of New World figs and their pollinating wasps. *Experientia* 45:637–47.

———. 1996. An overview of studies on a community of Panamanian figs. *J. Bioge-ography* 23:593–607.

Herre, E.A., C.A. Machado, E. Bermingham, J.D. Nason, et al. 1996. Molecular phylogenies of figs and their pollinator wasps. *J. Biogeography* 23:521–30.

Herrera, C.M., P. Jordano, J. Guitijn, and A. Traveset. 1998. Annual variability in seed production by woody plants and the masting concept: reassessment of principles and relationship to pollination and seed dispersal. *Am. Nat.* 152:576–94.

Heydon, M.J., and P. Bulloh. 1997. Mousedeer densities in a tropical rain forest: The impact of selective logging. *J. Appl. Ecol.* 34:484–96.

Hickman, C.J. 1979. The basic biology of plant numbers. In *Topics in Plant Population Biology*, edited by O.T. Solbrig, S. Jain, G.B. Johnson, and P.H. Raven, 232–63. New York: Columbia University Press.

Higgins, S.I., and D.M. Richardson. 1999. Predicting plant migration rates in a changing world: the role of long-distance dispersal. *Am. Nat.*153:464–75.

Hilty, S.L. 1980. Flowering and fruiting periodicity of a premontane rain forest in Pacific Colombia. *Biotropica* 12:292–306.

Holden, C. 2002. World's richest forest in peril. *Science* 295:963.

Hölldobler, B., and E.O. Wilson. 1990. *The Ants*. Cambridge: Harvard University Press.

Holt, R.D., and J.P. Lawton. 1993. Apparent competition and enemy-free space in insect host-parasitoid communities. *Am. Nat.* 142:623–45.

Holway, D.A. 1999. Competitive mechanisms underlying the displacement of native ants by the invasive Argentine ant. *Ecology* 80:238–51.

Hopkins, H.C. 1984. Floral biology and pollination ecology of the Neotropical species of *Parkia*. *J. Ecol.* 72:1–23.

Hubbell, S.P. 1980. Seed predation and the coexistence of tree species in tropical forests. *Oikos* 35:214–29.

———. 2001. *The Unified Neutral Theory of Biodiversity and Biogeography*. In Monog. Pop. Biol. 32. Princeton: Princeton Univ. Press.

Hubbell, S.P. and R.B. Foster. 1983. Diversity of canopy trees in a Neotropical forest and implications for conservation. In *Tropical Rain Forest: Ecology and Management*, edited by S.L. Sutton, T.C. Whitmore, and C. Chadwick, 25–41. Oxford: Blackwell Scientific Publications.

Hubbell, S.P., and L.K. Johnson. 1977. Competition and nest spacing in a tropical stingless bee community. *Ecology* 58:949–63.

———. 1978. Comparative foraging behavior of six stingless bee species exploiting a standardized resource. *Ecology* 59:1123–36.

Hulme, E.P., and K.M. Hunt. 1999. Rodent post-dispersal seed predation in deciduous woodland: predator response to absolute and relative abundance of prey. *J. Anim. Ecol.* 68:417–28.

Huppert, A., and L. Stone. 1998. Chaos in the Pacific's coral reef bleaching cycle. *Am. Nat.* 152:447–59.

Ichie, T., T. Hiromi, R. Yoneda, K. Kamiya, et al. 2004. Short-term drought causes synchronous leaf shedding and flushing in a lowland mixed dipterocarp forest, Sarawak, Malaysia. *J. Trop. Ecol.* 20:697–700.

Ickes, K., S.J. Dewalt, S. Appanah. 2001. Effects of native pigs (*Sus scrofa*) on woody understorey vegetation in a Malaysian lowland rain forest. *J. Trop. Ecol.* 17: 191–206.

Igarashi, Y., and N. Kamata. 1997. Insect predation and seasonal seedfall of the Japanese beech, *Fagus crenata* Blume, in northern Japan. *J. Applied Entomol.* 121:65–69.

Ihaka, R., and R. Gentleman. 1996. R: A language for data analysis and graphics. *J. Computational and Graphical Statistics* 5:299–314.

Illueca, J.E., and A.P. Smith. 1993. Exploring the upper tropical forest canopy. *Our Planet* 5:12–13.

Ims, R.A. 1990. On the adaptive value of reproductive synchrony as a predator-swamping strategy. *Am. Nat.* 136:485–98.

Ingle, N. 2003. Seed dispersal by wind, birds, and bats between Philippine montane rain forest and successional vegetation. *Oecologia* 134:251–61.

Inoue, T. 1997. Plant reproductive phenology and pollination mutualisms: implication from tropical studies. *Forest Science* 20:14–23 (in Japanese).

Inoue, T., and K. Nakamura. 1990. Physical and biological background for insect studies in Sumatra. In *Natural History of Social Wasps and Bees in Equatorial Sumatra*, edited by S.F. Sakagami, R. Ohgushi, and D.W. Roubik, 1–11. Sapporo: Hokkaido University Press.

Inoue, T., M. Kato, T. Kakutani, T. Suka, and T. Itino. 1990. Insect-flower relationship in the temperate deciduous forest of Kibune, Kyoto: an overview of the flowering phenology and the seasonal pattern of insect visits. *Contributions from the Biological Laboratory, Kyoto University.* 27:377–463.

Inoue, T., T. Nagamitsu, K. Momose, S.F. Sakagami, and A.A. Hamid. 1994. Stingless bees in Sarawak. In *Plant Reproductive Systems and Animal Seasonal Dynamics: Long-term Study of Dipterocarp Forests in Sarawak*, edited by T. Inoue and A.A. Hamid, 231–37. Otsu: Center for Ecological Research, Kyoto University.

Inoue, T., K. Nakamura, S. Salmah, and I. Abbas. 1993. Population dynamics of animals in unpredictably-changing tropical environments. *J. Biosciences* 18:425–55.

Inoue, T., S.F. Sakagami, S. Salmah, and N. Nukmal. 1984a. Discovery of successful absconding in the stingless bee *Trigona (Tetragonula) laeviceps*. *J. Apicultural Research* 23:136–42.

Inoue, T., S.F. Sakagami, S. Salmah, and S. Yamane. 1984b. The process of colony multiplication in the Sumatran stingless bee *Trigona (Tetragonula) laeviceps*. *Biotropica* 16:100–111.

Inoue, T., S. Salmah, S.F. Sakagami, S. Yamane, and M. Kato. 1990. An analysis of anthophilous insects in central Sumatra. In *Natural History of Social Wasps and Bees in Equatorial Sumatra*, edited by S.F. Sakagami, R. Ohgushi, and D.W. Roubik, 1–11. Sapporo: Hokkaido University Press.

Inoue, T., T. Yumoto, A.A. Hamid, H.S. Lee, and K. Ogino. 1995. Construction of a canopy observation system in a tropical forest of Sarawak. Selbyana 16:100–111.

Inouye, D.W. 1977. Species structure of bumblebee communities in North America and Europe. In *The Role of Arthropods in Forest Ecosystems*, edited by W.J. Mattson, 35–40. New York: Springer-Verlag.

Inui, Y., T. Itioka, K. Murase, R. Yamaoka, and T. Itino. 2001. Chemical recognition of partner plant species by foundress ant queens in *Macaranga-Crematogaster* myrmecophytism. *J. Chem. Ecol.* 27:2029–40.

Irvine, A.K., and J.E. Armstrong. 1990. Beetle pollination in tropical forests of Australia. In *Reproductive Ecology of Tropical Forest Plants*, edited by K.S. Bawa and M. Hadley, 135–50. Paris: Unesco.

Itino, T., S.J. Davies, H. Tada, Y. Hieda, et al. 2001a. Cospeciation of ants and plants. *Ecol. Res.* 16:787–93.

Itino, T., and T. Itioka. 2001. Interspecific variation and ontogenetic change in anti-herbivore defense in myrmecophytic *Macaranga* species. *Ecol. Res.* 16:765–74.

Itioka, T., and M. Yamamti. 2004. Severe drought, leafing phenology, leaf damage and lepidopteran abundance in the canopy of a Bornean aseasonal tropical rain forest. *J. Trop. Ecol.* 20:479–82.

Itino, T., T. Itioka, A. Hatada, A.A. Hamid. 2001b. Effects of food rewards offered by ant-plant *Macaranga* on the colony size of ants. *Ecol. Res.* 16:775–86.

Itioka, T., T. Inoue, H. Kaliang, M. Kato, et al. 2001a. Six-year population fluctuation of the giant honeybee *Apis dorsata* (Hymenoptera: Apidae) in a tropical lowland dipterocarp forest in Sarawak. *Ecol. Popul. Biol.* 94:545–49.

Itioka, T., M. Kato, H. Kaliang, M. Merdeck, et al. 2003. Insect responses to general flowering in Sarawak. In *Arthropods of Tropical Forests: Spatio-Temporal Dynamics and Resource Use in the Canopy*, edited by Y. Basset, G. Novotny, J. Kitching, and S. Miller, 126–34. Cambridge: Cambridge University Press.

Itioka, T., T. Nakashizuka, and L. Chong, eds. 2001b. *Proceedings of the International Symposium: Canopy Processes and Ecological Roles of Tropical Rain Forest*. Kuching, Sarawak, Malaysia: Forest Department of Sarawak, and Japan Science and Technology Agency.

Itioka, T., M. Nomura, Y. Inui, T. Itino, and T. Inoue. 2000. Difference in intensity of ant defense among three species of *Macaranga* myrmecophytes in a southeast Asian dipterocarp forest. *Biotropica* 32:318–26.

Itoh, A., T. Yamakura, K. Ogino, H.S. Lee, and P.S. Ashton. 1997. Spatial distribution patterns of two predominant emergent trees in a tropical rain forest in Sarawak, Malaysia. *Pl. Ecol.* 132:121–36.

Itoh, A., T. Yamakura, T. Ohkubo, M. Kanzaki, et al. 2003. Importance of topography and soil texture in the spatial distribution of two sympatric dipterocarp trees in a Bornean rain forest. *Ecol. Res.* 18:307–20.

Iversen, S.T. 1991, *The Usambara Mountains, NE Tanzania: Phytogeography of the vascular plant flora: Sybolae Botanicae Upsaliensis*, v. XXIX:3. Uppsala: Acta Universitatis Upsaliensis.

Janson, C.H., J. Terbourgh, and L.H. Emmons. 1981. Non-flying Mammals as Pollinating Agents in the Amazonian Forest. *Biotropica* 13: 1–6.

Janzen, D.H. 1966. Coevolution of mutualism between ants and acacias in Central America. *Evolution* 20:249–75.

———. 1967. Synchronization of sexual reproduction of trees within dry season in Central America. *Ecology* 21:620–37.

———. 1970. Herbivores and the number of tree species in tropical forests. *Amer. Nat.* 104:501–28.

———. 1971a. Euglossine bees as long distance pollinators of tropical plants. *Science* 171:203–205.

———. 1971b. Seed predation by animals. *Ann. Rev. Ecol. Syst.* 2:465–92.

———. 1974. Tropical black water rivers, animals and mast flowering by the Dipterocarpaceae. *Biotropica* 6:69–103.

———. 1976. Why bamboos wait so long to flower. *Ann. Rev. Ecol. Syst.* 7:347–91.

———. 1977. Promising directions of study in tropical animal-plant interactions. *Ann. Missouri Bot. Gard.* 64:706–36.

———. 1978. Seeding patterns of tropical trees. In *Tropical trees as living systems*, edited by P.B. Tomlinson and M.H. Zimmermann, 83–128. Cambridge: Cambridge University Press.

———. 1979. How to be a fig. *Ann. Rev. Ecol. Syst.* 10:13–51.

———. 1980. Specificity of seed-attacking beetles in a Costa Rican deciduous forest. *J. Ecol.* 68:929–52.

———. 1980. When is it coevolution? *Evolution* 34:611–12.

———. 1983. Seed and pollen dispersal by animals: convergence in the ecology of contamination and sloppy harvest. *Biol. J. Linn. Soc.* 20:103–13.

———. 1984. Two ways to be a tropical big moth: Santa Rosa saturniids and sphingids. *Oxford Surveys in Evolutionary Biology* 1:85–140.

———. 1985. On ecological fitting. *Oikos* 45:308–10.

———. 2003. How polyphagous are Costa Rican dry forest saturniid caterpillars? In *Arthropods of tropical forests: spatio-temporal dynamics and resource use in the canopy*, edited by Y. Basset, V. Novotny, S.E. Miller, and R.L. Kitching, 369–79. Cambridge: Cambridge University Press.

Johnson, L.K. 1981. Effect of flower clumping on defense of artificial flowers by aggressive stingless bees. *Biotropica* 13:151–57.

———. 1982. Foraging strategies and the structure of stingless bee communities in Costa

Rica. In *Social insects in the tropics,* edited by P. Jaisson, 31–58. Paris: University Paris-Nord.

Johnson, L.K., and S.P. Hubbell. 1974. Aggression and competition among stingless bees: field studies. *Ecology* 55:120–27.

———. 1975. Contrasting foraging strategies and coexistence of two bee species on a single resource. *Ecology* 56:1398–1406.

Johnson, S.D. 1992. Climatic and phylogenetic determinants of flowering seasonality in the Cape flora. *J. Ecol.* 81:567–72.

Joyce, C. 1991. A crane's eye view of tropical forests. *New Scientist* 21:30–31.

Justiniano, M.J., and T.S. Fredericksen. 2000. Phenology of tree species in Bolivian dry forests. *Biotropica* 32:276–81.

Kalko, E.K., E.A. Herre, and C.O. Handley, Jr. 1996. Relation of fig fruit characteristics to fruit-eating bats in the New and Old World tropics. *J. Biogeography* 23: 565–76.

Kalshoven, L.G.E. 1956. Notes on the habits and ecology of Indonesian forest insects of minor importance: Curculionidae. *Indonesian Forest Insect. Deel* 16:77–88.

Kameyama, T., R.D. Harrison, and N. Yamamura. 1999. Persistence of a fig wasp population and evolution of dioecy in figs: A simulation study. *Res. Popul. Ecol.* 41:243–52.

Kappeler, P., and E. Heymann. 1996. Nonconvergence in the evolution of primate life history and socio-ecology. *Biol. J. Linn. Soc.* 59:297–326.

Karban, R., and J.H. Myers. 1989. Induced plant responses to herbivory. *Ann. Rev. Ecol. Syst.* 20:331–48.

Karr, J.R., and F.C. James. 1975. Ecomorphological configurations and convergent evolution, In *Ecology and Evolution of Communities*, edited by M.L. Cody and J.M. Diamond, 258–91. Cambridge: Harvard University Press.

Karunaratne, W.A.I.P. 2004. Taxonomy and Ecology of Bees of Sri Lanka, Doctoral Thesis. Peradeniya, Sri Lanka: Peradeniya University.

Kato, M. 1996. Plant-pollinator interactions in the understory of a lowland mixed dipterocarp forest in Sarawak. *Am. J. Bot.* 86:732–43.

———. 2000. Anthophilous insect community and plant-pollinator interactions on Amami Islands in the Ryukyu Archipelago, Japan. *Contributions from the Biological Laboratory, Kyoto University* 29:157–252, pl. 2–3.

Kato, M., and T. Inoue. 1994. Origin of insect pollination. *Nature* 368:195.

Kato, M., and R. Miura. 1996. Flowering phenology and anthophilous insect community at a threatened natural lowland marsh at Nakaikemi in Tsuruga, Japan. *Contributions from the Biological Laboratory, Kyoto University* 29:1–48.

Kato, M., T. Inoue, A.A. Hamid, T. Nagamitsu, et al. 1995a. Seasonality and vertical structure of light-attracted insect communities in a dipterocarp forest in Sarawak. *Res. Popul. Ecol.* 37:59–79.

Kato, M., T. Inoue, and T. Nagamitsu. 1995b. Floral and pollination biology of *Gnetum* (Gnetaceae) in a lowland mixed dipterocarp forest in Sarawak. *Am. J. Bot.* 82:862–68.

Kato, M., T. Itino, and T. Nagamitsu. 1993a. Melittophily and ornithophily of long-tubed flowers in Zingiberaceae and Gesneriaceae in West Sumatra. *Tropics* 2:129–42.

Kato, M., T. Itioka, S. Sakai, K. Momose, et al. 2000. Various population fluctuation patterns of light-attracted beetles in a tropical lowland dipterocarp forest in Sarawak. *Population Ecol.* 42:97–104.

Kato, M., T. Kakutani, T. Inoue, and T. Itino. 1990. Insect-flower relationship in the primary beech forest of Ashu, Kyoto: An overview of the flowering phenology and the seasonal pattern of insect visits. *Contributions from the Biological Laboratory, Kyoto University* 27:309–75.

Kato, M., M. Matsumoto, and T. Kato. 1993b. Flowering phenology and anthophilous insect community in the cool-temperate subalpine forests and meadows at Mt. Kushigata in the central part of Japan. Contributions from the Biological Laboratory, Kyoto University 28:119–72, pl. 52.

Kelly, D. 1994. The evolutionary ecology of mast seeding. *Trends Ecol. Evol.* 9:465–70.

Kelly, D., M.J. Mckone, K.J. Batchelor, and J.R. Spence. 1992. Mast seeding of *Chinochloa* (Poaceae) and pre-dispersal seed predation by a specialist fly (*Diplotoxa*, Diptera: Chloropidae). *NZ J. Bot.* 30:125–33.

Kelly, D., and J.J. Sullivan. 1997. Quantifying the benefits of mast seeding on predator satiation and wind pollination in *Chionochloa pallens* (Poaceae). *Oikos* 78:143–50.

Kenta, T., K.K. Shimizu, M. Nakagawa, K. Okada, et al. 2002. Multiple factors contribute to outcrossing in a tropical emergent *Dipterocarpus tempehes*, including a new pollentube guidance mechanism for self-incompatibility. *Am. J. Bot.* 89: 60–66.

Kerdelhue, C., M.E. Hochberg, and J.-Y. Rasplus. 1997. Active pollination of *Ficus sur* by two sympatric fig wasp species in West Africa. *Biotropica* 29:69–75.

Kerdelhue, C., and J.Y. Rasplus. 1996. Non-pollinating Afrotropical fig wasps affect the fig-pollinator mutualism in *Ficus* within the subgenus *Sycomorus*. *Oikos* 75:3–14.

Kevan, P.G., ed. 1995. *The Asiatic Hive Bee: Apiculture, Biology, and Role in Sustainable Development in Tropical and Subtropical Asia.* Cambridge, Ontario: Enviroquest, Ltd.

Khoo, S.G. 1992. Foraging behaviour of stingless bees and honey bees. *Proceedings of the National IRPA (Intensification of Research in Priority Areas) Seminar (Agriculture Sector)* 2:461–62.

Kiew, R. 1993. Characteristics of Malaysian flowers visited by *Apis cerana*. In *Asian Apiculture,* edited by L.J. Connor, T. Rinderer, H.A. Sylvester, and S. Wongsiri, 361–68. Connecticut: Wicwas Press.

———. 1997. Analysis of pollen from combs of *Apis dorsata* in peninsular Malaysia. In *Proc. Int. Conf. on Tropical Bees and the Envrionment,* edited by M. Mardan, A. Sipat, K.M. Yusoff, H.M.S.R. Kew, and M.M. Abdullah, 174–86. Penang, Malaysia: Beenet Asia.

Kinnaird, M.F., T.G. O'Brien, and S. Suryadi. 1999. The importance of figs to Sulawesi's imperiled wildlife. *Trop. Biodiversity* 6:5–18.

Kiyono, Y., and Hastaniah. 1999. Six year observations on flowering and fruiting of 782 dipterocarp trees at Bukit Soeharto, East Kalimantan, Indonesia. In *Report on Forestry Research Overseas,* 15–20. Tsukuba: Forestry and Forest Products Research Institute.

Kjellberg, F., B. Doumesche, and J.L. Bronstein. 1988. Longevity of a fig wasp (*Blastophaga psenes*). *Proceedings of the Koninklijke Nederlandse Akademie Van Wetenschappen Series C Biological and Medical Sciences* 91:117–22.

Kjellberg, F., E. Jousselin, J.L. Bronstein, A. Patel, et al. 2001. Pollination mode in fig wasps: The predictive power of correlated traits. *Proc. Roy. Soc. Lond. B* 268:1113–21.

Kobayashi, K. 1974. Ecological study of seed production of kapur and keladan. *Trop. Forestry Quarterly J.* 34:16–20.

Kochmer, J.P., and S.N. Handel. 1986. Constraints and competition in the evolution of flowering phenology. *Ecol. Monog.* 56:303–25.

Kochummen, K.M. 1998. New species and varieties of Moraceae from Malaysia. *Gard. Bull. Sing.* 50:197–219.

Kochummen, K.M., and R. Go. 2000. Moraceae. In *Tree Flora of Sabah and Sarawak,* edited by E. Seopadmo and L.G. Saw, 181–334. Kuala Lumpur: Forest Research Institute Malaysia, Sabah Forestry Department Malaysia, Sarawak Forestry Department Malaysia.

Kochummen, K.M., J.V. LaFrankie, and N. Manokaran. 1990. Floristic composition of Pasoh Forest reserve, a lowland rain forest in Peninsular Malaysia. *J. Trop. For. Sci.* 3:1–13.

Koeniger, N., and G. Koeniger. 1980. Observations and experiments on migration and dance communication of *Apis dorsata* in Sri Lanka. *J. Apic. Res.* 19:21–34.

Koeniger, N., and G. Vorwohl. 1979. Competition for food among four sympatric species of Apini in Sri Lanka (*Apis dorsata, Apis cerana, Apis florea and Trigona iridipennis*). *J. Apic. Res.* 18:95–109.

Kokubo, J. 1987. Mortality factor of Dipterocarp seeds. *Trop. Forestry* 8:21–25.

Komai, F. 1992. Taxonomic revision of the genus *Andrioplecta* Obraztsov (Lepidoptera: Tortricidae). *Trans. Lepid. Soc. Japan.* 43:151–81.

Koptur, S., W.A. Haber, G.W. Frankie, and H.G. Baker. 1988. Phenological studies of shrub and treelet species in tropical cloud forests of Costa Rica. *J. Trop. Ecol.* 4:323–46.

Kress, W.J. 1994. Pollination of *Ravenala madagascariensis* (Strelitziaceae) by lemurs in Madagascar: evidence for an archaic coevolutionary system. *Am. J. Bot.* 81:542–51.

Kress, W.J., and J.H. Beach. 1994. Flowering plant reproductive systems. In *La Selva, Ecology and Natural History of a Neotropical Rain Forest,* edited by L.A. McDade, K.S. Bawa, H.A. Hespenheide, and G.S. Hartshorn, 161–82. Chicago: University of Chicago Press.

Kudo, G., and K. Kitayama. 1999. Drought effects on the summit vegetation on Mount Kinabalu by an El Niño event in 1998. *Sabah Parks Nat. J.* 2:101–10.

Kumpulainen, T., A. Grapputo, and J. Mappes. 2004. Parasites and sexual reproduction in psychid moths. *Evolution* 58:1511–20.

LaFrankie, J.V., and H.T. Chan. 1991. Confirmation of sequential flowering in *Shorea* (Dipterocarpaceae). *Biotropica* 23:200–203.

LaFrankie, J.V., S. Tan, and P.S. Ashton. 1995. Species list for the 52-ha forest dynamics research plot Lambir Hills National Park, Sarawak, Malaysia. Center for Tropical Forest Science Miscellaneous Internal Report, 2.9.95.

Laman, T.G. 1995. The ecology of strangler fig seedling establishment. *Selbyana* 16:223–29.

———. 1996a. *Ficus* seed shadows in a Bornean rain forest. *Oecologia* 107:347–55.

———. 1996b. Specialization for canopy position by hemiepiphytic *Ficus* species in a Bornean rain forest. *J. Trop. Ecol.* 12:789–803.

Laman, T.G., and G.D. Weiblen. 1999. Figs of Gunung Palung National Park (West Kalimantan, Indonesia). *Trop. Biodiversity* 5:245–97.

Lambert, F.R., and A.G. Marshall. 1991. Keystone characteristics of bird-dispersed *Ficus* in a Malaysian lowland rain forest. *J. Ecol.* 79:793–809.

Langenheim, J.H. 2003. *Plant resins: Chemistry, Evolution, Ecology and Ethnobotany.* Portland, Oregon: Timber Press.

Lee, H.S., P.S. Ashton, T. Yamakura, S. Davies, et al. 2003. The 52-ha forest dynamics plot at Lambir Hills, Sarawak, Malaysia: distribution maps, diameter tables and species documentation. Kuching, Sarawak, Malaysia: Sarawak Forest Department and CTFS-AA program.

Lee, H.S., S.J. Davies, J.V. LaFrankie, S. Tan, et al. 2002. Floristic and structural diversity of mixed dipterocarp forest in Lambir Hills National Park, Sarawak, Malaysia. *J. Trop. Forest Sci.* 14:379–400.

Legendre, P. and L. Legendre. 1998. *Numerical Ecology,* 2d edition. Amsterdam: Elsevier.

Leigh, E.G., Jr. 1999. *Tropical Forest Ecology: A View from Barro Colorado Island.* Oxford: Oxford University Press.

Leighton, M., and N. Wirawan. 1986. Catastrophic drought and fire in Bornean tropical rain forest associated with the 1982–83 El Niño Southern Oscillation event. In *Tropical Forests and the World Atmosphere,* edited by G.T. Prance, 75–102. Washington D.C.: American Association for the Advancement of Science.

Leith, H. 1974. *Phenology and Seasonality Modeling.* Berlin: Springer-Verlag.

Levey, D.J., W.R. Silva, and M. Galetti, eds. 2002. *Seed Dispersal and Frugivory: Ecology, Evolution and Conservation.* Wallingford: CABI Publishing.

Levin, D.A. 1976. The chemical defenses of plants to pathogens and herbivores. *Ann. Rev. Ecol. Syst.* 7:121–20.

Lieberman, D., and M. Lieberman. 1987. Forest tree growth and dynamics at La Selva, Costa Rica (1969-1982). *J. Trop. Ecol.* 3:347–58.

Lieberman, D., M. Lieberman, R. Peralta, and G.S. Hartshorn. 1996. Tropical forest structure and composition on a large-scale altitudinal gradient in Costa Rica. *J. Ecol.* 84:137–52.

Liechti, P., F.W. Roe, and N.S. Haile. 1960. The Geology of Sarawak, Brunei and the western part of North Borneo. British Territories in Borneo Geological Survey.

Lieftinck, M.A. 1956. Revision of some oriental anthophorine bees of the genus *Amegilla* Friese (Hymenoptera, Apoidea). *Zoologische Verhandelingen* 30:1–41.

Loiselle, B.A., and J.G. Blake. 1999. Dispersal of melastome seeds by fruit-eating birds of tropical forest understory. *Ecology* 80:330–36.

Lopez-Vaamonde, C., D.J. Dixon, J.M. Cook, and J.Y. Rasplus. 2002. Revision of the Australian species of *Pleistodontes* (Hymenoptera: Agaonidae) fig-pollinating wasps and their host-plant associations. *Zool. J. Linn. Soc.* 136:637–83.

Losos, E.C., and E.G. Leigh, Jr., eds. 2004. *Tropical Forest Diversity and Dynamism. Findings from a Large Scale Plot Network.* Chicago: University of Chicago Press.

Loveless, M.D. 2002. Genetic diversity and differentiation in tropical trees. In *Proceedings of the Symposium on Modelling and Experimental Research on Genetic Processes in Tropical and Temperate Forests*, edited by B. Degen, M.D. Loveless, and A. Kremer, 3–30. Belém, Brazil: Embrapa Amazônia Oriental.

Lumur, C. 1980. Rodent pollination of *Blakea* (Melastmataceae) in a Costa Rican cloud forest. *Brittonia* 32:512–17.

Lunau, K. 2000. The ecology and evolution of visual pollen signals. *Pl. Syst. Evol.* 222: 89–111.

MacArthur, R.H., and E.O. Wilson. 1963. An equilibrium theory of insular zoogeography. *Evolution* 17:373–87.

Machado, C.A., E. Jousselin, F. Kjellberg, S.G. Compton, and E.A. Herre. 2001. Phylogenetic relationships, historical biogeography and character evolution in fig-pollinating wasps. *Proc. Roy. Soc. Lond. B* 268:685–94.

Maeto, K. 1993. Acorn insects of *Quercus mongolica* var. *grosseserata* in Hitsujigaoka natural forest, Hokkaido: Life history of the principal species and their impacts on seed viability. *Transactions of the Meeting in Hokkaido Branch of the Japanese Forestry Society* 41:88–90.

Manokaran, N., and J.V. LaFrankie. 1990. Stand structure of Pasoh Forest Reserve, a lowland rain forest in Peninsular Malaysia. *J. Trop. For. Sci.* 3:14–24.

Manokaran, N., J.V. La Frankie, K.M. Kochummen, E.S. Quah, et al. 1990. Methodology for the fifty hectare research plot at Pasoh Forest Reserve. Forest Research Institute Malaysia, Kepong, Malaysia. Research Pamphlet Number 104. Kepong, Malaysia: Forest Research Institute Malaysia.

———. 1992. Stand table and distribution of species in the fifty hectare research plot at Pasoh Forest Reserve. Research Data, No. 1. Kuala Lumpur: Forest Research Institute of Malaysia.

Marquis, R.J. 1992. A bite is a bite is a bite? Constraints on response to folivory in *Piper arietinum* (Piperaceae). *Ecology* 73:143–52.

Marquis, R.J., and H.E. Baker. 1994. Plant-herbivore interactions: diversity, specificity, and impact. In *La Selva: Ecology and Natural History of a Neotropical Rain Forest,* edited by L.A. McDade, S.K. Bawa, A.H. Hespenheide, and S.G. Hartshorn, 261–81. Chicago: The University of Chicago Press.

Maschwitz, U., B. Fiala, S.J. Davies, and K.E. Linsenmair. 1996. A southeast Asian myrmecophyte with two alternative inhabitants: *Camponotus* or *Crematogaster* as partners of *Macaranga lamellata*. *Ecotropica* 2:29–40.

Mattson, J.W. 1978. The role of insects in the dynamics of cone production of red pine. *Oecologia* 33:327–49.

Mattson, W.J., J. Levieux, and C. Bernard-Dagan. eds. 1988. *Mechanisms of woody plant defenses against insects: search for pattern.* New York: Springer-Verlag.

Mawdsley, N.A., and N.E. Stork. 1997. Host-specificity and the effective specialization of tropical canopy beetles. In *Canopy Arthropods*, edited by N.E. Stork, J. Adis, and R.K. Didham, 104–30. London: Chapman and Hall.

McCune, B., and M.J. Mefford. 1999. *PC-ORD. Multivariate Analysis of Ecological Data.* Oregon: MjM Software Design.

McDade, L.A., K.S. Bawa, H.A. Hespenheide, and G.S. Hartshorn. 1994. *La Selva: Ecology and Natural History of a Neotropical Rain Forest.* Chicago: University of Chicago Press.

McGregor, G.R., and S. Nieuwolt. 1998. *Tropical Climatology.* 2d edition. West Sussex: John Wiley and Sons Ltd.

McKey, D., D.W. Davidson, and H. Gay. 1993. Ant-plant symbioses in Africa and the Neotropics: History, biogeography, and diversity. In *Biological Relationships Between Africa and South America*, edited by P. Goldblatt, 568–606. New Haven: Yale University Press.

Medway, L. 1972. Phenology of a tropical rain forest in Malaya. *Biol. J. Linn. Soc.* 4: 117–46.

Meehl, G.A. 1997. Pacific region climate change. *Ocean Coast. Manage.* 37:137–47.

Metcalfe, D.J., P.J. Grubb, and I.M. Turner. 1998. The ecology of very small-seeded shade-tolerant trees and shrubs in lowland rain forest in Singapore. *Pl. Ecol.* 134: 131–49.

Michaloud, G., and P.S. Michaloud. 1987. *Ficus* hemi-epiphytes (Moraceae) et arboles supports. *Biotropica* 19:125–36.

Michcner, C.D. 1974. *The Social Behavior of the Bees: A Comparative Study.* Cambridge: Harvard University Press.

———. 2000. *The Bees of the World.* Baltimore: The Johns Hopkins University Press.

Milton, K., D.M. Windsor, D.W. Morrison, and M.A. Estribi. 1982. Fruiting phenologies of two Neotropical *Ficus* species. *Ecology* 63:752–62.

Minckley, R.L., J.H. Cane, L. Kervin, and T.H. Roulston. 1999. Spatial predictability and resource specialization of bees (Hymenoptera: Apoidea) at a superabundant, widespread resource. *Biol. J. Linn. Soc.* 67:119–47.

Mitchell, A.W. 1982. *Reaching the Rain Forest Roof: A Handbook on Techniques of Access and Study in the Canopy.* Leeds: Leeds Philosophical and Literary Society.

Mitchell, A.W., K. Secoy, and T. Jackson. 2002. *Global Canopy Handbook.* Oxford: Global Canopy Network.

Molbo, D., C.A. Machado, J.G. Sevenster, L. Keller, and E.A. Herre. 2003. Cryptic species of fig-pollinating wasps: implications for the evolution of the fig-wasp mutualism, sex allocation, and precision of adaptation. *Proc. Nat. Acad. Sci. USA* 100: 5867–72.

Moles, S. 1994. Trade-offs and constraints in plant herbivore defense theory: a life history perspective. *Oikos* 71:3–12.

Momose, K., R. Ishii, S. Sakai, and T. Inoue. 1998a. Reproductive intervals and pollinators of tropical plants. *Proc. Roy. Soc. Lond.* 265:2333–39.

Momose, K., T. Nagamitsu, and T. Inoue. 1996. The reproductive ecology of an emergent dipterocarp in a lowland rain forest in Sarawak. *Pl. Sp. Biol.* 11:189–98.

———. 1998b. Thrips cross-pollination of *Popowia pisocarpa* (Annonaceae) in a lowland dipterocarp forest in Sarawak. *Biotropica* 30:444–48.

Momose, K., T. Yumoto, T. Nagamitsu, M. Kato, H. Nagamasu, et al. 1998c. Pollination biology in a lowland dipterocarp forest in Sarawak, Malaysia I. Characteristics of the plant-pollinator community in a lowland dipterocarp forest. *Am. J. Bot.* 85:1477–1501.

Morellato, P.C., D.C. Talora, A. Takahasi, C.C. Bencke, et al. 2000. Phenology of Atlantic rain forest trees: a comparative study. *Biotropica* 32:811–23.

Mori, S.A., and J.L. Brown. 1994. Report on wind dispersal in a lowland moist forest in central French Guiana. *Brittonia* 46:105–25.

Morley, R.J. 1998. Palynological evidence for Tertiary plant dispersals in SE Asian region in relation to plate tectonics and climate. In *Biogeography and Geological Evolution of Southeast Asia,* edited by R. Hall and J.D. Holloway, 177–200. Amsterdam: Backhuys Publishers.

———. 2000. *Origin and Evolution of Tropical Rain Forests.* New York: John Wiley and Sons.

Morton, C.M., S.A. Mori, G.T. Prance, K.G. Karol, et al. 1997. Phylogenetic relationships of Lecythidaceae: A cladistic analysis using rbcL sequence and morphological data. *Am. J. Bot.* 84:530–40.

Moure, J.S. 1961. A preliminary supra-specific classification of the Old World meliponine bees. *Studia Entomologica* 4:181–242.

Mulchay, D.H. 1979. The rise of the angiosperms: a genealogical factor. *Science* 206: 20–23.

Murali, K.S., and R. Sukumar. 1994. Reproductive phenology of a tropical dry forest in Mudumalai, southern India. *J. Ecol.* 82:759–67.

Murase, K., T. Itioka, Y. Inui, and T. Itino. 2002. Species specificity in settling-plant selection by foundress ant queens in *Macaranga–Crematogaster* myrmecophytism in a Bornean dipterocarp forest. *J. Ethol.* 20:19–24.

Murray, K.G., P. Feinsinger, W.H. Busby, Y.B. Linhart, et al. 1987. Evaluation of character displacement among plants in two tropical pollination guilds. *Ecology* 68:1283–93.

Murray, M.G. 1985. Figs (*Ficus* spp.) and fig wasps (Chalcidoidea, Agaonidae): hypotheses for an ancient symbiosis. *Biol. J. Linn. Soc.* 26:69–82.

Nagamasu, H., and K. Momose. 1997. Flora of Lambir Hills National Park, Sarawak, with special reference to the Canopy Biology Plot. In *General Flowering of Tropical Rain Forests in Sarawak,* edited by T. Inoue and A.A. Hamid, 20–67. Otsu: Center for Ecological Research, Kyoto University.

Nagamitsu, T. 1998. Community ecology of floral resource partitioning by eusocial bees in an Asian tropical rain forest. Doctoral Thesis, Kyoto University.

Nagamitsu, T., and T. Inoue, T. 1997a. Cockroach pollination and breeding system of *Uvaria elmeri* (Annonaceae) in a lowland mixed-dipterocarp forest in Sarawak. *Am. J. Bot.* 84:208–13.

———. 1997b. Aggressive foraging of social bees as a mechanism of floral resource partitioning in an Asian tropical rain forest. *Oecologia* 110:432–39.

———. 1998. Interspecific morphological variation in stingless bees (Hymenoptera, Meliponinae) associated with floral shape and location in an Asian tropical rain forest. *Entomol. Sci.* 1:189–94.

———. 2002. Foraging activity and pollen diets of subterranean stingless bee colonies in response to general flowering in Sarawak, Malaysia. *Apidologie* 33:303–14.

Nagamitsu, T., R.D. Harrison, and T. Inoue. 1999a. Beetle pollination of *Vatica parvifolia* (Dipterocarpaceae) in Sarawak, Malaysia. *Gardens' Bulletin Singapore* 51:43–54.

Nagamitsu, T., K. Momose, T. Inoue, and D.W. Roubik. 1999b. Preference in flower visits and partitioning in pollen diets of stingless bees in an Asian tropical Rain forest. *Res. Popul. Ecol.* 41:195–202.

Nakagawa, M., T. Itioka, K. Momose, T. Yumoto, et al. 2003. Resource utilization of insect seed predators during general flowering and seeding events in a Bornean dipterocarp rain forest. *Bull. Entomol. Res.* 93:455–66.

Nakagawa, M., K. Tanaka, T. Nakashizuka, T. Ohkuo, et al. 2000. Impact of severe drought associated with the 1997–1998 El Niño in a tropical forest in Sarawak. *J. Trop. Ecol.* 16:355–67.

Nason, J.D., E.A. Herre, and J.L. Hamrick. 1996. Paternity analysis of the breeding structure of strangler fig populations: Evidence for substantial long-distance wasp dispersal. *J. Biogeography.* 23:501–12.

———. 1998. The breeding structure of a tropical keystone plant resource. *Nature* 391: 685–87.

Natawiria, D., A.S. Kosasih, and A.D. Mulyana. 1986. Some insect pests of dipterocarp seeds (in East Kalimantan and Java). *Buletin Penelitian Hutan, Pusat Penelitian dan Pengembangan Hutan* 472:1–8.

Nefdt, R.J.C., and S.G. Compton. 1996. Regulation of seed and pollinator production in the fig-fig wasp mutualism. *J. Anim. Ecol.* 65:170–82.

Neumann, P., N. Koeniger, G. Koeniger, S. Tingek, et al. 2000. Home-site fidelity in migratory honeybees. *Nature* 406:474.

Newbery, D.M. and J. Proctor. 1984. Ecological studies in four contrasting lowland rain forests in Gunung Mulu National Park, Sarawak. IV. Associations between tree distributions and soil factors. *J. Ecol.* 72:475–93.

Newbery, D.M., E.J.F. Campbell, J. Proctor, and M.J. Still. 1996. Primary lowland dipterocarp forest at Danum Valley, Sabah, Malaysia: Species composition and patterns in the understorey. *Vegetatio* 122:193–220.

Newman, M.F., P.F. Burgess, and T.C. Whitmore. 1998a. *Borneo Island Light Hardwoods. Manuals of Dipterocarps for Foresters.* Edinburg/Jakarta: Royal Botanic Garden/CIFOR.

———. 1998b. *Borneo Island Medium and Heavy Hardwoods. Manuals of Dipterocarps for Foresters.* Edinburg/Jakarta: Royal Botanic Garden/CIFOR.

Newstrom, L.E., G.W. Frankie, and H.G. Baker. 1994a. A new classification for plant phenology based on flowering patterns in lowland tropical rain forest trees at La Selva, Costa Rica. *Biotropica* 26:141–59.

Newstorm, L.E., G.W. Frankie, H.G. Baker, and R.K. Colwell. 1994b. Diversity of long-term flowering patterns. In *La Selva: Ecology and Natural History of a Neotropical Rain Forest,* edited by L.A. McDade, S.K. Bawa, A.H. Hespenheide, and S.G. Hartshorn, 142–60. Chicago: University of Chicago Press.

Ng, F.S.P. 1977. Gregarious flowering of dipterocarps in Kepong, 1976. *Malaysian Forester* 40:126–36.

———. 1981. Vegetative and reproductive phenology of dipterocarps. *Malaysian Forester* 44:197–215.

Nilsson, G.S. 1985. Ecological and evolutionary interactions between reproduction of beech *Fagus silvatica* and seed eating animals. *Oikos.* 44:157–64.

Nilsson, G.S., and U. Woostljung. 1987. Seed predation and cross-pollination in mast-seeding beech (*Fagus sylvatica*) patches. *Ecology* 68:260–65.

Nomura, M., T. Itioka, and T. Itino. 2000. Difference in intensity of chemical defense among myrmecophytic and non-myrmecophytic sympatric species of *Macaranga* in a Southeast Asian dipterocarp forest. *Ecol. Res.* 15:1–11.

Nomura, M., T. Itioka, and K. Murase. 2001. Non-ant anti-herbivore defenses before plant-ant colonization in *Macaranga* myrmecophytes. *Popul. Ecol.* 43:207–12.

Normark, B.B., O.P. Judson, and N.A. Moran. 2003. Genomic signature of ancient asexual lineages. *Biol. J. Linn. Soc.* 79:69–84.

Numata, S., N. Kachi, T. Okuda, and N. Manokaran. 1999. Chemical defenses of fruits and mast-fruiting of dipterocarps. *J. Trop. Ecol.* 15:695–700.

Nykvist, N. 1996. Regrowth of secondary vegetation after the 'Borneo fire' of 1982–1983. *J. Trop. Ecol.* 12:307–12.

Odegaard, F. 2003. Taxonomic composition and host specificity of phytophagous beetles in a dry forest in Panama. In *Arthropods of Tropical Forests: Spatio-Temporal Dynamics and Resource Use in the Canopy,* edited by Y. Basset, V. Novotny, S.E. Miller, and R.L. Kitching, 220–36. Cambridge: Cambridge University Press.

Okuda, T., N. Manokaran, Y. Matsumoto, K. Niiyama, S.C. Thomas, and P.S. Ashton, eds. 2003. *Pasoh: Ecology of a Lowland Rain Forest in Southeast Asia.* Tokyo: Springer-Verlag.

Oldroyd, B.P., K.E. Osborne, and M. Mardan, 2000. Colony relatedness in aggregations of *Apis dorsata* Fabricius (Hymenoptera, Apidae). *Insectes soc.* 47:94–95.

Olesen, J.M., and P. Jordano. 2002. Geographic patterns in plant-pollinator mutualistic networks. *Ecology* 83:2416–24.

Oliveira, P.S., and A.T. Oliveira-Filho. 1991. Distribution of extrafloral nectaries in the woody flora of tropical communities in Western Brazil. In *Plant-Animal Interactions: Evolutionary Ecology in Tropical and Temperate Regions,* edited by P.W. Price, T.M. Lewinsohn, G.W. Fernandes, and W.W. Benson, 163–75. New York: John Wiley and Sons.

Ollerton, J., and A.J. Lack. 1992. Flowering phenology: an example of relaxation of natural selection? *Trends Ecol. Evol.* 7:274–76.

Ono, M., T. Igarashi, E. Ohno, and M Sakaki. 1995. Unusual thermal defence by a honeybee against mass attack by hornets. *Nature* 377:334–36.

Opler, P.A., G.W. Frankie, and H.G. Baker. 1976. Rainfall as a factor in the release, timing, and synchronization of anthesis by tropical trees and shrubs. *J. Biogeography* 3:231–36.

Osawa, N., and Y. Tsubaki. 2003. Seasonal variation and community structure of tropical bees in a lowland tropical forest of Peninsular Malaysia: the impact of general flowering. In *Pasoh: ecology of a lowland rain forest in Southeast Asia,* edited by T. Okuda, N. Manokaran, Y. Matsumoto, K. Niiyama, S.C. Thomas, and P.S. Ashton, 315–23. Tokyo: Springer-Verlag.

Otis, G.W. 1996. Distributions of recently recognized species of honey bees in Asia. *J. Kansas Entomol. Soc.* 69 (supplement):311–33.

Overdorff, D.J. 1992. Differential patterns in flower feeding by *Eulemur fulvus fulvus* and *Eulemur rubriventer* in Madagascar. *Am. J. Primatol.* 28:191–96.

Ozanne, C.M.P., D. Anhuf, S.L. Boulter, M. Keller, et al. 2003. Biodiversity meets the atmosphere: A global view of forest canopies. *Science* 301:183–86.

Paar, J., B.P. Oldroyd, and G. Kastberger. 2000. Giant honeybees return to their nest sites. *Nature* 406:475.

Page, R.D.M. 1993. *Component.* London: The natural history museum.

Palmiotto, P.A. 1998. The role of specialization in nutrient-use efficiency as a mechanism driving species diversity in a tropical rain forest. Doctoral Dissertation. New Haven: School of Forestry and Environmental Studies, Yale University.

Palmiotto, P., S.J. Davies, K.A. Vogt, P.M.S. Ashton, et al. In review. Soil-related habitat specialization in dipterocarp rain forest tree species in Borneo. *J. Ecol.*

Panayotou, T., and P.S. Ashton. 1992. *Not by Timber Alone.* Washington, D.C.: Island Press.

Patel, A. 1996a. Strangler fig-host associations in roadside and deciduous forest sites, South India. *J. Biogeography* 23:409–14.

———. 1996b. Variation in a mutualism: Phenology and the maintenance of gynodioecy in two Indian fig species. *J. Ecol.* 84:667–80.

Patel, A., M.C. Anstett, M. Hossaert McKey, and F. Kjellberg. 1995. Pollinators entering female dioecious figs: Why commit suicide? *J. Evol. Biol.* 8:301–13.

Pearce, F. 1999. Weather warning. *New Scientist* 164:36–39.

Pellmyr, O. 1997. Pollinating seed eaters: why is active pollination so rare? *Ecology* 78: 1655–60.

Penuelas, J., and I. Filella. 2001. Phenology: response to a warming world. *Science* 294: 793–95.

Perry, D.R. 1978. A method of access into the crowns of emergent and canopy trees. *Biotropica* 6:155–57.

———. 1984. The canopy of the tropical rain forest. *Scientific American* 251:114–22.

Peters, H.A. 2003. Neighbour-regulated mortality: the influence of positive and negative density dependence in species-rich tropical forests. *Ecol. Lett.* 6:757–65.

Peters, J.M., D.C. Queller, V.L. Imperatriz-Fonseca, D.W. Roubik, and J.E. Strassmann. 1999. Mate number, kin selection and social conflicts in stingless bees and honeybees. *Proc. R. Soc. Lond. B.* 266:379–84.

Plotkin, J.B., M.D. Potts, N. Leslie, N. Manokaran, J.V. LaFrankie, and P.S. Ashton. 2000. Species area curves, spatial aggregation, and habitat specialization in tropical forests. *J. Theor. Biol.* 207:81–99.

Potts, M.D. 2001. Species Spatial Patterning in Tropical Rainforests. Doctoral Thesis. Cambridge, Massachusetts: Harvard University.

———. 2003. Drought in an everwet rain forest. *J. Ecol.* 91:467–74.

Potts, M.D., P.S. Ashton, L.S. Kaufman, and J.B. Plotkin. 2002. Habitat patterns in tropical rain forests: A comparison of 105 plots in Northwest Borneo. *Ecology* 83: 2782–97.

Prance, G.T. 1980. A note on the probable pollination of *Combretum* by *Cebus* monkeys. *Biotropica* 12:239.

Primack, R.B. 1980. Phenological variation within natural populations: flowering in New Zealand montane shrubs. *J. Ecol.* 68:849–62.

———. 1987. Relationships among flowers, fruits, and seeds. *Ann. Rev. Ecol. Syst.* 18: 409–30.

Punchihewa, R.W.K. 1994. *Beekeeping for Honey Production in Sri Lanka: Management of Asiatic Hive Honeybee* Apis cerana *in its Natural Tropical Monsoonal Environment.* Sri Lanka: Sarvodaya Vishva Lekha.

Putz, F.E. 1979. Aseasonality in Malaysian tree phenology. *Malaysian Forester* 42:1–24.

Putz, F.E., and P. Chai. 1987. Ecological studies of lianas in Lambir National Park, Sarawak, Malaysia. *J. Ecol.* 75:523–31.

Quek, S.P., S.J. Davies, T. Itino, and N.E. Pierce. 2004. Codiversification in an ant-plant mutualism: the phylogeny of host use among *Crematogaster* (Formicidae) inhabitants of *Macaranga* (Euphorbiaceae). *Evolution* 58: 554–70.

Randall, G.M.M. 1986. The predation of predispersal *Juncus squarrosus* seeds by *Coleophora alticolella* (Lepidoptera) larvae over a range of altitudes in northern England. *Oecologia* 69:460–65.

Rasplus, J.Y. 1994. The one-to-one species specificity of the *Ficus*-Agaoninae mutualism: how casual? In *The Biodiversity of African Plants,* edited by L.J.G. van der Maesen, X.M. van der Burgt, and J.M. van Medenbach de Rooy, 639–49. Boston: Kluwer Academic Publishers.

Rathcke, B., and E.P. Lacey. 1985. Phenological patterns of terrestrial plants. *Ann. Rev. Ecol. Syst.* 16:179–214.

Reed, K.E., and J.G. Fleagle. 1995. Geographic and climatic control of primate diversity. *Proc. Nat. Acad. Sci. USA* 92:7874–76.

Rees, C.J.C. 1983. Microclimate and the flying Hemipteran fauna of a primary lowland rain forest in Sulawesi. In *Tropical Rain Forest: Ecology and Management*, edited by S.L. Sutton, T.C. Whitmore, and A.C. Chadwick, 121–36, Oxford: Blackwell Scientific Publications.

Regal, P.J. 1977. Ecology and evolution of flowering plant dominance. *Science* 196:622–27.

Rehr, S.S., P.P. Feeny, and D.H. Janzen. 1973. Chemical defence in Central American non-ant-acacias. *J. Anim. Ecol.* 42:405–16.

Reich, P.B. 1995. Phenology of tropical forests: patterns, causes and consequences. *Can. J. Bot.* 73:164–74.

Reich, P.B., and R. Borchert. 1982. Phenology and ecophysiology of the tropical tree, *Tabebuia neochrysantha* (Bignoniaceae). *Ecology* 63:294–99.

———. 1984. Water stress and phenology in a tropical dry forest in the lowlands of Costa Rica. *J. Ecol.* 72:61–71.

Renner, S.S., G. Clausing, and K. Meyer. 2001. Historical biogeography of Melastomataceae: the roles of Tertiary migration and long-distance dispersal. *Am. J. Bot.* 88: 1290–1300.

Rhoades, D.F. 1985. Offensive-defensive interactions between herbivores and plants: their relevance in herbivore population dynamics and ecological theory. *Am. Nat.* 125:205–38.

Richards, A.J. 1997. *Plant Breeding Systems.* 2d Ed. London: George Allen and Unwin.

Richards, P.W. 1996. *The Tropical Rain Forest: an Ecological Study.* Cambridge: Cambridge University Press.

Ricklefs, R.E. 2003. Genetics, evolution, and ecological communities. *Ecology* 84:588–91.

———. 2004. A comprehensive framework for global patterns in biodiversity. *Ecology Letters* 7:1–15.

Ricklefs, R.E., and D. Schluter, eds. 1993. *Species Diversity in Ecological Communities: Historical and Geographic Perspectives.* Chicago: University of Chicago Press.

Rincón, M., D.W. Roubik, B. Finegan, D. Delgado, and N. Zamora. 1999. Understory bees and floral resources in logged and silviculturally treated Cost Rican rain forest plots. *J. Kansas Entomol. Soc.* 72:379–93.

Rivera, G., and R. Borchert. 2001. Induction of flowering in tropical trees by a 30-min reduction in photoperiod: evidence from field observations and herbarium specimens. *Tree Physiol.* 21:201–12.

Robinson, S.K., and J. Terborgh. 1990. Bird communities of the Cocha Cashu Biological Station in Amazonian Peru. In *Four Neotropical Forests*, edited by A.H. Gentry, 199–236. New Haven: Yale University Press.

Rogstad, S.H. 1994. The biosystematics and evolution of the *Polyalthia hypoleuca* species complex (Annonaceae) of Malesia III. Floral ontogeny and breeding systems. *Am. J. Bot.* 81:145–54.

Rosen, D.E. 1975. A vicariance model for Caribbean biogeography. *Syst. Zool.* 24:431–64.

Rosenfeld, D. 1999. TRIMM observed first direct evidence of smoke from forest fires inhibiting rainfall. *Geophys. Res. Lett.* 26:3105–08.

Roubik, D.W. 1979. Africanized honey bees, stingless bees and the structure of tropical plant-pollinator communities. In *Proceedings IVth International Symposium on Pollination*, edited by D. Caron, 403–17. College Park, Maryland: Maryland Agricultural Experimental Station Miscellaneous Publication No. 1.

———. 1980. Foraging behavior of competing Africanized honeybees and stingless bees. *Ecology* 61:836–45.

———. 1982. The ecological impact of nectar-robbing bees and pollinating hummingbirds on a tropical shrub. *Ecology* 63:354–60.

———. 1983. Experimental community studies: time-series tests of competition between African and Neotropical bees. *Ecology* 64:971–78.

———. 1988. An overview of Africanized honey bee populations: reproduction, diet and competition. In *Proc. Intl. Conf. on Africanized Honeybees and Bee Mites*, edited by G. Needham, R. Page, and M. Delfinado-Baker, 45–54. Chichester, UK: E. Horwood Ltd.

———. 1989. *Ecology and Natural History of Tropical Bees.* Cambridge: Cambridge University Press.

———. 1990. Niche preemption in tropical bee communities: a comparison of neotropical and Malesian faunas. In *Natural History of Social Wasps and Bees in Equatorial Sumatra*, edited by S.F. Sakagami, R. Ohgushi, and D.W. Roubik, 245–57. Hokkaido: Hokkaido University Press.

———. 1992. Loose niches in tropical communities: Why are there so many trees and so few bees? In *Effects of Resource Distribution on Animal-Plant Interactions*, edited by M.D. Hunter, T. Ohgushi, and P.W. Price, 327–54. New York: Academic Press.

———. 1993. Tropical pollinators in the canopy and understory: field data and theory for stratum 'preferences'. *J. Insect Behav.* 6:659–73.

———. 1996a. Wild bees of Brunei Darussalam. In *Tropical Rain Forest Research:*

Current Issues, edited by D.S. Edwards, W.E. Booth, and S.C. Choy, 59–66. Dordrecht: Kluwer Academic Publishers.

———. 1996b. Measuring the meaning of honeybees. In *The Conservation of Bees*. edited by A. Matheson, S.L. Buchmann, C. O'Toole, P. Westrich, and I.H. Williams, 163–72. London: Academic Press, Ltd.

———. 1996c. Diversity in the real world: tropical forests as pollinator reserves. In *Tropical Bees and the Environment*, edited by M. Mardan, A. Sipat, K.M. Yusoff, H.M.S.R. Kew, and M.M. Abdullah, 174–86. Penang, Malaysia: Beenet Asia.

———. 1998. Grave-robbing by male *Eulaema* (Hymenoptera, Apidae): implications for euglossine biology. *J. Kansas Entomol. Soc.* 71:188–91.

———. 1999. The foraging and potential pollination outcrossing distances flown by African honey bees in Congo forest. *J. Kansas Entomol. Soc.* 72:394–401.

———. 2001. Ups and downs in pollinator populations: when is there a decline? *Conservation Ecology* 5:2. http://www.consecol.org/vol5/iss1/art2.

———. 2002. The value of bees to the coffee harvest. *Nature* 417:708.

Roubik, D.W., and P.E. Hanson. 2004. *Orchid Bees of Tropical America: Biology and Field Guide (Spanish/English edition)*. Heredia, Costa Rica: INBio Press (Editorial INBio).

Roubik, D.W., and J.E. Moreno 1990. Social bees and palm trees: What do pollen diets tell us? In *Social Insects and the Enrivonment*, edited by G.K. Veeresh, B. Mallik, and C.A. Viraktamath, 427–28. New Delhi: Oxford and IBH Publishing Co.

Roubik, D.W., and H. Wolda. 2001. Do competing honey bees matter? Dynamics and abundance of native bees before and after honey bee invasion. *Popul. Ecol.* 43:53–62.

Roubik, D.W., T. Inoue, and A.A. Hamid. 1995. Canopy foraging by two tropical honey bees: Bee height fidelity and tree genetic neighborhoods. *Tropics* 5:81–93.

Roubik, D.W., T. Inoue, and R.D. Harrison. 1999. Height communication by Bornean honey bees (Apiformes: Apidae; Apini). *J. Kansas Entomol. Soc.* 72:256–61.

Roubik, D.W., J.E. Moreno, C. Vergara, and D. Wittmann. 1986. Sporadic food competition with the African honeybee: projected impact on Neotropical social bees. *J. Trop. Ecol.* 2:97–111.

Roubik, D.W., S.F. Sakagami, and I. Kudo. 1985. A note on the nests and distribution of the Himalayan honey bee *Apis laboriosa* Smith in Nepal. *J. Kansas Entomol. Soc.* 58:746–49.

Roubik, D.W, S. Sakai, and F. Gattesco. 2003. Canopy flowers and certainty: loose niches revisited. In *Arthropods of Tropical Forests: Spatio-Temporal Dynamics and Resource Use in the Canopy*, edited by Y. Basset, V. Novotny, S. Miller, and R. Kitching, 360–68. Cambridge: Cambridge University Press.

Ruttner, F. 1988. *Biogeography and Taxonomy of Honeybees*. Berlin: Springer-Verlag.

Sakagami, S.F. 1975. Stingless bees (excl. *Tetragonula*) from the continental southeast Asia in the Collection of Bernice P. Bishop Museum, Honolulu (Hymenoptera, Apidae). *J. Facul. Sci., Hokkaido Univ. Zool.* 20:49–76.

———. 1978. *Tetragonula* stingless bees of the continental Asia and Sri Lanka (Hymenoptera, Apidae). *J. Fac. Sci., Hokkaido Univ. Zool.* 21:165–247.

———. 1982. Stingless bees. In *Social Insects*, edited by H.R. Hermann, 361–423. New York: Academic Press.

Sakagami, S.F., T. Inoue, and S. Salmah. 1990. Stingless bees of central Sumatra. In *Natural History of Social Wasps and Bees in Equatorial Sumatra*, edited by S.F. Sakagami, R. Ohgushi, and D.W. Roubik, 201–18. Sapporo: Hokkaido University Press.

Sakai, S. 2000. Reproductive phenology of gingers in a lowland dipterocarp forest in Borneo. *J. Trop. Ecol.* 16:337–54.

———. 2002. General flowering in lowland mixed dipterocarp forests of Southeast Asia. *Biol. J. Linn. Soc.* 75:233–47.

Sakai, S., and T. Inoue. 1999. A new pollination system: dung-beetle pollination discovered in *Orchidantha inouei* (Lowiaceae, Zingiberales) in Sarawak, Malaysia. *Am. J. Bot.* 86:56–61.

Sakai, S., and H. Nagamasu. 1998. Systematic studies of Bornean Zingiberaceae I. *Amomum* in Lambir Hills. *Edinburgh J. Bot.* 55:45–64.

Sakai, S., M. Kato, and T. Inoue. 1999a. Three population guilds and variation in floral characteristics of Bornean gingers (Zingiberaceae and Costaceae). *Am. J. Bot.* 86: 646–58.

Sakai, S., M. Kato, and H. Nagamasu. 2000. *Artocarpus* (Moraceae)-gall midge pollination mutualism mediated by a male-flower-parasitic fungus. *Am. J. Bot.* 87:440–45.

Sakai, S., K. Momose, T. Yumoto, M. Kato, and T. Inoue. 1999b. Beetle pollination of *Shorea parvifolia* (section *Mutica*, Dipterocarpaceae) in a general flowering period in Sarawak, Malaysia. *Am. J. Bot.* 86:62–69.

Sakai, S., K. Momose, T. Yumoto, T. Nagamitsu, et al. 1999c. Plant reproductive phenology over four years including an episode of general flowering in a lowland dipterocarp forest, Sarawak, Malaysia. *Am. J. Bot.* 86:1414–36.

Salafsky, N. 1994. Drought in the rain forest: effects of the 1991 El Niño Southern Oscillation event on a rural economy in west Kalimantan, Indonesia. *Climatic Change* 27:373–96.

———. 1998. Drought in the rain forest. Part II. An update based on the 1994 ENSO event. *Climatic Change* 39:601–603.

Salmah, S., T. Inoue, and S.F. Sakagami. 1990. An analysis of apid bee richness (Apidae) in central Sumatra. In *Natural History of Social Wasps and Bees in Equatorial Sumatra,* edited by S.F. Sakagami, R. Ohgushi, and D.W. Roubik, 139–74. Sapporo: Hokkaido University Press.

Sanford, R.L., P. Paaby, J.C. Luvall, and E. Phillips. 1994. Climate, geomorphology, and aquatic systems. In *La Selva: Ecology and Natural History of a Neotropical Rain Forest,* edited by L.A. McDade, S.K. Bawa, A.H. Hespenheide, and S.G. Hartshorn, 19–33. Chicago: The University of Chicago Press.

Sarmiento, G., and M. Monasterio. 1983. Life forms and phenology. In *Ecosystems of the world 14B: Tropical savannas,* edited by F. Bouliére, 219–40. Amsterdam: Elsevier Scientific.

Schemske, D.W. 1981. Floral convergence and pollinator sharing in two bee pollinated tropical herbs. *Ecology* 62:946–54.

Schultz, J.C. 1988. Many factors influence the evolution of herbivore diets, but plant chemistry is central. *Ecology* 69:896–97.

Schwarz, H.F. 1937. Results of the Oxford University Sarawak (Borneo) expedition: Bornean stingless bees of the genus *Trigona. Bull. Am. Museum Nat. Hist.* 73:281–328.

———. 1939. The Indo-Malayan species of *Trigona. Bull. Am. Museum Nat. Hist.*

Seal, J. 1957. Rainfall and sunshine in Sarawak. *Sarawak Museum J.* 8:500–544.

Seeley, T.D. 1985. *Honeybee Ecology: A Study of Adaptation in Social Life.* Princeton: Princeton University Press.

———. 1995. *The Wisdom of the Hive: the Social Physiology of Honeybee Colonies.* Cambridge: Harvard University Press.

Seeley, T.D., R.H. Seeley, and P. Akratanakul. 1982. Colony defense strategies of the honeybees in Thailand. *Ecol. Monogr.* 52:43–63.

Shanahan, M. 2000. *Ficus* seed dispersal guilds: ecology, evolution and conservation implications. Doctoral Thesis. Leeds: University of Leeds.

Shanahan, M., and S.G. Compton. 2001. Vertical stratification of figs and fig-eaters in a Bornean lowland rain forest: how is the canopy different? *Pl. Ecol.* 153:121–32.

Shanahan, M., and I. Debski. 2002. Vertebrates of Lambir Hills National Park, Sarawak, Malaysia. *Malayan Nat. J.* 56:103–18.

Shanahan, M., S.G. Compton, S. So, and R. Corlett. 2001a. Fig-eating by vertebrate frugivores: a global review. *Biol. Rev.* 76:529–72.

Shanahan, M., R.D. Harrison, R. Yamuna, W. Koen, and I.W.B. Thornton. 2001b. Colonisation of an island volcano, Long Island, Papua New Guinea, and an emergent island, Motmot, in its caldera lake. V. Figs (*Ficus* spp.), their dispersers and pollinators. *J. Biogeography* 28:1365–77.

Shapcott, A. 1999. Comparison of the population genetics and densities of five *Pinanga* palm species at Kuala Belalong, Brunei. *Mol. Ecol.* 8:1641–54.

Shibata, M., H. Tanaka, and T. Nakashizuka. 1998. Causes and consequences of mast seed production of four co-occurring *Carpinus* species in Japan. *Ecology* 79:54–64.

Silvertown, J.W. 1987. *Introduction to plant population ecology.* 2d ed. London: Longman.

———. 1980. The evolutionary ecology of mast seeding in trees. *Biol. J. Linn. Soc.* 14: 235–50.

Simoons, F.J. 1998. *Plants of life, Plants of Death.* Madison: University of Wisconsin Press.

Soepadmo, E., F.W. Fong, B.H. Kiew, M.T. Zakaria, et al. 1984. *An Ecological Survey of Lambir Hills National Park, Sarawak.* Kuala Lumpur: University of Malaya.

Sork, V.L. 1993. Evolutionary ecology of mast-seeding in temperate and tropical oaks (*Quercus* spp.). *Vegetatio* 107/108:133–47.

Spencer, H., G. Weiblen, and B. Flick. 1996. Phenology of *Ficus variegata* in a seasonal wet tropical forest at Cape Tribulation, Australia. *J. Biogeography* 23:467–75.

Sperens, U. 1997. Fruit production in *Sorbus aucuparia* L. (Rosaceae) and pre-dispersal seed predation by the apple fruit moth (*Argyresthia conjugella* Zell.). *Oecologia* 110: 368–73.

Stacy, E.A., and J.L. Hamrick. 2004. Using forest dynamics plots for studies of tree breeding structure: examples from Barro Colorado Island. In *Tropical Forest Diversity and Dynamism. Findings from a Large Scale Plot Network,* edited by E.C. Losos and E.G. Leigh, Jr., 264–78. Chicago: University of Chicago Press.

Starr, C.K. 1987. Queen or worker, which is the original honey bee? (Hymenoptera: Apidae). *Sociobiology* 13:287–93.

———. 1989. *Bombus folsomi* and the origin of Philippine bumble bees (Hymenoptera: Apidae). *Systematic Entomology* 14:411–15.

Starr, C.K., P.J. Schmidt, and J.O. Schmidt. 1987. Nest-site preferences of the giant honeybee, *Apis dorsata,* in Borneo. *Pan-Pacif. Entomol.* 63:37–42.

Steinberg, L.S.L., S.S. Mulkey, and S.J. Wright. 1989. Ecological interpretation of leaf carbon isotope ratios: influence of respired carbon dioxide. *Ecology* 70:1317–24.

Steven, D.D. 1983. Reproductive consequences of insect seed predation in *Hamamelis virginiana. Ecology* 64:89–98.

Stiles, F.G. 1977. Coadapted competitors: the flowering seasons of hummingbird-pollinated plants in a tropical forest. *Science* 198:1170–78.

———. 1978. Ecological and evolutionary implications of bird pollination. *Am. Zool.* 18:715–27.

Stork, N.E. 1987a. Canopy fogging, a method of collecting living insects for investigations of life history strategies. *J. Nat. Hist.* 21:563–66.

———. 1987b. Guild structure of arthropods from Bornean rain forest trees. *Ecol. Entomol.* 12:69–80.

———. 1988a. Insect diversity: facts, fiction and speculation. *Biol. J. Linn. Soc.* 35:321–37.

———. 1988b. Species number, species abundance and body length relationships of arboreal beetles in Bornean lowland rain forest trees. *Ecol. Entomol.* 13:25–37.

Strauss, S.Y. 1997. Floral characters link herbivores, pollinators, and plant fitness. *Ecology* 78:1640–45.

Summers, K., S. McKeon, J. Sellars, M. Keusenkothen, et al. 2003. Parasitic exploitation as an engine of diversity. *Biol. Rev.* 78:639–75.

Sun, C., B.A. Kaplin, K.A. Kristensen, V. Munyaligoga, et al. 1996. Tree phenology in a tropical montane forest in Rwanda. *Biotropica* 28:668–81.

Sutton, S.L., T.C. Whitmore, and A.C. Chadwick, eds. 1983. *Tropical Rain Forest: Ecology and Management.* Oxford: Blackwell Scientific Publications.

Tanaka, H. 1995. Seed demography of three co-occurring *Acer* species in a Japanese temperate deciduous forest. *J. Veg. Sci.* 6:887–96.

Tanaka, H., D.W. Roubik, M. Kato, F. Liew, and G. Gunsalam. 2001a. Phylogenetic position of *Apis nuluensis* of northern Borneo and phylogeography of *A. cerana* as inferred from mitochondrial DNA sequences. *Insectes soc.* 48:44–51.

Tanaka, H., T. Suka, D.W. Roubik, and M. Mohamed. 2001b. Genetic differentiation among geographic groups of three honeybee species, *Apis cerana, A. koschevnikovi* and *A. dorsata*, in Borneo. *Nature and Human Activities* 6:5–12.

Tanaka, H., T. Suka, S. Kahono, H. Samejima, et al. 2004. Mitochondrial variation and genetic differentiation in honeybees (*Apis cerana, A. koschevnikovi* and *A. dorsata*) of Borneo. *Tropics* 13:107–17.

Terborgh, J. 1986. Keystone plant resources in the tropical forest. In *Conservation Biology: the Science of Scarcity and Diversity,* edited by M.E. Soulé, 330–44. Sunderland: Sinauer.

Terborgh, J., N. Pitman, M. Silman, H. Schichter, and V.P. Nuñez. 2002. Maintenance of tree diversity in tropical forests. In *Seed Dispersal and Frugivory: Ecology, Evolution and Conservation,* edited by D.J. Levey, W.R. Silva, and M. Galetti, 1–17. Wallingford, UK: CABI Publishing.

Tevis, L. 1958. Germination and growth of ephemerals induced by sprinkling a sandy desert. *Ecology* 39:681–88.

Thien, L.B. 1974. Floral biology of *Magnolia. Am. J. Bot.* 61:1037–45.

———. 1980. Patterns of pollination of primitive angiosperms. *Biotropica* 12:1–13.

Thompson, J.N. 1994. *The Coevolutionary Process.* Chicago: University of Chicago Press.

Thomson, J.D. 1989. Germination schedules for pollen grains: implications for pollen selection. *Evolution* 43:220–23.

———. 2003. When is it mutualism? *Am. Nat.* (Suppl.).162:S1–S9.

Timmermann, A., J. Oberhuber, A. Bacher, M. Esch, et al. 1999. Increasing El Niño frequency in a climate model forced by future greenhouse warming. *Nature* 398:694–96.

Tissue, D.T., and S.J. Wright. 1995. Effects of seasonal water availability on phenology and the annual shoot carbohydrate cycle of tropical forest shrubs. *Func. Ecol.* 9:518–27.

Todzia, C. 1986. Growth habits, host tree species, and density of hemi-epiphytes on Barro Colorado Island, Panama. *Biotropica* 18:22–27.

Tollrian, R., C.D. Harvell, eds. 1999. *The Ecology and Evolution of Inducible Defenses.* Princeton: Princeton University Press.

Toma, T. 1999. Exceptional droughts and forest fires in the eastern part of Borneo Island. *Tropics* 9:55–72.

Toy, R.J. 1991. Interspecific flowering patterns in the Dipterocarpaceae in west Malaysia: implications for predator satiation. *J. Trop. Ecol.* 7:49–57.

Toy, R.J. and S.J. Toy. 1992. Oviposition preferences and egg survival in *Nanophyes shoreae* (Coleoptera, Apionidae), a weevil fruit-predator in Southeast Asian rain forest. *J. Trop. Ecol.* 8:195–203.

Traveset, A. 1991. Pre-dispersal seed predation in Central American *Acacia farnesiana*: factors affecting the abundance of cooccurring bruchid beetles. *Oikos* 87:570–76.

Turner, I.M. 1990. The seedling survivorship and growth of three *Shorea* spp. in a Malaysian tropical rain forest. *J. Trop. Ecol.* 6:469–78.

———. 1994. The taxonomy and ecology of the vascular plant flora of Singapore: A statistical analysis. *Bot. J. Linn. Soc.* 114:215–27.

———. 2001. An overview of the plant diversity of South-East Asia. *Asian J. Trop. Biol.* 4:1–16.

Uriarte, M., R. Condit, C.D. Canham, and S.P. Hubbell. 2004. A spatially explicit model of sapling growth in a tropical forest: does the identity of neighbors matter? *J. Ecol.* 92:348–360.

Valencia, R., R. Condit, R. Foster, K. Romoleroux, et al. 2004a. Yasuní Forest Dynamics Plot, Ecuador. In *Forest Diversity and Dynamism: Findings from a Network of Large-scale Tropical Forest plots,* edited by E.C. Losos and E.G. Leigh 609–20. Chicago: University of Chicago Press.

Valencia, R., R.B. Foster, G. Villa, R. Condit, J.C. Svenning, C. Hernández, K. Romeroleroux, E. Losos, E. Magård, and H. Balslev. 2004b. Tree species distributions and local habitat variation in the Amazon: large forest plot in eastern Ecuador. *J. Ecol.* 92:214–229.

Valencia, R., R. Condit, K. Romoleroux, R. Foster, et al. 2004c. Tree Species Diversity and Distribution in a Forest Plot at Yasuní National Park, Amazonian Ecuador. In *Forest Diversity and Dynamism: Findings from a Network of Large-scale Tropical Forest plots,* edited by E.C. Losos and E.G. Leigh 107–18. Chicago: University of Chicago Press.

van Schaik, C.P. 1986. Phenological changes in a Sumatran rain forest. *J. Trop. Ecol.* 2: 327–47.

van Schaik, C.P., J.W. Terborgh, and S.J. Wright. 1993. The phenology of tropical forests: adaptive significance and consequences for primary consumers. *Ann. Rev. Ecol. Syst.* 24:353–77.

Verkerke, W. 1988. Syconium morphology and its influence on the flower structure of *Ficus sur* (Moraceae). *Proceedings of the Koninklijke Nederlandse Akademie Van Wetenschappen Series C Biological and Medical Sciences* 91:319–44.

Villanueva, R., and D.W. Roubik. 2004. Why are African honey bees and not European bees invasive? Pollen diet diversity in community experiments. *Apidologie* 35:481–91.

Visser, M.E., and L.J. Holleman. 2000. Warmer springs disrupt the synchrony of oak and winter moth phenology. *Proc. Roy. Soc. Lond. B.* 268:15–23.

Wainhouse, D., D.J. Cross, and R.S. Howell. 1990. The role of lignin as a defence against the spruce bark beetle *Dendroctonus micans*: effect on larvae and adults. *Oecologia* 85:257–65.

Walker, E.H. 1976. *Flora of Okinawa and the Southern Ryukyu Islands.* Washington, D.C.: Smithsonian Institution Press.

Wallace, H.M., and S.J. Trueman. 1995. Disperal of *Eucalyptus torelliana* seeds by resin-collecting stingless bees, *Trigona carbonaria. Oecologia* 104:12–16.

Walsh, R.P.D. 1996. Drought frequency changes in Sabah and adjacent parts of northern Borneo since the late nineteenth century and possible implications for tropical rain forest dynamics. *J. Trop. Ecol.* 12:385–407.

Ward, P.S. 1991. Phylogenetic analysis of psedomyrmecine ants associated with domatia-bearing plants. In *Ant-Plant Interactions,* edited by C.R. Huxley and D.F. Cutler, 335–52. Oxford: Oxford University Press.

Ware, A.B., and S.G. Compton. 1994a. Dispersal of adult female fig wasps: 1. Arrivals and departures. *Entomologia Experimentalis et Applicata* 73:221–29.

———. 1994b. Dispersal of adult female fig wasps: 2. Movements between trees. *Entomologia Experimentalis et Applicata* 73:231–38.

Watson, H. 1985. Lambir Hills National Park: Resource inventory with management recommendations. Kuching, Sarawak, Malaysia: National Parks and Wildlife Office, Forest Department.

Webster, P.J., and T.N. Palmer. 1997. The past and the future of El Niño. *Nature* 390: 562–64.

Weiblen, G.D. 2000. Phylogenetic relationships of functionally dioecious *Ficus* (Moraceae) based on ribosomal DNA sequences and morphology. *Am. J. Bot.* 87:1342–57.

———. 2002. How to be a fig wasp. *Ann. Rev. Entomol.* 47:299–330.

West, S.A., E.A. Herre, D.M. Windsor, and P.R.S. Green. 1996. The ecology and evolution of the New World non-pollinating fig wasp communities. *J. Biogeography* 23: 447–58.

West, S.A., M.G. Murray, C.A. Machado, A.S. Griffin, and E.A. Herre. 2001. Testing Hamilton's rule with competition between relatives. *Nature* 409:510–13.

West-Eberhard, M.J.W. 2003. *Developmental Plasticity and Evolution.* Oxford: Oxford University Press.

Wheelwright, N.T. 1985. Competition for dispersers, and the timing of flowering and fruiting in a guild of tropical trees. *Oikos* 44:465–77.

Whitmore, T.C. 1981. Paleoclimate and vegetation history. In *Wallace's line and plate tectonics,* edited by T.C. Whitmore, 36–42. Oxford: Claradon Press.

———. 1984. *Tropical Rain Forests of the Far East.* Oxford: Claradon Press.

———. 1998. *An Introduction to Tropical Rain Forests.* Oxford: Oxford University Press.

Wiens, D., C.L. Calvin, C.A. Wilson, C.I. Davern, et al. 1987. Reproductive success, spontaneous embryo abortion, and genetic load in flowering plants. *Oecologia* 71: 501–509.

Wille, A. 1979. Phylogeny and relationships among the genera and subgenera of the stingless bees (Meliponinae) of the world. *Revista de Biologia Tropical* 27:241–77.

Willmer, P.G., and S.A. Corbet. 1981. Temporal and microclimatic partitioning of the floral resources of *Justicia aurea* amongst a concourse of pollen vectors and nectar robbers. *Oecologia* 51:67–78.

Willson, M.F. and N. Burley. 1983. Mate choice in plants: tactics, mechanisms, and consequences. *Monographs in Population Biology.* Princeton: Princeton University Press.

Wilms, W., V.L. Imperatriz-Fonseca, and W. Engels. 1996. Resource partitioning between highly eusocial bees and possible impact of the introduced Africanized honey bee on native stingless bees in the Brazilian Atlantic rain forest. *Studies on Neotropical Fauna and Environment* 31:137–51.

Wilms, W., and B. Wiechers. 1997. Floral resource partitioning between native *Melipona* bees and introduced Africanized honeybee in the Brazilian Atlantic rain forest. *Apidologie* 28:339–55.

Wong, K.M. 1998. Patterns of plant endemism and rarity in Borneo and the Malay peninsula. In *Academica Sinica Monograph Series 16: Rare, threatened and endangered floras of Asia and the Pacific Rim,* edited by C.I. Peng and P.P. Lowry II, 139–69. Taiwan: Institute of Botany.

Wong, K.M., and A. Phillipps. 1996. *Kinabalu: Summit of Borneo.* Kota Kinabalu, Sabah, Malaysia: The Sabah Society.

Wood, G.H.S. 1956. Dipterocarp flowering season in Borneo, 1955. *Malaysian. Forester* 19:193–201.

Woods, P. 1989. Effects of logging, drought, and fire on structure and composition of tropical forests in Sabah, Malaysia. *Biotropica* 21:290–98.

Wright, S.J. 1991. Seasonal drought and the phenology of understory shrubs in a tropical moist forest. *Ecology* 72:1643–57.

———. 2002. Plant diversity in tropical forests: a review of mechanisms of species coexistence. *Oecologia* 130:1–14.

Wright, S.J., and O. Calderón. 1995. Phylogenetic patterns among tropical flowering phenologies. *J. Ecol.* 83:937–48.

Wright, S.J., and F.H. Cornejo. 1990a. Seasonal drought and leaf fall in a tropical forest. *Ecology* 71:1165–75.

———. 1990b. Seasonal drought and the timing of flowering and leaf fall in a Neotropical forest. In *Reproductive Ecology of Tropical Forest Plants*, edited by K.S. Bawa and M. Hadley, 49–61. Paris: UNESCO.

Wright, S.J., C. Carrasco, O. Calderón, and S. Paton. 1999. The El Niño Southern Oscillation, variable fruit production, and famine in a tropical forest. *Ecology* 80:1632–47.

Wyatt-Smith, J. 1963. *Manual of Malayan Silviculture for Inland Forest*. Kuala Lumpur: Forest Department.

———. 1995. *Manual of Malayan Silviculture for Inland Forest*. 2d ed. Kuala Lumpur: Forest Research Institute Malaysia.

Wycherley, P.R. 1973. The phenology of plants in the humid tropics. *Micronesica* 9:75–96.

Xu, Z.F., M. Liu Hong, Q. Chen Gui, and Y. Cui Jing. 1996. Ethnobotanical culture of fig trees in Xishuangbanna. *J. Plant Resources and Environment* 5:48–52.

Yamada, T., T. Yamakura, M. Kanzaki, A. Itoh, et al. 1997. Topography-dependent spatial pattern and habitat segregation of sympatric *Scaphium* species in a tropical forest at Lambir, Sarawak. *Tropics* 7:67–80.

Yamakura, T., M. Kanzaki, A. Itoh, T. Ohkubo, et al. 1995. Topography of a large-scale research plot established within a tropical rain forest at Lambir, Sarawak. *Tropics* 5:41–56.

———. 1996. Forest structure of a tropical rain forest at Lambir, Sarawak with special reference to the dependency of its physiognomic dimensions on topography. *Tropics* 6:1–18.

Yap, S.K. 1982. The phenology of some fruit tree species in a lowland dipterocarp forest. *Malaysian Forester* 45:21–35.

Yap, S.K., and H.T. Chan. 1990. Phenological behavior of some *Shorea* species in Peninsular Malaysia. In *Reproductive ecology of tropical forest plants*, edited by K.S. Bawa and M. Hadley, 21–35. Paris: UNESCO.

Yasuda, M., J. Matsumoto, N. Osada, S. Ichikawa, et al. 1999. The mechanism of general flowering in Dipterocarpaceae in the Malay Peninsula. *J. Trop. Ecol.* 15:437–49.

Yasunari, T. 1995. Characteristics of rainfall variability in the humid tropics. In *Humid Tropical Environment,* edited by T. Tamura, S. Shimada, H. Kadomura, and M. Umitsu, 26–41. Tokyo: Asakura (in Japanese).

Young, H.J. 1986. Beetle pollination of *Dieffenbachia longispatha* (Araceae). *Am. J. Bot.* 73:931–44.

———. 1988. Neighborhood size in a beetle pollinated tropical aroid. Effects of low density and asynchronous flowering. *Oecologia* 76:461–66.

Yumoto, T. 1987. Pollination systems in a warm temperate evergreen broad-leaved forest on Yaku Island. *Ecol. Res.* 2:133–45.

———. 2000. Bird-pollination of three *Durio* species (Bombacaceae) in a tropical rain forest in Sarawak. *Am. J. Bot.* 87:1181–88.

Yumoto, T., T. Inoue, and A.A. Hamid. 1996. Monitoring and inventorying system in canopy observation system in Canopy Biology Program in Sarawak, Malaysia. In *Biodiversity and the Dynamics of Ecosystems, vol. 1, The international network for DIVERSITAS in Western Pacific and Asia (DIWPA)*, edited by I.M. Turner, C.H. Diong, S.S.L. Lim, and P.K.L. Ng, 203–15. Otsu: Center for Ecological Research, Kyoto University.

Yumoto, T., T. Itino, and H. Nagamasu. 1997. Pollination of hemiparasites (Loranthaceae) by spiderhunters (Nectariniidae) in the canopy of a Bornean tropical rain forest. *Selbyana* 18:51–60.

Yumoto, T., K. Momose, and H. Nagamasu. 2000. A new pollination syndrome-Squirrel pollination in a tropical rain forest in Lambir Hills National Park, Sarawak, Malaysia. *Tropics* 9:147–51.

Zeuner, F.E., and F.J. Manning. 1976. A monograph on fossil bees (Hymenoptera: Apoidea). *Bull. Brit. Mus. Nat. Hist. (Geol.)* 27:151–268.

Zimmerman, J.K., D.W. Roubik, and J.D. Ackerman. 1989. Asynchronous phenology of a Neotropical orchid and its euglossine bee pollinator. *Ecology* 70:1192–95.

Index

Ecological Studies

Volumes published since 1998